AF545030

Hybridization Techniques for Electron Microscopy

Edited by

Gérard Morel, Ph.D., D. es Sc.
Research Director
National Scientific Research Center
Claude Bernard University
School of Medicine Lyon-Sud
Lyon, France

CRC Press
Boca Raton Ann Arbor London Tokyo

Library of Congress Cataloging-in-Publication Data

Hybridization techniques for electron microscopy / edited by Gérard Morel.
p. cm.
Includes bibliographical references and index.
ISBN 0-8493-4414-X
1. In situ hybridization. 2. Electron microscopy. I. Morel, Gérard.
QH452.8.H93 1993
574.87′322--dc20 92-27129
CIP

Direct all inquiries to CRC Press, Inc., 2000 Corporate Blvd., N.W., Boca Raton, Florida 33431.

International Standard Book Number 0-8493-4414-X

Library of Congress Card Number 92-27129

Printed in the United States of America 2 3 4 5 6 7 8 9 0

Printed on acid-free paper

PREFACE

For several years, I have taught a graduate level course in *in situ* hybridization for electron microscopy. At every turn, I was confronted by the "myth" of complexity that surrounds this technique.

The objective of this book is to eliminate this "myth" by (1) presenting a succinct summary of the basic aspects of hybridization and labeling methods; (2) showing the transition of *in situ* hybridization from light microscopy, using confocal microscopy and semi-thin sections; (3) introducing the three most useful *in situ* hybridization techniques: pre-embedding, ultrathin frozen section, and post-embedding in hydrophilic resin methods; and (4) presenting interesting applications, such as detection of viral nucleic acid and RNA in plants.

Each chapter describes concisely the precise methodology recommended. The authors, specialists within their respective fields of study, have selected what they consider to be relevant and important technical points, most of which are presented in the format of a protocol.

In the future, molecular biology and ultrastructural *in situ* hybridization techniques will become even more sophisticated, making an even greater contribution to the understanding of cell biology.

THE EDITOR

Gérard Morel, Ph.D., D. es Sc., is Research Director at the National Scientific Research Centre at the Claude Bernard University, School of Medicine Lyon-Sud, at Lyon, France.

Dr. G. Morel received his baccalauréat from Lalande College, Bourg en Bresse in 1967. He obtained his M.S. and Ph.D. degrees in 1973 and 1976, respectively, from the Department of Physiology. He was appointed an assistant of Histology at the same university in 1974 and was Doctor es Science in 1980. He was appointed by CNRS in 1981 and became Research Director in 1989.

Dr. G. Morel is a member of the American Endocrine Society, International Society of Neuroendocrinology, Société Française de Microscopie Electronique (is serving on the council), Société de Biologie Cellulaire de France, Société de Neuroendocrinologie Expérimentale de France, Society for Neuroscience, and Society for Cell Biology.

Dr. G. Morel has been the recipient of research grants from INSERM (National Institute of Health and Medical Research) and private industry.

Dr. G. Morel is the author of more than 100 papers and 10 chapters and has been the author of two books. His current major research interests include the localization of receptor molecules and regulation of gene expression.

CONTRIBUTORS

J. G. J. Bauman
I.T.R.I. — T.N.O.
Rijswiju, The Netherlands

G. J. Brakenhoff, Ph.D.
Department of Molecular Cell Biology
University of Amsterdam
The Netherlands

Judy Brangeon, Ph.D.
Institut de Recherches sur les Plantes
CNRS
Université Paris-Sud
Orsay, Cedex, France

Pierre Couble, Ph.D.
Centre de Genetique Moleculaire et Cellulaire
Université Lyon
Villeurbanne, Cedex, France

Francoise Escaig-Haye, Ph.D.
INSERM
Hopital St. Vincent de Paul
Paris, France

Jean-Guy Fournier, Ph.D.
INSERM
Hopital St. Vincent de Paul
Paris, France

Lucien Frappart, M.D., Ph.D.
Laboratoire d'Anatomie Pathologique
Hopital Edouard Herriot
Lyon, Cedex, France

Dominique Le Guellec, Ph.D.
Cytologie Moleculaire
CNRS
Villeurbanne, Cedex, France

Gerard Morel, Ph.D.
Histology Laboratory
CNRS
School of Medicine Lyon-Sud
Lyons, France

Christine Pechoux, Ph.D.
Laboratory of Anatomy and Pathology
Edouard Herriot Hospital
Lyon, France

Chris W. Perrett, Ph.D.
Department of Medicine
Keele University
Keele, U.K.

Francine Puvion-Dutilleul, Ph.D.
Laboratoire de Biologie et Ultrastructure du Noyau
CNRS
Villejuif, Cedex, France

Lucienne Sossountzov, Ph.D.
Laboratoire de Physiologie du Developpement des Plantes
CNRS
Université Pièrre et Marie Curie
Paris, Cedex, France

Alain Trembleau, Ph.D.
Laboratoire de Neurobiologie des Signaux Intercellulaires
CNRS
Université Pièrre et Marie Curie
Paris, Cedex, France

H. van Dekken, M.D., Ph.D.
Department of Pathology
Erasmus University Rotterdam
Rotterdam, The Netherlands

TABLE OF CONTENTS

INTRODUCTION

Gérard Morel

The *in situ* hybridization method is based on the technique used for studying *in vitro* pairing reactions between labeled RNA or DNA molecules in solution (called "probe") and complementary nucleic acid either in solution[1-3] or bound to membrane[4-6] (called "target nucleic acid"). In these cases, hybrid molecules are detected by the use of a radioactive counter or an autoradiographic technique (for radioactive labels) or an immunological stained reaction (for nonradioactive labels). In the *in situ* molecular hybridization method, the target nucleic acid is maintained in its original cellular location. It is, thus, theoretically possible to obtain hybrids with a sequence complementary to the labeled probe used. DNA/DNA, DNA/RNA, and RNA/RNA hybrid formation reactions are basically similar and may be considered together.

The principle of the molecular hybridization method is as follows: the two strands of native DNA molecules are held together by hydrogen bonds of nucleotide pairs. Disruption of the hydrogen bonds causes the separation of the two strands and denaturation of DNA, whereas native RNA occurs in single strands, and thus does not need to be denatured before hybridization. Denaturation of DNA can be brought about by increasing the temperature, raising the concentration of Na^+ in the buffer, or using substances such as formamide ($NCONH_2$) which compete with the bases for the formation of hydrogen bonds. The so-called "melting temperature" corresponds to the temperature of 50% of denaturation.

Renaturation of the nucleic acid target and the probe to give hybrids is obtained at about 15°C below the melting temperature, at high ionic strength, for decreasing electrostatic repulsion between molecules. Addition of formamide to the hybridization buffer allows hybridization at a lower temperature (0.65°C/% of formamide).[7] Pretreatment to permit access of the probe to nucleic acid targets is sometimes necessary and is obtained by proteolysis.

After hybridization for several hours, noncomplexed or incompletely complexed probes are removed by enzymatic digestion and by washes in decreasing concentrations of ionic buffer and increasing washing temperature.

Hybrids are revealed, on the one hand, using an autoradiographic method if the probe is labeled with radioactive nucleotides, and, on the other hand, using an immunocytological reaction, if the probe is labeled with antigen-conjugated nucleotides.

The general procedure for *in situ* hybridization is the following:

0-8493-4414-X/93/$0.00 + $.50

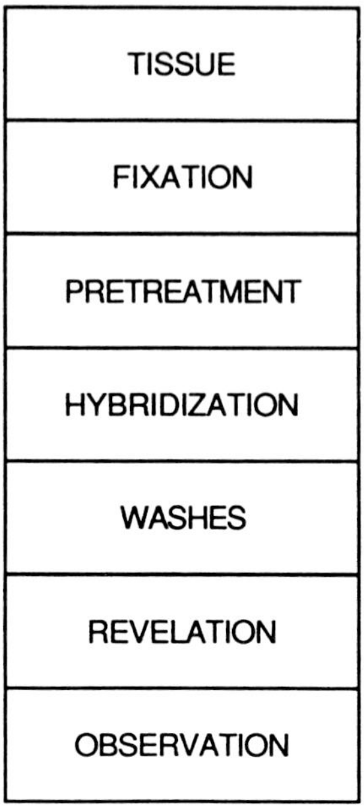

TISSUE
FIXATION
PRETREATMENT
HYBRIDIZATION
WASHES
REVELATION
OBSERVATION

Ultrastructural *in situ* hybridization is much more recent than its counterpart in light microscopy. The pioneering work done by Gall and Pardue, and by John et al., starting from 1969, showed that it was possible to reveal RNA/DNA hybrids formed *in situ*.[8-9] However, the results of such experiments, as reviewed by Henning in 1973, are still open to interpretation.[10]

Recent developments in the techniques of molecular biology (genetic recombination, gene amplification, oligonucleotide synthesis, or the polymerase chain reaction), in addition to what they have contributed to the knowledge of the genome, have provided morphologists with tools which are both effective and diverse. Until 1990, the only probe which was used for detecting target nucleic acid, whether DNA or RNA, was DNA.[11] Troxler et al.[12] were the first to use a single-stranded RNA probe, which had the advantages of greater sensitivity, higher hybridization efficiency, and the absence of competing strands, as in the case of light microscopy.[12-13] Wenderoth and Eisenberg[14] compared *in situ* hybridization using biotinylated riboprobes on ultrathin frozen sections to sections of LR White embedded tissue. At present, the results obtained using oligonucleotide probes, whether single or in combination, are similar to those obtained previously. The revelation of the final product of protein synthesis by immunocytological techniques is no longer the only possible way of revealing the activity of this synthesis, whose initial product, mRNA, can also be detected. Similarly, all viral genomes can be revealed without antibodies being produced.

In the same way, the development of techniques of probe labeling, with the availability of the new labeled nucleotides, gives numerous possibilities

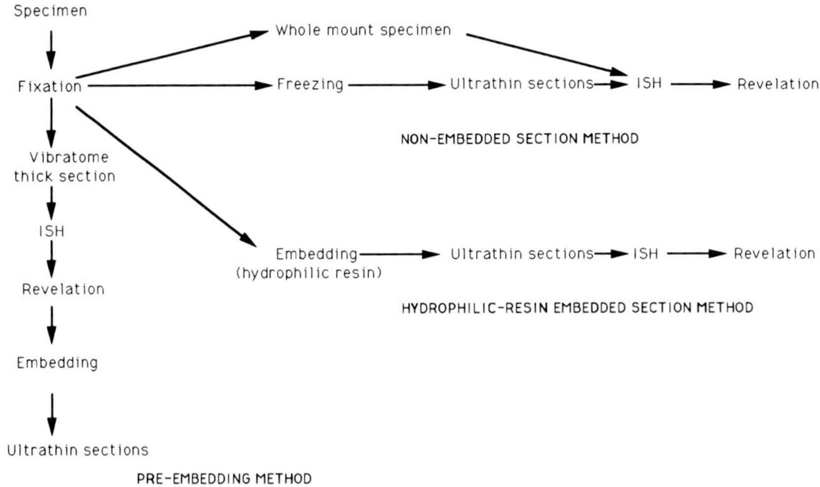

FIGURE 1. Principles of ultrastructural *in situ* hybridization methods.

for autoradiographic detection as well as for immunocytological detection. At first, the nucleotides used were tritium-labeled, which, though, provided excellent cellular localization, required prolonged exposure times to visualize the signal due to their low energy emission.[15-21] In order to increase the specific activity of the probe (10^9 dpm/μg vs. up to 10^8 dpm/μg with tritiated probe), ^{125}I labeling of nucleic acid molecules was introduced,[22-24] but was not widely used at the electron microscopical level.[25-26] The main isotope currently used is sulfur 35, which was a greater activity,[20-21] and is commercially available as a labeled nucleotide; and more recently phosphorus 33 has been introduced. But the development of commercial nonradioactive nucleotides (biotin- or digoxigenin-nucleotides) has improved the resolution of the detected signal and reduced the revelation time.[19,27-29]

Though many studies have been done using *in situ* hybridization with light microscopy, this is not yet the case for electron microscopy. The first results published in 1971 by Jacob et al. revealed the presence of ribosomal RNA using a tritiated probe.[15] Then the first detections of viral DNA were reported,[16-18] at the same time as the characterization of repetitive sequences in the chromosomes.[30] Later, the messenger RNA was visualized.[21,29] However, the current phase of development of this technique only began in the 1980s, with the utilization of nonradioactive probes.[19]

Now three methods have been developed (Figure 1): two methods on sections (nonembedded[19,21,31-34] and hydrophilic-resin embedded[28,35-36] tissue) and another, carried out before embedding (pre-embedding *in situ* hybridization).[16,37-39] The basic principles of these methods are always the same as described above.

Fixation is the first step, with aldehydic fixatives being the most widely used, in the form of paraformaldehyde with or without the addition of glutaraldehyde. The concentration of glutaraldehyde is normally less than 0.5%, but results have been described with 2 or 2.5% of glutaraldehyde to detect rRNA.[15-19,30,40-42] No data have been produced after osmium tetroxide fixation.[21]

Only minor modifications have been introduced into the tissue embedding process for *in situ* hybridization. In the ultrathin frozen section method, the cryoprotective agent is sucrose, and the freezing procedure must avoid the production of large ice crystals.[43] In the embedding in hydrophilic resin method, all the most common resins have been successfully used for *in situ* hybridization, e.g., glycerol metacrylate, Lowicryl K4M or K11M, LR White, and Spurr. In the pre-embedding method large section, about 50 to 100 μm obtained using a vibratome are incubated with the probe, without difficulties of penetration.[44] The pretreatment of tissue for *in situ* hybridization is always limited as far as possible in order to preserve cytology.[34,38,45] The conditions of hybridization and washing are the steps which will be modified in the future. Incubation periods varying from 5[15] to 16[17] to 45 h[18] have been used for electron microscopic experiments. The shortest hybridization period currently used is 3 to 4 h,[21,33,34] since it seems that hybrid formation, as DNA renaturation, is very rapid.

After hybridization, the sections are washed with SSC. The stringency of this step (temperature and concentration of this buffer) seems to be reduced, compared to the first studies. Moreover, it is further reduced before immunocytological detection, in comparison to autoradiographic detection.[34]

Revelation, the last step of *in situ* hybridization, depends on the probe label: autoradiography is carried out with radioactive-labeled probes, immunocytology with nonradioactive-labeled probes. No difference is observed with classical detection methods between isotopic molecules and antigens. The preparation of nuclear photographic emulsion membrane, exposure time, and quantification remains the essential problem of autoradiography.[46] The penetration of antibodies limits the use of nonradioactive-labeled probes in the pre-embedding *in situ* hybridization method. The combination of these revelation methods, or the use of two sizes of gold particles, makes it possible to detect two different nucleic acids simultaneously, or an antigen and a nucleic acid.[33-34,41]

The comparison of these methods showed that sensitivity and ultrastructural preservation were mutually opposed. *In situ* hybridization on ultrathin frozen sections was the most sensitive method, but the pre-embedding method preserved ultrastructure. Embedding in lowicryl or LR White seemed to give similar results (Table 1).[47-49]

The areas of application of *in situ* hybridization at the ultrastructural level have greatly expanded. Historically, reiterated nucleotide sequences were the first to be visualized,[9] then ribosome RNA,[15] along with virus[16-17] and messenger RNA.[21,29] Since these studies were carried out, the methods have been refined.

TABLE 1
Chief Characteristics of Ultrastructural Nonradioactive *in situ* Hybridization Methods: Pre-Embedding Vs. Frozen Sections Vs. Post-Embedding

Method	Pre-embedding	Frozen section	Post-embedding
Storage	Not possible	After freezing	After embedding
Pretreatment	No[a]	No	No
Probe concentration	10 pmol/ml	4 pmol/ml	100 pmol/ml
Detection system	Enzymatic	Colloidal gold	Colloidal gold
Ultrastructure preservation	High	Low	Medium
Resolution	Low	High	High
Sensitivity	Medium	High	Low

[a] This step is necessary if colloidal gold particles are used.

From Le Guellec, D., Trembleau, A., Pechoux, C., Gossard, F., and Morel, G., *J. Histochem. Cytochem.*, 40, 979, 1992.

In situ hybridization has been applied on:

- Non-embedded materials including whole mount chromosomes,[19-20,50] ultrathin frozen sections,[28,47-49,51] or triton-extracted whole mounts of single cells[32-33,52]
- Material embedded in hydrophilic resin: lowicryl K4M[28,45,53-54] or K11M,[55] LR White,[14,56] glycerol methacrylate[15,37]
- The embedding method, followed by embedding in araldide,[39,41] or Spurr,[58] or Epon,[44] after osmium tetroxide fixation.

The main applications have been detection of:

- Virus
- Ribosomic nucleic acid
- Messenger RNA
- Genes on chromosomes
- Messenger RNA in plants

Viral nucleic acid in infected cells seems to be the best material for *in situ* hybridization studies, since the target nucleic acid is present in large amounts in the cells. The simplification of specimen processing and the development of highly specific hybridization are the most advanced of the *in situ* hybridization methods both on sections of cells embedded in lowicryl or LR White[12] and on ultrathin frozen sections.[59]

Depending on the protocol used, related DNA sequences (both double-stranded and single-stranded segments),[45,60] single-stranded DNA portions only,[60] related single-stranded nucleic acids (both DNA and RNA),[61-62] and RNA molecules only[61-63] can be specifically labeled at the electron microscopic level, at the surface of lowicryl K4M sections, using biotinylated viral DNA probes and immunodetection of hybrids. The data clearly demonstrate that post-embedding *in situ* hybridization is a suitable tool for studying the functional compartmentalization of the nucleus[45,60] and the establishment of structure-function relationships.[64-65]

The intranuclear development of DNA viruses, such as herpes simplex virus type 1 (HSV-1), adenovirus type 5 (Ad 5), and cytomegalovirus[66] (SV 40)[18,40] presents special problems (large amounts of viral DNA-binding proteins and portions of single-stranded DNA).[63-64] Therefore, the optimal protocols for detecting viral nucleic acid sequences in infected cells differ from those used for the detection of cellular nucleic acid, as revealed by comparison with protocols allowing the detection of ribosomal genes, and those used for the study of the maturation of preribosomal RNA.[67-69]

Following the molecular hybridization of ribosomal RNA on ultrathin frozen sections in 1971 by Jacob et al.,[15] several attempts have been made to localize the expression of ribosomal genes in animal cells. Nucleoli are highly ordered nuclear components which provide an opportunity to study the mode of integration of several complexed functions within restricted nuclear territories.[70] They are sites of ribosome biogenesis, which involves transcription of ribosomal RNA genes and multiple steps of maturation of the preribosomal subunits.[18,47,50,53-54,67-69,71,74-77]

Localization of messenger RNA at the light microscopic level has become a widely used tool to examine tissue distribution and temporal expression. It lacks the resolution necessary to determine the spatial relationship between mRNA, and the protein product encoded by that message, but electron microscopic *in situ* hybridization has been used to elucidate the mechanisms involved in protein genesis and maintenance. All three methods of *in situ* hybridization have been used in this way in the field of endocrinology[12,21,29,34,36-39,41,44,49,55,79-82] as well as for determining the distribution of cytoskeletal mRNA[14,32-33,48] associated or not with antigen detection.[32-34,41]

Two methods for *in situ* hybridization to whole mount chromosome preparations at the electron microscopic level have been described: one using radioactive-labeled probes[19,20,31] and the other based on the use of a biotinylated probe.[19,83] All these methods may constitute useful additions to standard light microscopic *in situ* methods, where higher resolution of fine structure in the mapping of genes on chromosome is desired. In the case of plant tissue, the use of this technique is still limited.[29,84,85] In general, it provides a useful method for relating cellular and/or intracellular ultrastructure to molecular aspects of plant gene expression. In particular, the *in situ* hybridization approach can reveal accumulations of specific mRNA in discrete cellular compartments.[58]

The future applications of *in situ* hybridization at the electron microscopic level are as difficult to determine as were the applications of immunocytology 20 years ago. But these techniques should provide a tool for studying the location of nucleic acid in the cell, and such morphological knowledge, along with the functional findings of molecular biology, should contribute greatly to the understanding of cell biology.

REFERENCES

1. **Marmur, J. and Doty, P.,** Thermal renaturation of deoxyribonucleic acids, *J. Mol. Biol.,* 3, 585, 1961.
2. **Hall, B. D. and Spiegelman, S.,** Sequence complementarity of T2-specific RNA, *Proc. Natl. Acad. Sci. U.S.A.,* 47, 137, 1961.
3. **Yankofski, S. A. and Spiegelman, S.,** Distinct cistrons for two ribosomal RNA components, *Proc. Natl. Acad. Sci. U.S.A.,* 49, 538, 1963.
4. **Nygaard, A. P. and Hall, B. D.,** A method for detection of RNA-DNA complexes, *Biochem. Biophys. Res. Commun.,* 12, 98, 1963.
5. **Gillespie, D. and Spiegelman, S.,** A quantitative assay for DNA-RNA hybrids with DNA immobilized on a membrane, *J. Mol. Biol.,* 12, 829, 1965.
6. **McCarthy, B. J. and Church, R. B.,** The specificity of molecular hybridization reactions, *Annu. Rev. Biochem.,* 39, 131, 1970.
7. **Marmur, J. and Ts'o, P. O.,** Denaturation of deoxyribonucleic acid by formamide, *Biochim. Biophys. Acta,* 51, 32, 1961.
8. **Gall, J. C. and Pardue, M. L.,** Molecular hybridization of radioactive DNA to the DNA of cytological preparations, *Proc. Natl. Acad. Sci. U.S.A.,* 64, 600, 1969.
9. **John, H. A., Birnstiel, M. L., and Jones, K. W.,** RNA-DNA hybrids at the cytological level, *Nature (London),* 223, 582, 1969.
10. **Henning, W.,** Molecular hybridization of DNA and RNA *in situ, Int. Rev. Cytol.,* 36, 1, 1973.
11. **Haase, A., Brahic, M., Stowring, L., and Blum, H.,** Detection of viral nucleic acids by in situ hybridization, in *Methods Virol.,* Vol. 7, Academic Press, London, 1985.
12. **Troxler, M., Pasamontes, L., Egger, D., and Bienz, K.,** *In situ* hybridization for light and electron microscopy: a comparison of methods for the localization of viral RNA using biotinylated DNA and RNA probes, *J. Virol. Methods,* 30, 1, 1990.

12a. **Cox, K. H., DeLeon, D. V., Angerer, L. M., and Angerer, R. C.,** Detection of mRNAs in sea urchin embryos by *in situ* hybridization using assimetric RNA probes, *Dev. Biol.,* 101, 485, 1984.

13. **Couderc, T., Christodoulou, C., Kopecka, H., Marsden, S., Taffs, L. F., Crainic, R., and Horaud, F.,** Molecular pathogenesis of neuronal lesions induced by poliovirus type I, *J. Gen. Virol.,* 70, 2907, 1989.
14. **Wenderoth, M. P. and Eisenberg, B. R.,** Ultrastructural distribution of myosin heavy chain mRNA in cardiac tissue: a comparison of frozen and LR White embedment, *J. Histochem. Cytochem.,* 39, 1025, 1991.
15. **Jacob, J., Todd, K., Binstiel, M. L., and Bird, A.,** Molecular hybridization of ^{3}H-labelled ribosomal RNA in ultrathin sections prepared for electron microscopy, *Biochim. Biophys. Acta,* 228, 761, 1971.

16. **Orth, G., Jeanteur, P., and Croissant, O.,** Evidence for and localization of vegetative viral DNA replication by autoradiographic detection of RNA-DNA hybrids in sections of tumors induced by the Shope papillomas virus, *Proc. Natl. Acad. Sci. U.S.A.,* 68, 1876, 1971.
17. **Croissant, O., Dauguet, C., Jeanteur, P., and Orth, G.,** Application de la technique d'hybridation moléculaire in situ à la mise en évidence au microscope électronique, de la réplication végétative de l'ADN viral dans les papillomes provoqués par le virus de Shope chez le Lapin cottontail, *C. R. Acad. Sci. Paris,* 274, 614, 1972.
18. **Geuskens, M. and May, E.,** Ultrastructural localization of SV40 viral DNA in cells, during lytic infection, by in situ molecular hybridization, *Exp. Cell Res.,* 87, 175, 1974.
19. **Hutchison, N., Langer-Safer, P., Ward, D., and Hamkalo, B.,** *In situ* hybridization at the electron microscopic level: hybrid detection by autoradiography and colloidal gold, *J. Cell Biol.,* 95, 609, 1982.
20. **Li, C.-B., Wu, M., Margitich, I. S., and Davidson, N.,** Gene mapping on human metaphase chromosomes by *in situ* hybridization with 3H, 35S, and 32P labeled probes and transmission electron microscopy, *Chromosoma (Berlin),* 93, 305, 1986.
21. **Morel, G., Dubois, P. M., and Gossard, F.,** Détection ultrastructurale des ARN messagers codant pour l'hormone de croissance dans l'antéhypophyse du rat par hybridization *in situ, C. R. Acad. Sci. Paris,* 302, 479, 1986.
22. **Prensky, W., Steffensen, D. M., and Hughues, W. L.,** The use of iodinated RNA for gene location, *Proc. Natl. Acad. Sci. U.S.A.,* 70, 1860, 1973.
23. **Commerford, S. L.,** Iodination of nucleic acid *in vitro, Biochemistry,* 10, 1993, 1972.
24. **Getz, M. J., Altenburg, L. C., and Saunders, G.,** The use of RNA labelled *in vitro* with iodine-125 in molecular hybridization experiments, *Biochim. Biophys. Acta,* 287, 485, 1972.
25. **Gautier, A.,** Ultrastructural localization of DNA in ultrathin tissue sections, *Int. Rev. Cytol.,* 44, 175, 1976.
26. **Seinert, G., Thomas, C., and Brachet, J.,** Localization by *in situ* hybridization of amplified ribosomal DNA during *Xenopus laevis* oocyte maturation (a light and electron microscopy study), *Proc. Natl. Acad. Sci. U.S.A.,* 73, 833, 1976.
27. **Binder, M.,** *In situ* hybridization at the electron microscope level, *Scan. Microsc.,* 1, 331, 1984.
28. **Binder, M., Tourmente, S., Roth, J., Renaud, M., and Gehring, W. J.,** *In situ* hybridization at the electron microscope level: localization of transcripts on ultrathin sections of Lowicryl K4M-embedded tissue using biotinylated probes and protein A-gold complexes, *J. Cell Biol.,* 102, 1646, 1986.
29. **Harris, N. and Croy, R. R. D.,** Localization of mRNA for pea legumin: *in situ* hybridization using a biotinylated cDNA probe, *Protoplasma,* 130, 57, 1986.
30. **Eckhardt, R. C.,** Cytological localization of repeated DNAs, in *Handbook of Genetics,* Vol. 5, *Molecular Genetics,* King, R. C., Ed., Plenum Press, New York, 1976, 31.
31. **Hutchison, N. J.,** Hybridisation histochemistry: *in situ* hybridisation at the electron microscope level, in *Immunolabelling for Electron Microscopy,* Polak and Varndell, Eds., Elsevier, New York, 1984, 341.
32. **Singer, R. H., Langevin, G. L., and Lawrence, J. B.,** Ultrastructural visualization of cytoskeletal messenger RNAs and their associated proteins using double label *in situ* hybridization, *J. Cell Biol.,* 108, 2343, 1989.
33. **Singer, R. H., Lawrence, J. B., Silva, F., Langevin, G. L., Pomeroy, M., and Billings-Gagliardi, S.,** Strategies for ultrastructural visualization of biotinated probes hybridized to messenger RNA *in situ,* in *Current Topics in Microbiology and Immunology,* Vol. 143, Springer-Verlag, Berlin, 1989, 55.
34. **Morel, G.,** Hybridization *in situ* sur coupes ultrafines de tissus congelés: sondes radioactives, sondes non-radioactives, double marquage, in *Techniques en Microscopie Electronique: Cryométhodes, Immunocytologie, Autoradiographie, Hybridation in situ,* Morel, G., Ed., Editions INSERM, Paris, 1991, 519.

35. **Steiner, G., Felsani, A., Kettmann, R., and Brachet, J.,** Presence of rRNA in the heavy bodies of sea urchin eggs. An *in situ* hybridization study with the electron microscope, *Exp. Cell Res.*, 154, 203, 1984.
36. **Webster, H. F., Lamperth, L., Favilla, J. T., Lemke, G., Tesin, D., and Manuelidis, L.,** Use of a biotinylated probe and *in situ* hybridization for light and electron microscopic localization of mRNA in myelin-forming Schwann cells, *Histochemistry,* 86, 441, 1987.
37. **Trembleau, A., Fevre-Montange, M., and Calas, A.,** Localisation ultrastructurale d'ARNm codant pour l'ocytocine par hybridation *in situ.* Etude par radioautographie à haute résolution à l'aide d'une sonde oligonucléotidique tritiée, *C. R. Acad. Sci. Paris,* 307, 869, 1988.
38. **Trembleau, A., Fevre-Montange, M., Landry, M., and Calas, A.,** Hybridation *in situ* ultrastructurale avant inclusion, in *Techniques en Microscopie Electronique: Cryométhodes, Immunocytologie, Autoradiographie, Hybridation in situ,* Morel, G., Ed., Editions INSERM, Paris, 1991, 503.
39. **Guitteny, A. F. and Bloch, B.,** Ultrastructural detection of the vasopressin messenger RNA in the normal and Brattleboro rat, *Histochemistry,* 92, 277, 1989.
40. **Geuskens, M.,** Autoradiographic localization of DNA, in nonmetabolic conditions, in *Principles and Techniques of Electron Microscopy, Biological Applications,* Vol. 7, Hayat, M. A., Ed., Van Nostrand Reinhold Company, 1977, 163.
41. **Tong, Y., Zhao, H., Simard, J., Labrie, F., and Pelletier G.,** Electron microscopic autoradiographic localization of prolactin mRNA in a rat pituitary, *J. Histochem. Cytochem.*, 37, 567, 1989.
42. **Wachtler, F., Schöfer, C., Mosgöller, W., Weipoltshammer, K., Schwarzacher, H. G., Guichaoua, M., Hartung, M., Stahl, A., Berge-Lefranc, J. L., Gonzalez, I., and Sylvester, J.,** Human ribosomal RNA gene repeats are localized in the dense fibrillar component of nucleoli: light and electron microscopic *in situ* hybridization in human sertoli cells, *Exp. Cell Res.*, 198, 135, 1992.
43. **Morel, G.,** Coupes semi-fines et ultrafines de tissus congelés: congélation, cryo-ultramicotomie, contraste, in *Techniques en Microscopie Electronique: Cryométhodes, Immunocytologie, Autoradiographie, Hybridation in situ,* Morel, G., Ed., Editions INSERM, Paris, 1991, 31.
44. **Trembleau, A., Calas, A., and Fevre-Montange, M.,** Ultrastructural localization of oxytocin mRNA in the rat hypothalamus by *in situ* hybridization using a synthetic oligonucleotide, *Mol. Brain Res.*, 8, 37, 1990.
45. **Puvion-Dutilleul, F. and Puvion, E.,** Ultrastructural localization of viral DNA in thin sections of herpes-simplex virus type I infected cells by *in situ* hybridization, *Eur. J. Cell. Biol.*, 49, 99, 1989.
46. **Hejblum, G., Mary, J. Y., Morice, V., and Aurengo, M.,** Interpretation of electron microscope autoradiographs: correction for cross-fire when membrane components are highly labelled, *J. Microsc.*, 166, 219, 1992.
47. **Geraud, M.-L., Herzog, M., and Soyer-Gobillard, M.-O.,** Nucleolar localization of rRNA coding sequences in Prorocentrum micans Ehr. (dinomastigote, kingdom Protoctist) by *in situ* hybridization, *BioSystem,* 26, 61, 1991.
48. **Wenderoth, M. A. and Eisenberg, B. R.,** Ultrastructural distribution of myosin heavy chain mRNA in cardiac tissue: a comparaison of frozen and LR white embedment, *J. Histochem. Cytochem.*, 39, 1025, 1991.
49. **Le Guellec, D., Trembleau, A., Pechoux, C., and Morel, G.,** Ultrastructural nonradioactive *in situ* hybridization of GH mRNA in rat pituitary gland: pre-embedding vs. ultra-thin frozen sections vs. post-embedding, *J. Histochem. Cytochem.*, 40, 979, 1992.
50. **Manuelidis, L., Langer-Safer, P. R., and Ward, D. C.,** High resolution mapping of satellite DNA using biotin-labeled DNA probes, *J. Cell Biol.*, 95, 619, 1982.
51. **Pollock, J. A., Ellisman, M. H., and Benzer, S.,** Subcellular localization of transcripts in Drosophilia photoreceptor neurons: chaotic mutants have an aberrant distribution, *Genes Dev.*, 4, 806, 1990.

52. **Silva, F. G., Lawrence, J. B., and Singer, R. H.,** Progress toward ultrastructural identification of individual mRNAs in thin section: myosin heavy chain in developing myotubes, in *Techniques in Immunocytochemistry,* Bullock, G. and Petrusz, P., Eds., Vol. 4, Academic Press, London, 1989, 77.
53. **Escaig-Haye, F., Grigoriev, V., and Fournier, J.-G.,** Détection ultrastructurale d'ARN ribosomal par hybridation *in situ* à l'aide d'une sonde biotinylée sur coupes ultrafines de cellule animale en culture, *C. R. Acad. Sci. Paris,* 309, 429, 1989.
54. **Thiry, M. and Thiry-Blaise, L.,** *In situ* hybridization at the electron microscope level. An improved method for precise localization of ribosomal DNA and RNA, *Eur. J. Cell Biol.,* 50, 235, 1989.
55. **Le Guellec, D., Frappard, L., and Wilhems, R.,** Ultrastructural localization of fibronectin mRNA in chick embryo by *in situ* hybridization using ^{35}S or biotin labeled cDNA probes, *Cell Biol.,* 70, 159, 1990.
56. **Brangeon, J., Priul, J. L., and Forchioni, A.,** Localization of mRNAs for the small and large subunits of rubisco using electron microscopic *in situ* hybridization, *Plant Physiol.,* 86, 990, 1988.
57. **Jamrich, M. K. A., Mahon, K. A., Gavis, E. R., and Gall, J. G.,** Histone RNA in amphibian oocytes visualized by *in situ* hybridization to methacrylate-embedded tissue sections, *EMBO J.,* 3, 1939, 1984.
58. **Brangeon, J., Nato, A., and Forchioni, A.,** Ultrastructural detection of ribulose-1,5-biphosphate carboxylase protein and its subunit mRNAs in wild type and holoenzyme-deficient Nicotiana using immuno-gold and *in-situ* hybridization techniques, *Planta,* 177, 151, 1989.
59. **Pautrat, G., Morel, G., Dihl, F., Allasia, C., Rey, F., Spire, B., Sire, J., Kourilsky, F., and Chermann, J. C.,** Distribution polaire des particules virales à la surface des cellules d'une lignée lymphocytaire infectées *"in vitro"* par le virus HIV: détection intracellulaire du génome viral, *Rétrovirus,* 1, 17, 1989.
60. **Puvion-Dutilleul, F., Picard, E., Laithier, E., and Puvion, E.,** Cytochemical study of parental Herpes simplex virus type I DNA during initial infection of the cell, *Eur. J. Cell Biol.,* 50, 187, 1989.
61. **Puvion-Dutilleul, F. and Puvion, E.,** Sites of transcription of adenovirus type 5 genomes in relation to early viral DNA replication in infected HeLa cells. A high resolution *in situ* hybridization and autoradiographical study, *Bio. Cell,* 71, 135, 1991.
62. **Puvion-Dutilleul, F., Roussev, R., and Puvion, E.,** Distribution of viral RNA molecules during the adenovirus type 5 infectious cycle in Hela cells, *J. Struct. Biol.,* 108, 209, 1992.
63. **Puvion-Dutilleul, F. and Puvion, E.,** Ultrastructural localization of defined sequences in viral RNA and DNA by *in situ* hybridization of biotinylated DNA probes on sections of Herpes simplex virus type 1 infected cells, *J. Electron. Microsc. Tech.,* 18, 336, 1991.
64. **Puvion-Dutilleul, F. and Puvion, E.,** Replicating single-stranded adenovirus type 5 DNA molecules accumulate within well-delimited intranuclear areas of lytically infected Hela cells, *Eur. J. Cell Biol.,* 52, 379, 1990.
65. **Puvion-Dutilleul, F. and Puvion, E.,** Analysis by *in situ* hybridization and autoradiography of sites of replication and storage of single and double-stranded adenovirus type 5 DNA in lytically infected HeLa cells, *J. Struct. Biol.,* 103, 280, 1990.
66. **Beals, T. F.,** Ultrastructure of *in situ* hybridization, *Ultrastruct. Pathol.,* 16, 87, 1992.
67. **Puvion-Dutilleul, F., Bachellerie, J. P., and Puvion, E.,** Nucleolar organization of HeLa cells as studied by *in situ* hybridization, *Chromosoma,* 100, 395, 1991.
68. **Puvion-Dutilleul, F., Mazan, S., Nicoloso, M., Christensen, M. E., and Bachellerie, J. P.,** Localization of U3 RNA molecules in nucleoli of Hela and mouse 3T3 cells by high resolution *in situ* hybridization, *Eur. J. Cell Biol.,* 56, 178, 1991.

69. **Puvion-Dutilleul, F., Mazan, S., Nicoloso, M., Christensen, M. E., Bachellerie, J. P., and Puvion E,** Alteration of nucleolar ultrastructure and ribosome biogenesis by actinomycin D. Implications for U3 snRPN function, *Eur. J. Cell Biol.*, 58, 149, 1992.
70. **Hadjiolov, A. A.,** The nucleolus and ribosome biogenesis, in *Cell Biology Monographs,* Vol. 12, Alfert, M., Beermann, W., Goldstein, L., Porter, K. R., and Sitte, P., Eds., Springer, Berlin, 1985.
71. **Escaig-Haye, F., Grigoriev, V., Peranzi, G., Lestienne, P., and Fournier, J.-G.,** Analysis of human mitochondrial transcripts using electron microscopic *in situ* hybridization, *J. Cell Sci.*, 100, 851, 1991.
72. **Manuelidis, L.,** Different central nervous system cell types display distinct and nonrandom arrangements of satellite DNA sequences, *Proc. Natl. Acad. Sci. U.S.A.*, 81, 3123, 1984.
73. **Manuelidis, L.,** Indications of centromere movement during interphase and differentiation, *Ann. N.Y. Acad. Sci.*, 450, 205, 1985.
74. **Thiry, M., Scheer, U., and Goessens, G.,** Localization of nucleolar chromatin by immunocytochemistry and *in situ* hybridization at the electron microscopic level, *Electron Micosc. Rev.*, 4, 85, 1990.
75. **Thiry, M., Schoonbroodt, S., and Goessens, G.,** Cytochemical distinction of various nucleolar components in insect cells, *Bio. Cell,* 72, 133, 1991.
76. **Thiry, M. and Thiry-Blaise, L.,** Locating transcribed and non-transcribed rDNA spacer sequences within the nucleolus by in situ hybridization and immunoelectron microscopy, *Nucleic Acid Res.*, 19, 11, 1990.
77. **Wachtler, F., Mosgöller, W., and Schwarzacher, H. G.,** Electron microscopic *in situ* hybridization and autoradiography: localization and transcription of rDNA in human lymphocyte nucleoli, *Exp. Cell Res.*, 187, 346, 1990.
78. **Gosh, S. and Paweletz, N.,** Localization of ribosomal cistrons at the ultrastructural level by *in situ* hybridization technique, *Cell Biol. Inter. Rep.*, 14, 521, 1990.
79. **Morel, G., Chabot, J.-G., Gossard, F., and Heisler, S.,** Is atrial natriuretic peptide synthesized and internalized by gonadotrophs?, *Endocrinology,* 124, 1703, 1989.
80. **Morel, G., Dihl, F., and Gossard, F.,** Ultrastructural distribution of GH mRNA and GH intron I sequences in rat pituitary gland: effects of GH releasing factor and somatostatin, *Mol. Cell. Endocrinol.*, 65, 81, 1989.
81. **LeGuellec, D., Frappart, L., and Desprez, P. Y.,** Ultrastructural localization of mRNA encoding for the EGF receptor in human breast cell cancer line BT20 by *in situ* hybridization, *J. Histochem. Cytochem.*, 39, 1, 1991.
82. **Pomzroy, M. E., Lawrence, J. B., Singer, R. H., and Billings-Gagliardi,** Distribution of myosin heavy chain mRNA in embryonic muscle tissue visualized by ultrastructural *in situ* hybridization, *Dev. Biol.*, 143, 58, 1991.
83. **Manning, J. E., Hershey, N. D., Broker, T. R., Pellegrini, M., Mitchell, H. K., and Davidson, N.,** A new method of *in situ* hybridization, *Chromosoma,* 53, 107, 1975.
84. **Martineau, B. and Taylor, W. C.,** Cell specific photosynthetic gene expression in maize determined using cell separation. Techniques and *in situ* hybridization, *Plant Physiol.*, 80, 613, 1986.
85. **Aayagi, K. and Chua, N.-H.,** Cell-specific expression of pyruvate, Pidikinase. *In situ* mRNA hybridization and immunolocalization labeling of protein in wheat seed, *Plant Physiol.*, 86, 364, 1988.

Chapter 1

THE MOLECULAR PRINCIPLES OF *IN SITU* HYBRIDIZATION

Chris W. Perrett

TABLE OF CONTENTS

0-8493-4414-X/93/$0.00 + $.50

I. INTRODUCTION

Over the last 20 years several molecular biological techniques have had an increasing impact on biology and medicine. Thus, the development of recombinant DNA technology and sensitive methods for the detection of specific DNA and RNA sequences by molecular hybridization (and, more recently, the polymerase chain reaction) has opened up new fields of research.[1-3] Nucleic acid hybridization has generally been carried out in solution, on solid supports (nitrocellulose or nylon membranes), and on tissue sections or cell preparations (*in situ* hybridization). Hybridization techniques using solid supports involve the immobilization of extracted and electrophoretically separated (or individually cloned) DNA fragments to a membrane, followed by hybridization to a specific DNA probe (dot blotting, Southern blotting).[2] Similar methods can be applied to RNA analysis (Northern blotting).[2] Hybridization can also be performed directly to DNA and RNA in agarose gels, thus speeding up the procedure and making it more sensitive in certain circumstances.[4] However, although such methods provide valuable information on the structure of different classes of DNA and RNA in tissues and cell populations, they reveal little about the distribution of specific sequences in individual cells. *In situ* hybridization is a technique which allows the morphological demonstration of specific DNA/RNA sequences in individual cells in tissue sections, single cells, or chromosome preparations and is an especially important method for studying DNA and RNA in heterogeneous cell populations. This chapter aims to present the general molecular principles underlying this technique.

II. NUCLEIC ACID STRUCTURE

Although a detailed treatment of the chemistry and biochemistry of the nucleic acids is not relevant some of the important features of these biopolymers are summarized here. A more extensive treatment of nucleic acid structure has been covered in several excellent recent texts.[5,6]

DNA and RNA are linear biological polymers consisting of nucleotide monomers. Each nucleotide has three constituent parts:

- A sugar (pentose) structure containing five carbon atoms numbered 1′ to 5′. The sugar is deoxyribose in DNA and ribose (containing a 2′ OH group) in RNA.
- Chemical bases attached to carbon C_1'; the bases consist of single heterocyclic rings (pyrimidines, Py) or fused rings (purines, Pu). Cytosine (C) and thymine (T) (uracil, U) are the pyrimidines in DNA (RNA), guanine (G), and adenine (A) the purines in both DNA and RNA.
- A phosphate group.

Monomers are joined by a phosphodiester bond between C_5' of one nucleotide and C_3' of the preceding one. This makes the polymer asymmetric and gives it direction (from 5′ to 3′; Figure 1A). The phosphate group makes the molecule acidic under physiological conditions and this is neutralized *in vivo* by basic proteins (e.g., histones) and *in vitro* by various cations in the medium (e.g., Na^+, Mg^{2+}). DNA polymers are generally double-stranded *in vivo*. In double-stranded molecules (''duplexes'') the individual strands align in opposite (antiparallel) directions and the structure is given thermodynamic stability by hydrogen bonding between the bases on separate strands, as well as hydrophobic interactions between adjacent bases on the same strand. The bases, which can adopt different tautomeric structures, form hydrogen bonds between the keto and amino forms (Figure 1B); two bonds are formed between A and T (A and U), three between G and C. When such bonding occurs throughout the molecule the strands are said to be ''complementary''. *In vivo* mature RNA molecules generally remain single-stranded although intramolecular hydrogen bonding can occur, giving rise to hairpin loop structures. Under certain conditions *in vitro,* however, complementary RNA:RNA double strands and DNA:RNA hybrids can form and these are thermodynamically more stable than DNA:DNA duplexes (Section VII). Denaturation (''melting'') of the double strands is generally reversible and can be carried out by increasing the temperature or using alkaline conditions (although RNA is hydrolyzed under alkaline conditions). The joining of complementary strands is known as ''hybridization'' or ''re-annealing''. The temperature required to denature duplexes containing G-C sequences is greater than that for the corresponding A-T (A-U) sequence, since an extra hydrogen bond has to be disrupted.

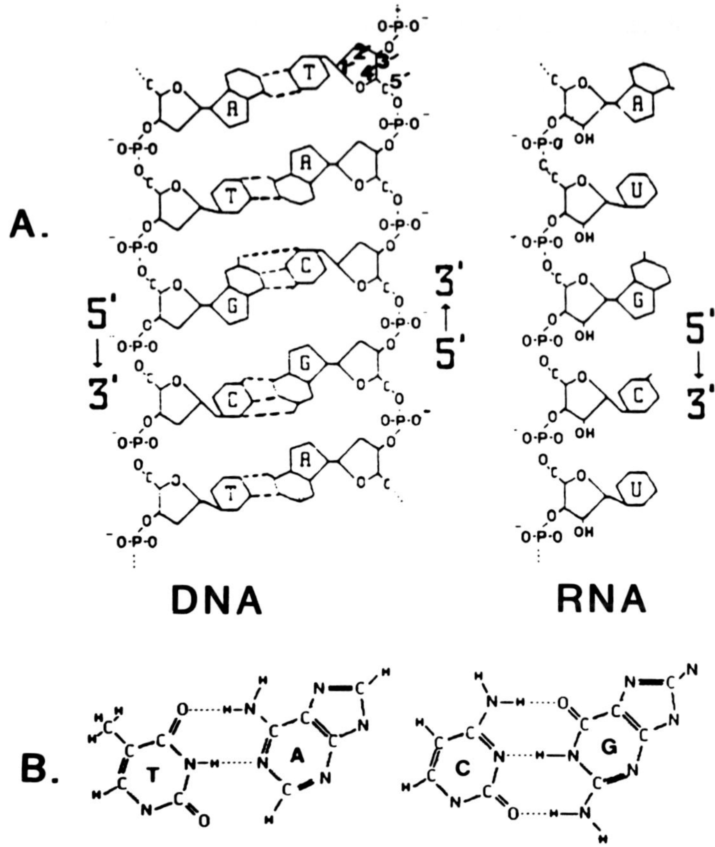

FIGURE 1. The chemical structure of DNA and RNA. A, double-stranded DNA and single-stranded RNA, showing strand direction; B, hydrogen bonding between complementary constituent bases in DNA.

III. MOLECULAR CLONING

A. PRINCIPLES

Molecular cloning is a technique which amplifies an identified DNA segment in large amounts. In a typical gene cloning experiment the DNA sequence to be amplified is either from a complex population of cDNA molecules, generated from each of the respective mRNAs, or genomic DNAs. In cDNA cloning, cDNA copies of mRNAs are synthesized by reverse transcription *in vitro* and then inserted into cloning vehicles known as "vectors" to allow amplification.[2] The vectors themselves possess genetic functions that allow selective amplification in hosts (usually bacteria) and selection of recombinants (i.e., cells containing vector + "foreign" DNA insert). For the cloning of genomic sequences the extracted DNA is usually mechanically sheared or digested with specific enzymes (type II restriction endonucleases, Section IV) to generate suitably sized fragments, which are introduced into the cloning vehicles.[2] After introduction and amplification of recombinants, detection of the desired sequence is achieved using biochemical, hybridization, or immunological methods.[2] There are now several classes of vector in common use, each having properties desirable for specific cloning experiments. They include plasmids, bacteriophages (viruses having bacteria as hosts, e.g., λ and M13), and vectors used for cloning large DNA inserts such as cosmids, bacteriophage Pl, and yeast artificial chromosomes (YACs).[7]

Having prepared a recombinant vector-and-insert combination, a bacterial cell is infected by the recombinant. Amplification proceeds via replication of the recombinant vector in the bacterium and transmission to daughter cells. Bacterial selection of recombinant is then done on the basis of drug resistance, color testing, or the presence of lysed bacteria (''plaques''). The stocks used in *in situ* hybridization experiments are either supplied as transformed bacteria (containing recombinant plasmids) or solutions containing recombinant phage or plasmid DNA. Quite often it is more convenient to transfer the insert from the vector used for cloning to a more suitable vector used for probe generation (Section V). This technique involves removal of the foreign insert from one vector (using a suitable restriction endonuclease) and insertion and joining (using the enzyme DNA ligase) into a second vector; the method is known as subcloning.

B. PLASMID VECTORS

Plasmids are replicating DNA molecules generally found extra-chromosomally in bacteria. They are double-stranded, closed, circular DNA molecules of varying size and contain genes which are often advantageous to the bacterial host, e.g., antibiotic resistance. A widely used plasmid is pBR322 (Figure 2A) which was originally synthesized by combining fragments of several naturally occurring plasmids.[8] pBR322 has several features typical of all useful plasmid cloning vectors:

- The ability to autoreplicate in the bacterial host. This is generally initiated by the host's own replicating enzymes unwinding the DNA at the origin of replication (''ori'') sequence; pBR322 has a copy number of 20 but this may be increased by plasmid amplification.[9]
- The existence of selectable markers. pBR322 contains genes giving resistance to the antibiotics tetracycline and ampicillin. The proteins synthesized from these respective genes (1) prevent tetracycline from entering the cell and (2) catalyze the hydrolysis of the B-lactam ring of ampicillin. Thus, the host bacterium is allowed to grow in the presence of such antibiotics.[10,11]

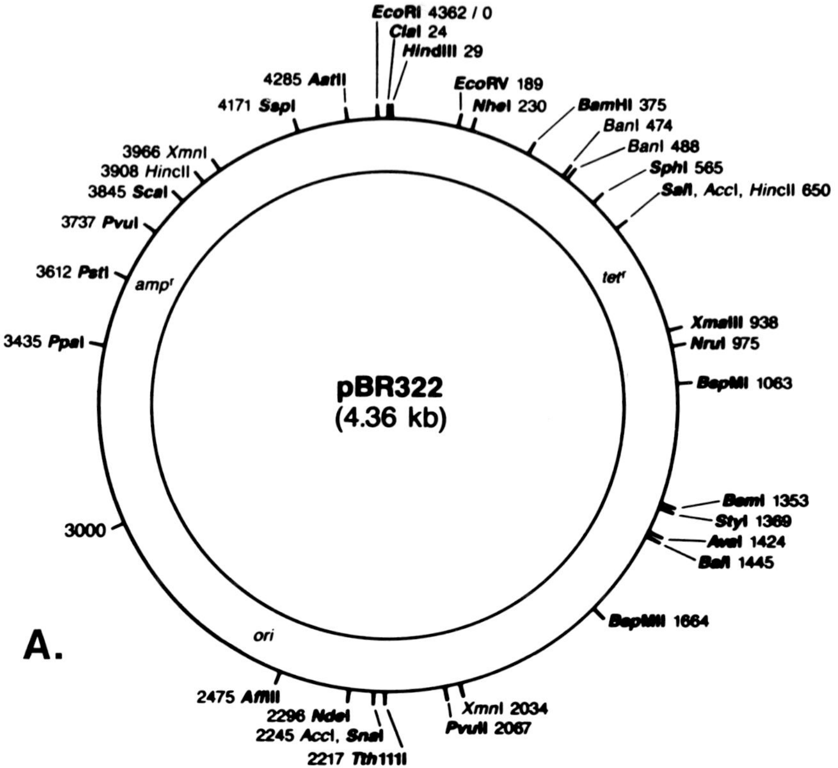

FIGURE 2. Some vectors used in molecular cloning. A, plasmid pBR322; B, plasmids pUC18/19; C, plasmids pGEM-3/4; D, bacteriophages M13mp18/19.

- The ability to select recombinant plasmids. This is done by inserting the foreign DNA into one of the selectable markers, thus disrupting the gene. Such a procedure involves cutting the plasmid DNA at a specific site within the gene using a restriction endonuclease (Section IV) (several such sites are shown in Figure 2A, e.g., *Bam* HI and *Sal* I in the tetracycline resistance gene). The foreign DNA (with compatible ends, Section IV) is now inserted into this cleaved region and covalently linked with DNA ligase. Bacteria harboring recombinant plasmids would be ampicillin resistant but sensitive to tetracycline; those harboring nonrecombinant plasmids would be both ampicillin and tetracycline resistant.
- A known plasmid map and sequence. The sites for restriction endonuclease digestion (''restriction map'') and complete nucleotide sequence are now well established for several useful plasmids (see reference 2 and catalogs of molecular biology suppliers). It is necessary to know this to predict the enzymes required for DNA insertion and removal from the plasmid.

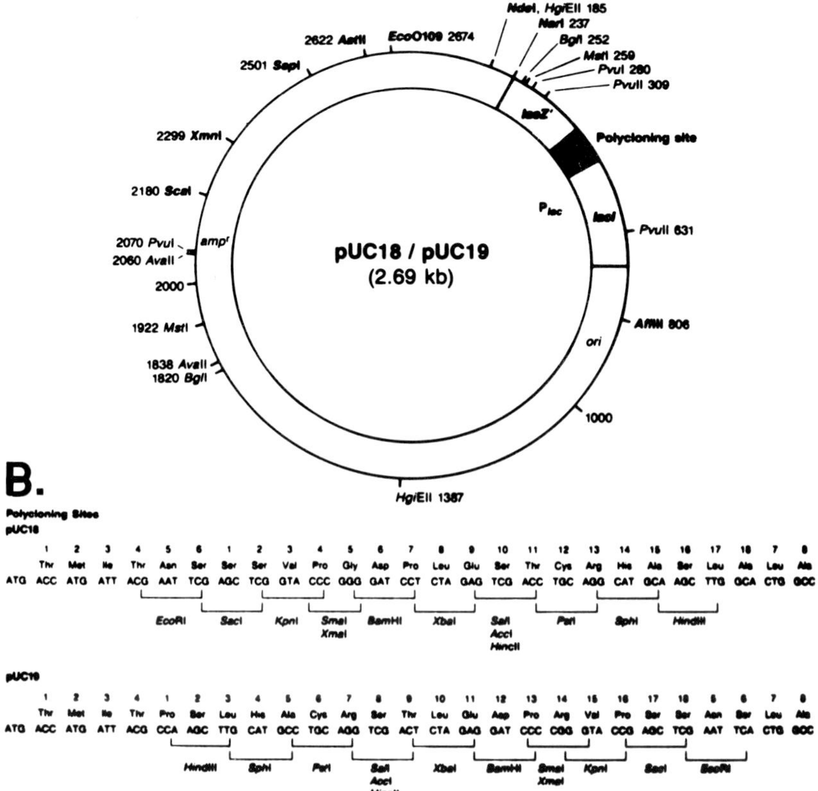

FIGURE 2 (continued).

Most other useful plasmid vectors have been derived from pBR322 by the introduction of functional stretches of DNA which increases the plasmid's versatility as a cloning vehicle.[8] Two examples are shown in Figures 2B and 2C. Plasmids pUC18/pUC19 are similar to pBR322 but carry two additional features. By rearranging the nucleotide sequence, all useful unique restriction endonuclease sites are clustered together in a short segment of the vector (the "polycloning site", "multiple cloning site" (MCS), or "polylinker"). pUC18 and pUC19 differ in the respective orientation of this polycloning site: in PUC18 the *Eco* RI site lies immediately downstream from the lacZ′ gene; in PUC19 it is the *Hin*d III site. Furthermore, the polycloning site, which accepts the foreign DNA fragment, is found in the lacZ′ gene, which synthesizes the amino-terminal fragment of β-galactosidase.[12] Bacteria harboring pUC plasmids are recognized by their ability to metabolize the chromogenic substrate X-gal (5-bromo-4-chloro-3-indolyl-β-D-galactoside) and appear as blue colonies. Bacteria containing recombinant plasmids, with a disrupted lacZ′ gene,

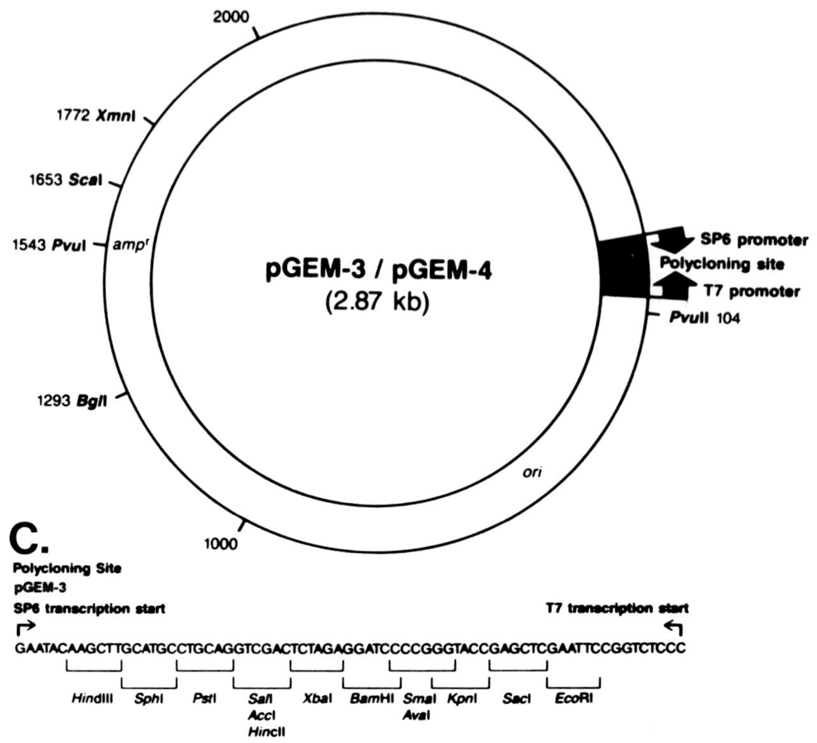

FIGURE 2 (continued).

appear as white colonies. Other plasmid vectors have been designed which are derivatives of pBR322 that carry promoter sequences recognized by specific RNA polymerases, most notably those obtained from bacteriophages T3, T7, and SP6. These promoters are adjacent to the polycloning site and therefore either side of the inserted DNA. As such, these vectors permit the synthesis *in vitro* of RNA strands complementary to one or other of the insert strands (Section V). Examples of such vectors include the pSP series, Bluescript, and Gemini (pGEM) series. Vectors pGEM-3 and pGEM-4 are shown in Figure 2C. They differ only in the orientation of the polycloning site; in pGEM-3 the *Hin*d III site is adjacent to the SP6 promoter, in pGEM-4 it is adjacent to the T7 promoter. Again, by a suitable choice of enzyme, it would be possible to clone a foreign DNA fragment in one particular vector (e.g., pBR322) and then remove it and subclone into a new vector (e.g., pGEM) for production of a specific (RNA) probe.

C. FILAMENTOUS BACTERIOPHAGE VECTORS

The *E. coli*-specific filamentous bacteriophages, in particular M13 and its derivatives, are useful cloning vectors since they can produce single-

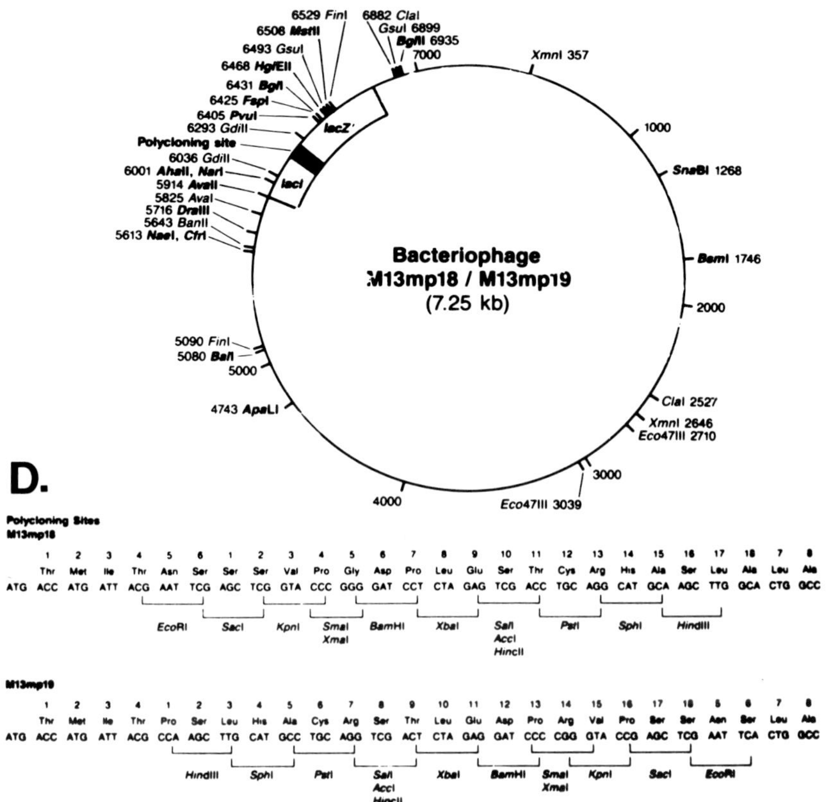

FIGURE 2 (continued).

stranded circular DNA during infection of the host.[13] As a result, by inserting foreign DNA, single-stranded recombinants are produced which can then be used to manufacture probes (Section V). The double-stranded foreign DNA is inserted into the vector when the latter is at the double-stranded ("replicative form") of its life cycle in the host. M13-derived vectors are shown in Figure 2D; again, they differ in the orientation of the polycloning site (with respect to the lacZ' gene).

IV. ENZYMES USED IN DNA MANIPULATION

In recent years, with an increased understanding of the enzymatic reactions of molecular biology, a wide range of enzymes has become available for cutting, synthesizing, and modifying DNA and RNA sequences.

A. RESTRICTION AND DNA METHYLATION ENZYMES

Type II restriction/modification enzymes are binary systems consisting of a restriction endonuclease that cleaves a specific DNA sequence at the phosphodiester linkage and a separate methylase that modifies the same sequence. A large number of type II restriction endonucleases/enzymes has now been isolated (see references 14, 15, and the catalogs of most molecular biology suppliers). The vast majority of such enzymes recognize specific sequences that are four, five, or six nucleotides in length and show a twofold axis of rotational symmetry. These enzymes are used *in vivo* by bacteria as a defense against viral DNA; the bacterial DNA is modified by the endogeneous methylase and is therefore protected. Restriction enzymes are generally named after the host organism from which they have been isolated, e.g., *Escherichia coli* (strain R) = *Eco* R, *Thermus aquaticus* = *Taq*. All type II restriction enzymes require Mg^{2+} in order to function although each differs in its NaCl requirement. A single buffer (potassium glutamate buffer — KGB) has now been devised in which virtually all restriction enzymes work.[16] The incubation temperature is generally 37°C unless the enzyme has been derived from an unusual source. Thus, *Taq* I is obtained from bacteria found in hot springs and functions optimally at 65°C.

1. Restriction Site

The restriction site is often indicated by the 5′ to 3′ sequence of nucleotides recognized by the enzyme on one strand of DNA. Thus, *Eco* RI recognizes GAATTC, *Fin* I GTCCC and *Xho* II PuGATCPy. The full sequences would therefore be

5′ ..... GAATTC ..... 3′ 3′ ..... CTTAAG ..... 5′	5′ ..... GTCCC ..... 3′ 3′ ..... CAGGG ..... 5′	5′ ..... PuGATCPy ..... 3′ 3′ ..... PyCTAGPu ..... 5′
Eco RI	*Fin* I	*Xho* II

Most sites have a twofold axis of rotational symmetry (e.g., *Eco* RI) but not all (e.g., *Fin* I). Sites for such enzymes as Xho II may or may not have such symmetry, depending upon the base sequence (e.g., AGATCT does, GGATCT does not). The location of a restriction enzyme site is now well established for several vector sequences[2] and for cloned genomic and cDNA sequences on computer packages such as the Wisconsin package.[17]

Cutting frequency is a function of the number of constituent bases. A tetranucleotide sequence should occur once every $4^4 = 256$ nucleotides, a hexanucleotide sequence every $4^6 = 4096$ nucleotides, and an octanucleotide sequence (recognized by "rare cutting" enzymes such as *Not* I, GCGGCCGC) once every 65,536. As such these latter enzymes are used to generate large DNA fragments in genome analysis.[1] The above calculations assume that the nucleotides are present in equal amounts (GC = 50%) and ordered in a random fashion. In practice neither of these assumptions is valid so that fragment size for a particular restriction enzyme and DNA segment can vary greatly.

2. Cutting and Joining DNA

In molecular cloning and subcloning experiments restriction enzymes are used to specifically cleave vector DNA at unique sites prior to the insertion of foreign DNA.

a. Compatible Cohesive Ends

Some restriction enzymes cleave each strand of DNA on the 5′ side of the axis of dyad symmetry. The resulting staggered break creates DNA fragments carrying protruding 5′ cohesive termini. The phosphodiester cleavage always leaves the phosphate group on the 5′ end of the DNA fragment:

$$\begin{array}{lcl} \downarrow & & \\ 5' \ldots\ldots \mathrm{GpApApTpTpC} \ldots\ldots 3' & \xrightarrow{\textit{Eco}\ \mathrm{RI}} & 5' \ldots\ldots \mathrm{G_{OH}}3' \\ 3' \ldots\ldots \mathrm{CpTpTpApApG} \ldots\ldots 5' & & 3' \ldots\ldots \mathrm{CpTpTpApAp}5' \\ \phantom{3' \ldots\ldots \mathrm{CpTpTpApAp}}\uparrow & & + \ 5'\mathrm{pApApTpTpC} \ldots\ldots 3' \\ & & 3'_{\mathrm{OH}}\mathrm{G} \ldots\ldots 5' \end{array}$$

During digestion hydrogen bonding between cohesive termini is not favored and the two fragments separate. However, if incubated under conditions which favor hydrogen bonding re-annealing can occur and the phosphodiester bonds can be resealed with a DNA ligase (e.g., that derived from bacteriophage T4).

$$\begin{array}{lcr} 5' \ldots\ldots \mathrm{G_{OH}}3' & & 5'\mathrm{pApApTpTpC} \ldots\ldots 3' \\ & + & \\ 3' \ldots\ldots \mathrm{CpTpTpApAp}5' & & 3'_{\mathrm{HO}}\mathrm{G} \ldots\ldots 5' \end{array}$$

$$\xrightarrow[\text{ligase}]{\text{DNA}} \quad \begin{array}{l} 5' \ldots\ldots \mathrm{GpApApTpTpC} \ldots\ldots 3' \\ 3' \ldots\ldots \mathrm{CpTpTpApApG} \ldots\ldots 5' \end{array}$$

Since all DNA fragments created by *Eco* RI cleavage carry the same 5′ protruding ends, they can be joined in novel combinations, e.g., joining compatible vector and foreign DNA ends. Other enzymes cleave each DNA strand on the 3′ side of the axis of symmetry. This creates protruding 3′ termini which can be similarly resealed with DNA ligase:

```
                ↓
5′ ..... CpTpGpCpApG ..... 3′   Pst I   5′ ..... CpTpGpCpA_OH3′
                               ──────→
3′ ..... GpApCpGpTpC ..... 5′           3′ ..... Gp5′
         ↑
                                                   5′pG ..... 3′
                                  +
                                        3′_HOApCpGpTpC ..... 5′
```

Again, different fragments carrying *Pst* I cohesive ends can be rejoined in a similar fashion during cloning.

If both DNA strands are cleaved at the axis of symmetry, "blunt" ("flush")-ended fragments are produced:

```
           ↓
5′ ..... GpGpCpC ..... 3′   Hae III  5′ ..... GpG_OH3′        5′pCpC ..... 3′
                            ──────→                      +
3′ ..... CpCpGpG ..... 5′            3′ ..... CpCp5′       3′_HOGpG ..... 5′
           ↑
```

Blunt ends from differently generated fragments can be rejoined by DNA ligase, although the reaction efficiency is lower than joining cohesive ends.

Finally, not all restriction enzymes cleave within the recognition sequence, some cleave at a distance away from this sequence:

```
                                    ↓
5′ ..... GpApCpGpCpNpNpNpNpNpN .................. 3′
3′ ..... CpTpGpCpGpNpNpNpNpNpNpNpNpNpNpN .... 5′
                                            ↑
```

Hga I (5/10) (N = A,G,C, or T)

b. Isochizomers

In general, different restriction enzymes recognize different sequences although some enzymes, isolated from different sources, cleave within the same target sequences; these are known as "isoschizomers" (e.g., *Bam* HI and *Bst* I both recognize the sequence GGATCC). In addition, several enzymes have been discovered which recognize tetranucleotide sequences within the hexanucleotide sequences of other enzymes. Thus, *Sau* 3AI recognizes the sequence ↓GATC whereas *Bam* HI recognizes the sequence G↓GATCC. In each case an identical protruding tetranucleotide terminus is created. Thus, DNA fragments obtained by digestion with *Sau* 3AI can be joined to vector DNA cleaved with *Bam* HI (e.g., pBR322, having a *Bam* HI site in the tetracycline resistance gene; Figure 2A). However, only one in every sixteen

Sau 3AI/*Bam* HI fusions would regenerate a *Bam* HI site, so the DNA fragment would be more efficiently recovered by *Sau* 3AI digestion. Some restriction enzymes are sensitive to base methylation within the target sequence. Thus, *Mbo* I also recognizes the sequence GATC but does not cleave when A is methylated (by the *dam* methylase of *E. coli*). In order to use this enzyme, mammalian DNA should be used as the substrate, since A is not methylated in this position, or the molecular cloning should be done in *dam*$^-$ *E. coli*. *Sau* 3AI is insensitive to the methylation of the GATC sequence but *Dpn I*, which also recognizes GATC, unusually requires methylation in order to function. In a few cases, ligation of fragments generated by one restriction enzyme to those generated by a second enzyme results in hybrids that are recognized by neither of the original enzymes. Thus, when *Sal* I fragments (G↓TCGAC) are ligated to *Xho* I fragments (C↓TCGAG) the resulting target is not cleaved by either enzyme:

$$\begin{array}{lcr} 5' \ldots\ldots G_{OH}3' & & 5'pTpCpGpApG \ldots\ldots 3' \\ & + & \\ 3' \ldots\ldots CpApGpCpTp5' & & 3'_{HO}C \ldots\ldots 5' \end{array}$$

$$\longrightarrow \quad \begin{array}{l} 5' \ldots\ldots GpTpCpGpApG \ldots\ldots 3' \\ 3' \ldots\ldots CpApGpCpTpC \ldots\ldots 5' \end{array}$$

B. DNA/RNA POLYMERASES AND MODIFYING ENZYMES

The synthesis of DNA or RNA using labeled nucleotides is the general technique for making DNA and RNA probes. DNA polymerases are used in a variety of reactions for preparing labeled DNA probes, including nick translation, primer extension, end filling, and end labeling reactions. RNA polymerases are used to generate labeled single-stranded RNA probes. By using strong promoters and stable, purified RNA polymerases several cycles of transcription can occur from the same DNA template (often eight moles of transcript/mole of template) resulting in the production of large amounts of probe (up to 10 μg RNA/μg plasmid). Three RNA polymerases have been used extensively (Table 1), obtained from the *Salmonella typhimurium* bacteriophage SP6 and *E. coli* bacteriophages T3 and T7.

The holoenzyme of *E. coli* DNA polymerase I is used extensively in nick translation (Section V) due to its polymerase and 5′ to 3′ exonuclease activity. Proteolytic cleavage of the holoenzyme removes this latter activity and creates the Klenow (large) fragment, which contains polymerase and 3′ to 5′ exonuclease activity. As such the Klenow fragment is used in reactions such as specific and random primer extension, end filling (where staggered ends created by restriction enzyme digestion are "filled in" with labeled nucleotides to produce blunt ends) and replacement synthesis (where nucleotides are

TABLE 1
Some Properties of Useful DNA and RNA Polymerases

Enzyme	5′ → 3′ Polymerase activity	5′ → 3′ Exonuclease activity	3′ → 5′ Exonuclease activity	Applications	Ref.
E. coli DNA polymerase I (holoenzyme)	ss[a] DNA template + DNA/RNA primer with 3′OH	ds DNA or RNA/DNA hybrids, proceeds from 5′p-end	ds DNA/ss DNA with 3′OH; blocked by polymerase activity	Nick translation	18
Klenow fragment	Same	None	Same	Primer extension, end filling, replacement synthesis	19
T4 DNA polymerase	Same	None	200 × activity of Klenow fragment	End filling, replacement synthesis	20
T7 DNA polymerase	Same	None	1000 × activity of Klenow fragment	Primer extension, end filling, replacement synthesis	21
Taq DNA polymerase	Same	ds DNA from 5′p-end	None	Polymerase chain reaction	22
Terminal transferase	ss DNA substrate, ds DNA with protruding 3′OH	None	None	3′ End labeling	23
SP6 RNA polymerase	ds DNA with specific promoter	None	None	ss RNA probes	24
T3 RNA polymerase	Same	None	None	Same	25
T7 RNA polymerase	Same	None	None	Same	26

[a] Abbreviations used: ss, single-stranded; ds, double-stranded.

removed by the 3′ to 5′ exonuclease activity of the enzyme and replaced by polymerase action with labeled nucleotide). Since both bacteriophage T4 and T7 DNA polymerases have greater 3′ to 5′ exonuclease activity compared to the Klenow fragment they are often used in replacement synthesis reactions. Other polymerases (Table 1) include terminal transferase which, as it requires no primer, is used to (3′) end-label single-stranded oligonucleotides and *Taq* polymerase, a heat-stable polymerase used in the polymerase chain reaction (PCR; Section V).

Other useful DNA-modifying enzymes used in probe preparation include the endonuclease DNase I, which introduces random nicks into DNA (and is used during "nick translation") and bacteriophage T4 polynucleotide kinase,

TABLE 2
Some Properties of Useful DNA Modifying Enzymes

Enzyme	Property	Applications	Ref.
DNase I	Degrades ss[a]DNA/dsDNA by removing 5′p-oligonucleotides (1–4) having cut 5′ to Py, creating "nicks"	Nick translation	27
T4 Polynucleotide kinase	Phosphorylates 5′OH group of ds DNA/ss DNA/RNA	5′ End-labeling	28

[a] Abbreviations used: ss, single-stranded; ds, double-stranded; Py, pyrimidine.

which is used to (5′) end-label nucleic acid sequences, in particular oligonucleotides. These are shown in Table 2.

V. CONSTRUCTION OF PROBES

Both DNA and RNA probes can be used to localize DNA and mRNA during *in situ* hybridization. The optimal probe length to ensure efficient tissue penetration and hybridization is 50 to 300 nucleotides, although shorter oligonucleotide probes are now being increasingly used. For *in situ* chromosomal hybridization, longer probes are preferred.

A. DNA PROBES

1. Double-Stranded DNA Probes

Labeled double-stranded DNA probes are used extensively in *in situ* hybridization studies. The probes are biologically stable (since DNases, which require Mg^{2+} to function, can be efficiently inhibited by EDTA), although hybridization is limited by complementary strands re-annealing during the hybridization reaction. Such probes are usually labeled by nick translation or random primer extension and must be denatured before use (generally heat-denatured since alkaline denaturation will destroy some, e.g., biotinylated, nucleotides).

a. Nick Translation

This method utilizes the combined action of DNA polymerase I (holoenzyme) and DNase I.[29] DNA polymerase I adds labeled nucleotides to the 3′OH end created by DNase I "nicking" and then, using its 5′ to 3′ exonuclease activity, removes nucleotides from the 5′ side of the nick (adding them to the 3′ side of the nick). This results in a movement sideways of the nick (nick translation) along the DNA (Figure 3). Using highly radioactive nucleotides (Section VI), it is possible to prepare labeled DNA of specific activity 5×10^8 cpm/μg probe. Nick translation can be performed using either whole plasmid or insert alone. If the insert is to be labeled it is excised

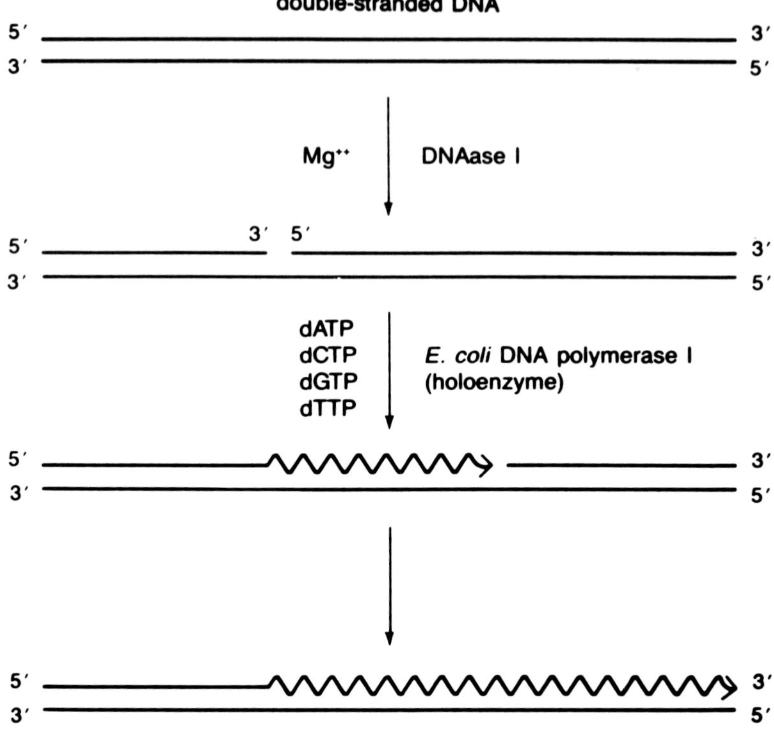

FIGURE 3. Labeling double-stranded DNA by nick translation.

from the plasmid (using the appropriate restriction enzymes) and separated and purified using agarose or polyacrylamide gels.[2] In heterologous probing, where probe and sequence to be identified come from different species, insert labeling is generally preferred since the stringency of the hybridization conditions (Section VII) often has to be lowered to identify such sequences (owing to base mismatching) and this may result in increased background labeling due to plasmid sequences.

b. Random Primer Extension ("Random Priming")

Primer extension reactions using random heterogeneous sequences of primers containing 6 to 12 nucleotides are used for labeling double-stranded probes to high specific activity (5×10^9 cpm/μg probe).[30] It is necessary to linearize the recombinant DNA at a remote restriction site to the insert and then denature. The primers will form hybrids at many sites along the linearized strands and, as a result, most nucleotides of the template sequences will be copied (except the 5′ terminus), using the Klenow fragment as polymerase. This technique has the advantages over nick translation of high probe specific activity and the requirement for small amounts of probe (10 to 50 ng minimum). However, since only a little product is made during each reaction the technique is not generally used for *in situ* hybridization studies where large numbers of tissue sections, requiring high amounts of probe, are to be analyzed. As the insert is not usually removed from the plasmid hybridization conditions should be adjusted to prevent background labeling. The probe length by random priming is generally comparable to that produced by nick translation.[2]

2. Single-Stranded DNA Probes

Production of single-stranded DNA probes has become possible with the use of M13 cloning vectors[13] and, recently, PCR.[2] They have the advantage that probe reassociation is not a drawback during hybridization.

a. Use of M13 Vectors

Having created a single-stranded vector containing foreign DNA insert, the single-stranded probe is synthesized as the complementary strand to the insert using the Klenow fragment. A primer is chosen (the "universal" primer) which is complementary to the vector sequence adjacent to the polylinker; the most commonly chosen region is the lacZ′ gene region in M13mp vectors (Figure 2D).[13] Probe specific activity generated in this way is generally 10^9 cpm/μg probe. Although these probes would be well suited to *in situ* hybridization experiments, they have seen restricted use, probably because probe synthesis requires the tedious separation of newly synthesized probes from unlabeled template DNA by polyacrylamide gel electrophoresis.[2] In an alternative procedure, complementary labeled nucleotides are synthesized on vector template, leaving insert single-stranded. In this case the probe can be used directly without separation.[31]

b. Polymerase Chain Reaction (PCR)

PCR is a technique for amplifying target genomic DNA or cDNA sequences, including those derived from paraffin sections, a millionfold.[2,22,32,33] The technique utilizes several rounds (usually 30) of denaturation of the double-stranded target sequence, annealing of oligonucleotide primers to known flanking sequences of the target DNA and primer extension using the heat-stable *Taq* polymerase (Figure 4). Although generally used to amplify double-stranded DNA, by suitably adjusting the primer ratio, it is possible to produce amplified labeled single-stranded target DNA of high specific activity.[34] The technique will be used increasingly to produce single-stranded probes from foreign DNA in vector sequences by choosing primers flanking the insert. It also has the advantage that vector sequences are not amplified.

B. SINGLE-STRANDED RNA PROBES

Single-stranded RNA probes ("riboprobes") are constructed from vectors containing SP6, T3, or T7 polymerase promoters (Figure 5). The vector is

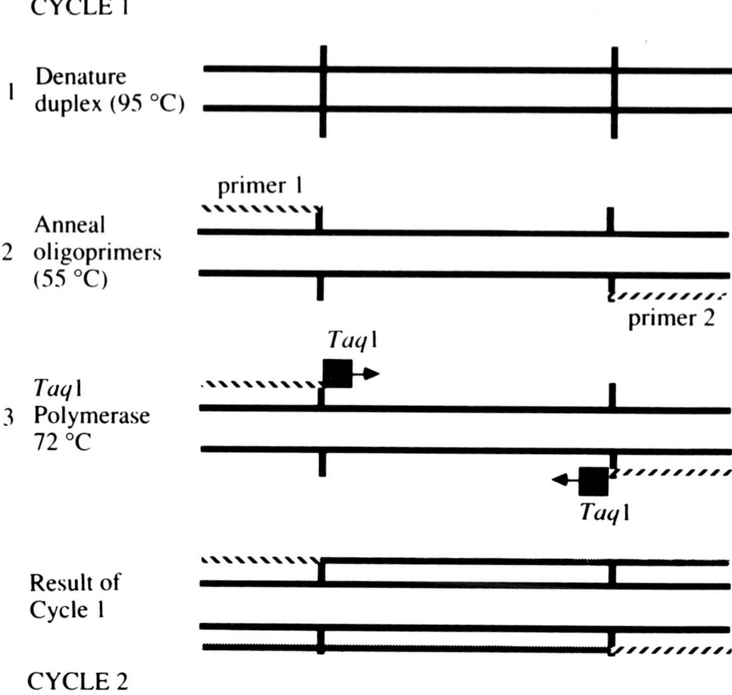

FIGURE 4. One cycle of the polymerase chain reaction. Usually 30 to 40 cycles are used.

linearized (in the multiple cloning site) at a site distal to the insert and chosen promoter, so that probes of defined length, containing no vector sequences, are transcribed. The correct orientation of the insert in the vector must be known. This can be deduced from an analysis of the restriction map of recombinant vector on agarose gels. Once orientation is known a suitable RNA polymerase can be chosen to generate labeled "sense" RNA probes (equivalent in nucleotide sequence to mRNA) and "antisense" (cRNA) probes from the appropriate promoter. The cRNA probes are used to detect nucleic acid sequences during *in situ* hybridization, and sense RNA probes are suitable negative controls. Riboprobes have a number of advantages over double-stranded DNA probes: they generally do not require denaturation before use, as they are single-stranded (although probe denaturation is often routinely done to remove secondary structure caused by intramolecular base-pairing), and, as there is no risk of re-annealing in solution, a comparatively greater amount of probe is available for hybridization (compared to double-stranded probes), thus resulting in strong signals.[35] This is especially true since cRNA hybrids are more stable than DNA: DNA duplexes (Section VII). However,

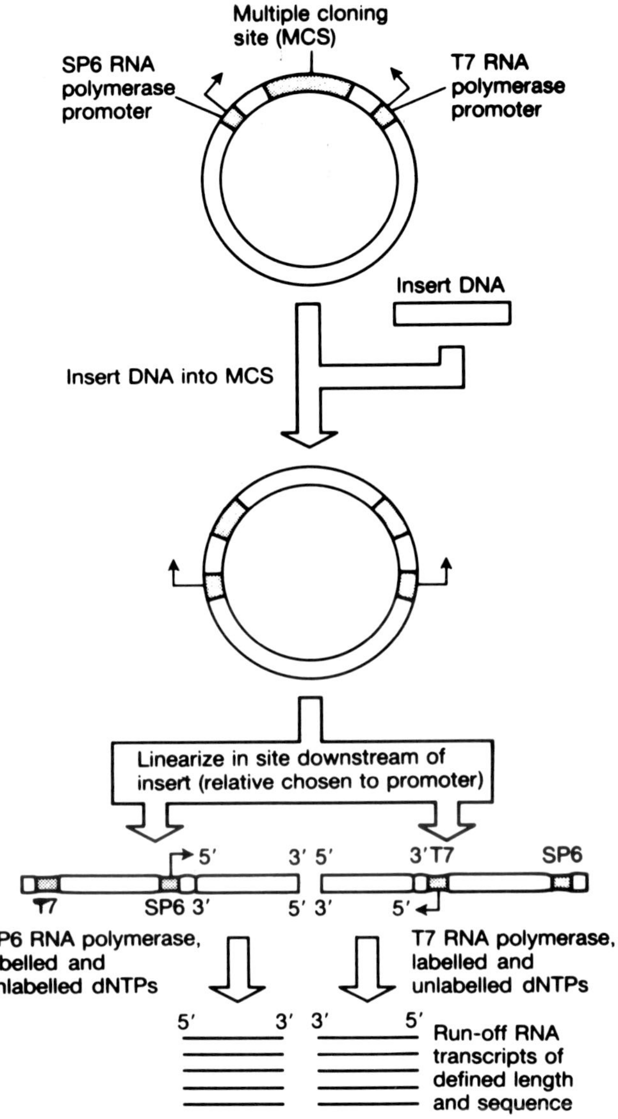

FIGURE 5. Construction of single-stranded sense and antisense (cRNA) probes using SP6 and T7 polymerases and a plasmid vector.

such factors may also result in the strong, nonspecific background often found with cRNA probes; this can be circumvented by the use of ribonuclease A in the washing steps. Ribonuclease A cleaves single-stranded RNA but not hybrids.[36] Because of this, RNA probes should be prepared in RNase-free conditions, using placental RNase inhibitor (1000 units/ml) in the incubation mix.

C. OLIGONUCLEOTIDE PROBES

1. Advantages of Oligonucleotide Probes

The use of oligonucleotide probes is limited to cases where the DNA, RNA (or amino acid) sequence is known. As well as being conveniently prepared using oligonucleotide synthesizers, without recourse to molecular cloning, labeled oligonucleotides have several advantages over some conventional probes:

- They are single-stranded, so there is no competitive hybridization in solution.
- They are small and therefore effectively penetrate tissues.
- Labeling (with polynucleotide kinase or terminal transferase) only requires a small amount of oligonucleotide (10 pmoles of a 20 base sequence, equivalent to about 50 ng). However, although the specific activity of such probes can be as high as the labeled nucleotide itself, typical specific activities are 10^6 cpm/μg probe.
- Using a suitable choice of probes it is possible to discriminate between the sequences of structurally similar genes.[37]
- Hybridization specificity can be confirmed by constructing probes complementary to different regions of the same target sequence (although, in some cases, not all target sequences have been shown to be equivalent for probing).[38]

2. Oligonucleotide Probe Design

A primary requirement for any oligonucleotide probe is that it must hybridize specifically to the target sequence and not to unrelated sequences. Statistically, shorter probes have an increased probability of finding perfect matches, by chance alone, in a given target genome. The K value for a given probe is defined as the number of perfect matches expected in a random assortment of nucleotides of length C, when the probe and target have a G + C content of 50%:

$$K = (0.25)^L \cdot 2C \tag{1}$$

where L is the probe length. As the oligonucleotide length increases, the chance that it will find an exact match decreases. For a probe to be useful the K value must certainly not exceed 1 and should at the most be 0.1. Thus, C for an mRNA population (or cDNA library) is approximately 10^7 and for a mammalian genome 10^9. From these values the minimum probe length can be computed to be 14 and 17 nucleotides, respectively. However, the distribution of nucleotides in coding sequences is random, and, as a result, longer oligonucleotides are usually used to increase the specificity of hybridization.[39] By increasing the length of the probe the concomitant increase in mismatches which occurs will not be significant, since mispaired sequences are generally

TABLE 3
Some Merits of Radioactive and Nonradioactive Probes

Label	Resolution	Sensitivity	Exposure (d)	Stability (weeks)
^{32}P	+	+ + +	3–7	0.14[a]–0.5
^{35}S	+ +	+ +	10	6
^{3}H	+ + +	+	14	>30
Biotin	+ + +	+ +	0.16	>52
Digoxigenin	+ + +	+ +	0.16	>52

[a] For RNA probes; Reference 42.

unstable in short oligonucleotides (less than 20). In longer oligonucleotides, however, or in short oligonucleotides where the mismatch occurs at the end of the molecule, hybridization may well occur under the experimental conditions employed.[40] Thus, the generally chosen length for oligonucleotide probes is 16 (for mRNA hybridization) and 20 (for hybridization against genomic DNA). Recently, oligonucleotide probe design has been extended further by inserting 20 to 70 nucleotide sequences into riboprobe vectors to synthesize cRNA oligonucleotides; these have been used successfully in *in situ* hybridization experiments.[41]

VI. RADIOACTIVE AND NONRADIOACTIVE LABELS

DNA and RNA probes can be modified directly by chemical means, introducing haptens such as biotin and digoxigenin into the nucleotide and producing a nonradioactive probe. Alternatively, labeled nucleotides (radioactive or nonradioactive) can be introduced by enzymatic methods. Radioactive probes have the advantages that they generally have a structure close to the natural structure of the "cold" substrate for the enzyme (whereas use of haptenized probes may radically change synthesis and hybridization conditions), and they can be labeled to high specific activity (and the efficiency of probe synthesis can be easily monitored[2]); this therefore results in sensitive autoradiographic detection methods. However, such probes suffer from the disadvantages of safety problems and waste disposal, reduced probe stability, and speed of visualization. Some of the merits of using radioactive and nonradioactive probes are shown in Table 3.

A. RADIOACTIVE PROBES

All the radioactive probes used for *in situ* hybridization are β-emitters and detection is by the aggregation of silver grains in emulsions or films. This aggregation is either caused directly by radiation, or indirectly by secondary emission from scintillants (in the case of weak β-emitters) or from intensifying screens.[2] Radioactive precursors have a specific activity in Ci/mmol or Bq/mmol where:

$$1\text{Ci} = 2.22 \times 10^{12}\ \text{dpm} = 37\ \text{GBq} \tag{2}$$

Some of the commonly used radionuclides are considered below:

1. *Phosphorus-32*. A high energy β-emitter (1.71 MeV) with relatively low half-life (14.3 days) produced commercially from ^{32}S. The maximum specify activity of a solution "carrier-free" of ^{32}P (where all phosphorus atoms are radioactive) is 338 TBq/mmol; so the maximum commercial preparations (222 TBq/mmol, 6000 Ci/mmol) correspond to two radioactive nucleotides in three. Radioactive nucleotides can be ^{32}P-labeled at α- or γ-phosphate groups (Figure 6A). [α-^{32}P]-labeled nucleotides are used in the *in vitro* synthesis of DNA and RNA probes and the 3′-end-labeling of oligonucleotides. If only 1 type of radioactive nucleotide is used out of the 4 precursors, this corresponds to 2 labels in 12 and a maximum specific activity of 5×10^9 cpm/μg (using 222 TBq/mmol nucleotides). Such labeling is achieved with random primed DNA probes. Thus, ^{32}P labeling is very sensitive (detection limit, with intensifying screen, 50 dpm/cm^2) although the isotope suffers from poor resolution (due to the high energy and range of the β-particles) as well as from probe decay (to ^{32}S) and radioactive covalent bond breakage. [γ-^{32}P]-labeled nucleotides are used for 5′ end-labeling of oligonucleotides.
2. *Phosphorus-33*. An attractive alternative to ^{32}P because of lower energy (0.25 MeV), and therefore greater resolution, and longer half-life (28 d). This isotope has not yet been used widely because of its expensive commerical production (from ^{33}S).
3. *Sulfur-35*. A weaker β-emitter than ^{32}P (0.167 MeV) and therefore less sensitive (detection limit 400 dpm/cm^2), but greater autoradiographic resolution and, as the half-life is longer (87.4 d), the probes are more stable. The isotope is prepared efficiently from ^{35}Cl and this results in high specific activity. In ^{35}S-labeled precursors, the radioactive S atom is substituted for O in the α- or γ-phosphate position (Figure 6B) and radioactive disintegration of S into Cl does not disrupt this bond. The maximum specific activity of ^{35}S-containing nucleotides is 55 TBq/mmol, and commercially available preparations have two labeled nucleotides in three (37 TBq/mmol, 1500 Ci/mmol). The wide utility and availability of ^{35}S-labeled nucleotides has been limited by difficult preparation chemistry, although new preparation procedures may improve this situation.[43]

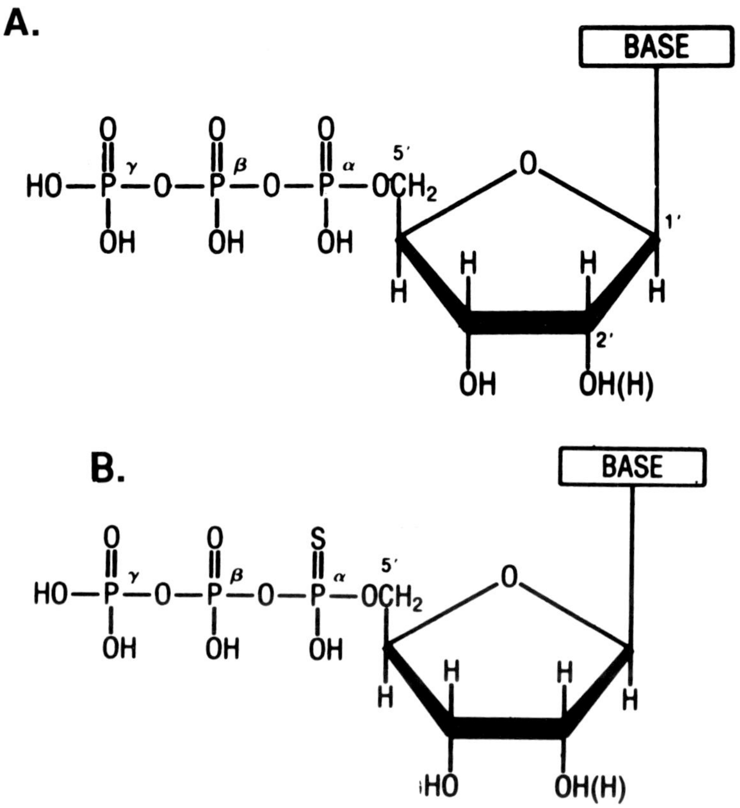

FIGURE 6. Radioactive and nonradioactive precursors used for probe preparation. A, ribo/deoxyribonucleotide triphosphate; B, [α-^{35}S]-ribo/deoxyribonucleotide; C, 5-iodocytidine (5-iododeoxycytidine) triphosphate; D, biotinylated hapten (X = 5 to 18 C atoms); E, digoxigenin hapten.

4. *Iodine-125.* This isotope can only be introduced into pyrimidines where it is added to the C_5 ring position (Figure 6C) with little effect on hybridization.[44] The isotope has low energy (0.035 MeV) and relatively long half-life (60 d), but the general use of ^{125}I has been limited by nucleotide availability.
5. *Hydrogen-3 (Tritium).* Tritium is very stable, having a long half-life (12.35 years), and the emission of weak β-particles (0.018 MeV) results in the best resolution. ^{3}H can be substituted in one or several hydrogen atoms in both the base and pentose ring. Commercially available nucleotides have from one (e.g., [5-^{3}H] UTP) to four ([1′, 2′, 2, 8-^{3}H] dATP) labeled atoms and this results in specific activities of 1 to 4 TBq/mmol. The drawback of the isotope is the long exposure time and low sensitivity (detection limit 8000 dpm/cm^2 by fluorography).

C.

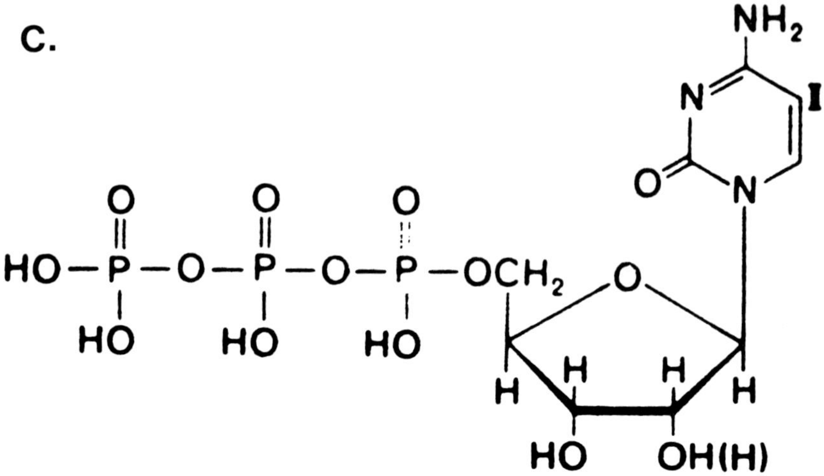

FIGURE 6 (continued).

D.

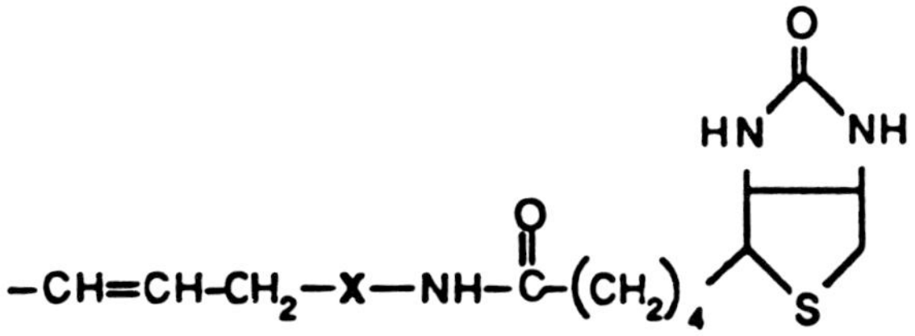

FIGURE 6 (continued).

B. NONRADIOACTIVE PROBES

Nonradioactive probes have chemical structures (haptens) inserted into the base for detection after hybridization. There are now several nonradioactive probes available, including mercurated and sulfonated probes.[45] However, the two in widest use are probes labeled with biotin and digoxigenin.

1. Biotinylated Probes

Biotin is a vitamin (Figure 6D). It can be enzymatically introduced into the DNA or RNA probe using biotinylated precursor nucleotides, including dUTP-7-biotin, dUTP-11-biotin, and UTP-21-biotin. The biotin is attached to C_5 of the uracil ring. The numbering system reflects the number of carbon atoms in the spacer joining the base to the biotin. The function of the spacer is to limit the steric hindrance of the nucleotide during the enzymatic preparation of the probe and to allow biotin access to its affinity molecule. Alternatively, biotin can be chemically introduced into the probe using reactive

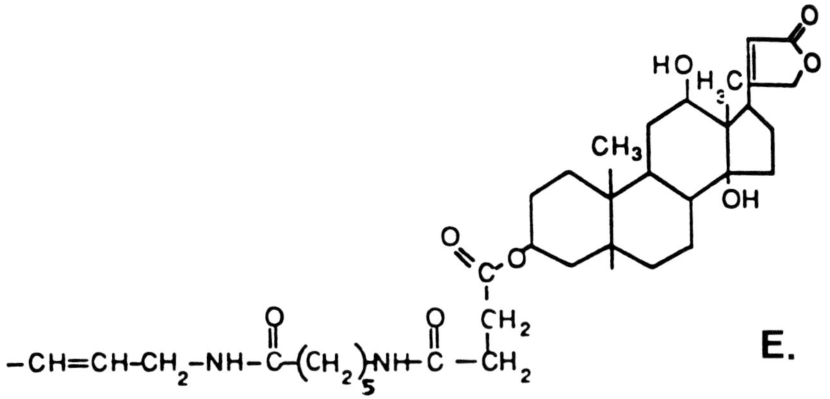

FIGURE 6 (continued).

biotin analogs such as photobiotin.[46] After hybridization, biotinylated probes can be detected either by using labeled monoclonal or polyclonal antibodies to biotin or, more usually, using streptavidin. Avidin and streptavidin both have an exceedingly high affinity for their natural ligand, biotin (Kd $= 10^{-15}$, equivalent to a covalent bond). Bound streptavidin can be detected by using attached signal molecules; these include visible markers (e.g., alkaline phosphatase), fluorescent markers (e.g., peroxidase), or colloidal gold.

2. Digoxigenin-Labeled Probes

Digoxigenin, a steroid extract, can be similarly attached to precursor nucleotides to make labeled probes (Figure 6E). The nonradioactive probe can be detected by signaled antibody binding.

VII. NUCLEIC ACID HYBRIDIZATION

Hybridization is the joining of complementary strands of DNA:DNA, DNA:RNA, or cRNA:RNA. In *in situ* hybridization experiments, unlike immunohistochemistry, the hybridization conditions can be adjusted to suit the experiment. Thus only complementary sequences will remain stable (''high stringency'') or, by suitable adjustment, mismatched sequences can be tolerated (''low stringency''). The factors which determine the extent of hybridization reactions are hybrid stability and hybridization kinetics.

A. HYBRID STABILITY

Hybrid stability is generally regulated by the cation concentration and the temperature, measured by the melting temperature (Tm). Tm is the temperature at which 50% of the duplex is single-stranded. It is dependent upon various factors including: the proportion of G + C in the probe, the length of probe (L), the concentration of monovalent (M) cations (usually Na^+), and the amount of formamide in the reaction mixture (%F). For DNA in solution the equation tying up these factors is

$$Tm = 81.5°C + 16.6\log_{10} M + 0.41\ (\%\ G + C) - 0.63\ (\%F)\ \frac{-600}{L} \qquad (3)$$

for probes of greater than 100 nucleotides.[47]

The equation is valid for [M] = 0.01 to 0.4 *M* and for DNAs whose GC content is 30 to 70%. It predicts the Tm less accurately in solutions of higher monovalent ionic strength. For mismatched double-stranded DNA in solution the Tm decreases by 1 to 1.5°C for every 1% decrease in homology (complementary bases) although it is sequence dependent, since GC-rich mismatches reduce the Tm by a greater amount compared to AT-rich sequences.[48] The mismatch effect is also dependent upon the position of mismatches, since mismatches all in one region will not have such a marked effect compared to alternative base mismatches.[49] Similar equations have been derived for RNA probes hybridizing to immobilized RNA[50]

$$Tm = 79.8°C + 18.5\log_{10} M + 0.58\ (\%\ G + C) + 11.8\ (\%\ G + C)^2 - 0.35\ (\%F) - \frac{820}{L} \qquad (4)$$

and DNA:RNA hybrids[51]

$$Tm = 79.8°C + 18.5\log_{10} M + 0.58\ (\%\ G + C) + 11.8\ (\%\ G + C)^2 - 0.50\ (\%F) - \frac{820}{L} \qquad (5)$$

A comparison of these equations shows that RNA:RNA hybrids are the most stable and DNA:DNA duplexes the least stable. In aqueous solution the Tm of a DNA:DNA duplex is approximately 10°C lower than the equivalent cRNA:RNA hybrid and these differences become more marked in solutions containing formamide.[51] For shorter (e.g., oligonucleotide) probes an equation estimating Tm is[52]

$$Tm\ oligo < 18\ bases \simeq 2°\ (A + T) + 4°\ (G + C) \qquad (6)$$

This estimate is independent of the salt concentration but tends to overestimate the Tm of longer probes. An alternative equation (Equation 3 above) has also been found to be accurate for probes of 14 to 70 bases.[2] For oligonucleotide probes, as with longer probes, there is a 1 to 1.5°C decrease in the Tm for every 1% change in homology although, as above, this depends upon GC content and position of the mismatch. Mismatches in the middle of a short probe have a much more marked effect on hybrid stability compared to those at the end.[40] If hybridizations are carried out in quaternary alkylammonium salts, the Tm of the hybrid is then independent of base composition.[53]

For a given salt concentration the hybridization of single strands is carried out at a temperature less than, but close to, the Tm. Thus, DNA of 50% G + C content has a Tm of 87°C in 0.4 *M* NaCl; under these conditions a temperature of Tm − 20/25°C is chosen to generate stable and matched hybrids. At higher temperatures the strands dissociate and at lower temperatures mismatched hybrids (as well as looped structures in RNA probes) become stabilized. Oligonucleotide probes present a special case of this. From Equation 3 as the length of the probe L decreases the Tm is lowered to a point where it is often impracticable to carry out the hybridization at Tm − 20/25°C. Thus, Tm − 5/10°C is often chosen in this case but, since such a temperature also favors hybrid dissociation, the conditions are essentially reversible. Thus, rapid posthybridization washes are necessary with such probes. Often, however, the Tm arrived at using the above considerations is still high (55 to 70°C), especially for *in situ* work, where consideration of the retainment of cell structure and adhesion of the section to a slide is necessary. In such cases formamide is often included in the hybridization solution, destabilizing the duplexes and lowering the Tm (from Equations 3 to 5). Thus, for DNA of 50% G + C content in 0.4 *M* NaCl containing 50% formamide the Tm is 54.5°C. Under such conditions the hybridization temperature chosen is in the range 37 to 42°C.

B. HYBRIDIZATION KINETICS

An analysis of probe hybridization kinetics is known in detail for reactions in solution.[54] In contrast, hybridization onto solid supports has been less extensively studied although some of the factors affecting the hybridization reaction rate onto immobilized sequences are considered below.

1. Probe Concentration and Length

As with solution hybridization, reassociation of double-stranded probes greatly affects the reaction kinetics, although there is no such problem with single-stranded probes. Generally, as the double-stranded probe concentration increases the number of filter bound probe sequences increases, although not dramatically. Similarly, the longer the probe the slower the rate of reassociation. These factors have been brought together in an empirical equation which determines the approximate hybridization time for such probes to immobilized sequences.[2] The probe concentration used is generally up to 300 ng/ml for *in situ* studies. The concentration of the probe can be artificially increased by introducing inert polymers of high molecular weight into the hybridization solution, at a concentration of 10% w/v. Both dextran sulfate and polyethylene glycol[55] have been used; they increase the rate about tenfold and also enhance the signal owing to the phenomenon of "networking".[48]

2. Temperature and Ionic Strength

The hybridization reaction rate curve is typically bell-shaped and has a maximum rate at Tm − 20/25°C, close to the temperature for optimal matched hybrid stability.[48] The reaction rate increases with monovalent ionic strength and this is particularly noticeable up to 0.1 *M*, where a twofold increase in concentration results in an increased rate of five- to tenfold. Above 0.1 *M*, the rate dependence is less but is still marked up to about 1.5 *M*.[56] Since high salt concentrations stabilize mismatched duplexes (from Equation 3), then the washing step is carried out under more stringent conditions. Typically, hybridization is done in 6× SSC, SSPE buffers, and washing in 0.1 to 2× such buffers.[2] Other reagents are also included to minimize nonspecific background binding of the probe; these include Denhardt's reagent, heparin, BLOTTO, heterologous DNA (e.g., from salmon sperm), and poly A/poly C sequences.[2,48]

C. *IN SITU* HYBRIDIZATION

All of the above considerations have been formulated for solution and filter hybridization. Little work has been carried out on the factors affecting *in situ* hybridization although it does seem that temperature, probe concentration, and length and salt concentration have a similar qualitative effect to that observed for filter hybridization. Other factors also play a part when using tissue sections. These include accessibility of the probe to the sequence (since proteins, especially structural proteins, may interact with the probe) and the effect of tissue fixation on hybrid stability (since, for instance, formalin fixation raises the Tm of matched sequences[57]). In practice, conditions are adjusted empirically to maximize the signal-to-noise ratio.

REFERENCES

1. **Davies, K. E. and Read, A. P.,** *Molecular Basis of Inherited Disease,* IRL Press, Oxford, 1988.
2. **Sambrook, J., Fritsch, E. F., and Maniatis, T.,** *Molecular Cloning: A Laboratory Manual,* 2nd ed., Cold Spring Harbor Press, Cold Spring Harbor, NY, 1989.
3. **Primrose, S. B.,** *Molecular Biotechnology,* 2nd ed., Blackwell Scientific, Oxford, 1991.
4. **Ehtesham, N. Z. and Hasnain, S. E.,** Direct in-gel hybridization, without blotting, using nick-translated cloned DNA probe, *Biotechniques,* 11, 718, 1991.
5. **Watson, J. D., Hopkins, N. H., Roberts, J. W., Steitz, J. A., and Weiner, A. M.,** *Molecular Biology of the Gene,* Vol. 1, Benjamin/Cummings, Menlo Park, CA, 1987.

6. **Darnell, J., Lodish, H., and Baltimore, D.,** *Molecular Cell Biology,* 2nd ed., Scientific American, New York, 1990.
7. **Wicking, C. and Williamson, B.,** From linked marker to gene, *Trend. Genet.,* 7, 288, 1991.
8. **Old, R. W. and Primrose, S. B.,** *Principles of Gene Manipulation: An Introduction to Genetic Engineering,* 4th ed., Blackwell Scientific, Oxford, 1989.
9. **Clewell, D. B.,** Nature of Col El plasmid replication in *Escherichia coli* in the presence of chloramphenicol, *J. Bacteriol.,* 110, 667, 1972.
10. **Backman, K. and Boyer, H. W.,** Tetracycline resistance determined by pBR322 is mediated by one polypeptide, *Gene,* 26, 197, 1983.
11. **Sykes, R. B. and Mathew, M.,** The beta-lactamases of gram-negative bacteria and their role in resistance to beta-lactam antibiotics, *J. Antimicrob. Chemother.,* 2, 115, 1976.
12. **Vieira, J. and Messing, J.,** The pUC plasmids, an M13mp7-derived system for insertion mutagenesis and sequencing with synthetic universal primers, *Gene,* 19, 259, 1982.
13. **Messing, J.,** New M13 vectors for cloning, *Methods Enzymol.,* 101, 20, 1983.
14. **McClelland, M. and Nelson, M.,** The effect of site-specific DNA methylation on restriction endonucleases and DNA modification methyltransferases — a review, *Gene,* 74, 291, 1988.
15. **Roberts, R. J.,** Restriction enzymes and their isoschizomers, *Nucleic Acids Res.* (Supp.), 16, r271, 1988.
16. **McClelland, M., Hanish, J., Nelson, M., and Patel, Y.,** KGB: A single buffer for all restriction endonucleases, *Nucleic Acids Res.,* 16, 364, 1988.
17. **Devereaux, J., Haeberli, P., and Smithies, O.,** A comprehensive set of sequence analysis for the VAX, *Nucleic Acids Res.,* 12, 387, 1984.
18. **Kelley, W. S. and Stump, K. H.,** A rapid procedure for isolation of large quantities of *Escherichia coli* DNA polymerase I utilizing a λ*polA* transducing phage, *J. Biol. Chem.,* 254, 3206, 1979.
19. **Steitz, T. A., Freemont, P. S., Ollis, D. L., Joyce, C. M., and Grindley, J. M.,** Functional implications of the Klenow fragment structure, *Biochem. Soc. Trans.,* 14, 205, 1986.
20. **Nossal, N. G.,** DNA synthesis on a double-stranded DNA template by the T4 bacteriophage DNA polymerase and T4 gene 32 DNA unwinding protein, *J. Biol. Chem.,* 249, 5668, 1974.
21. **Tabor, S., Huber, H. E., and Richardson, C. C.,** *Escherichia coli* thioredoxin confers processivity on the DNA polymerase activity of the gene 5 protein of bacteriophage T7, *J. Biol. Chem.,* 262, 16212, 1987.
22. **Saiki, R. K., Gelfand, D. H., Stoffel, S., Scharf, S. J., Higuchi, R., Horn, G. T., Mullis, K. B., and Erlich, H. A.,** Primer-directed enzymatic amplification of DNA with a thermostable DNA polymerase, *Science,* 239, 487, 1988.
23. **Chang, L. M. S. and Bollum, F. J.,** Molecular biology of terminal transferase, *CRC Crit. Rev. Biochem.,* 21, 27, 1986.
24. **Melton, D. A., Krieg, P. A., Rebagliati, M. R., Maniatis, T., Zinn, K., and Green, M. R.,** Efficient *in vitro* synthesis of biologically active RNA and RNA hybridization probes from plasmids containing a bacteriophage SP6 promoter, *Nucleic Acids Res.,* 12, 7035, 1984.
25. **Morris, C. E., Klement, J. F., and McAllister, W. T.,** Cloning and expression of the bacteriophage T3 RNA polymerase gene, *Gene,* 41, 193, 1986.
26. **Tabor, S. and Richardson, C. C.,** A bacteriophage T7 RNA polymerase/promoter system for controlled exclusive expression of specific genes, *Proc. Natl. Acad. Sci. U.S.A.,* 82, 1074, 1985.
27. **Melgar, E. and Goldthwait, D. A.,** Deoxyribonucleic acid nucleases. II. The effects of metals on the mechanism of action of deoxyribonuclease I, *J. Biol. Chem.,* 234, 4409, 1968.

28. **Richardson, C. C.,** Bacteriophage T4 polynucleotide kinase, in *The Enzymes,* Vol. 14, 3rd ed., Boyer, P. D., Ed., Academic Press, New York, 1981, 299.
29. **Rigby, P. W. J., Dieckmann, M., Rhodes, C., and Berg, P.,** Labeling deoxyribonucleic acid to high specific activity in vitro by nick translation with DNA polymerase I, *J. Mol. Biol.,* 113, 237, 1977.
30. **Feinberg, A. P. and Vogelstein, B.,** A technique for radiolabeling DNA restriction endonuclease fragments to high specific activity, *Anal. Biochem.,* 132, 6, 1983.
31. **Arrand, J. E.,** Preparation of nucleic acid probes, in *Nucleic Acid Hybridisation: A Practical Approach,* Hames, B. D. and Higgins, S. J., Eds., IRL Press, Oxford, 1985, 17.
32. **Wright, P. A. and Wynford-Thomas, D.,** The polymerase chain reaction: Miracle or Mirage? A critical view of its uses and limitations in diagnosis and research, *J. Pathol.,* 162, 99, 1990.
33. **Wright, D. K. and Manos, M. M.,** Sample preparation from paraffin-embedded tissues, in *PCR Protocols: A Guide to Methods and Applications,* Innis, M. A., Gelfand, D. H., Sninsky, J. J., and White, T. J., Eds., Academic Press, New York, 1990, 153.
34. **McCabe, P. C.,** Production of single-stranded DNA by asymmetric PCR, in *PCR Protocols: A Guide to Methods and Applications,* Innis, M. A., Gelfand, D. H., Sninsky, J. J., and White, T. J., Eds., Academic Press, New York, 1990, 76.
35. **Cox, K. H., DeLeon, D. V., Angerer, L. M., and Angerer, R. C.,** Detection of mRNAs in sea urchin embryos by *in situ* hybridization using asymmetric RNA probes, *Dev. Biol.,* 101, 485, 1984.
36. **Adams, R. L. P., Knowler, J. T., and Leader, D. P.,** *The Biochemistry of the Nucleic Acids,* 11th ed., Chapman and Hall, London.
37. **Jameson, J. L., Klibanski, A., Black, P. McL., Zervas, N. T., Lindell, C. M., Hsu, D. W., Ridgway, E. C., and Habener, J. F.,** Glycoprotein hormone genes are expressed in clinically nonfunctioning pituitary adenomas, *J. Clin. Invest.,* 80, 1472, 1987.
38. **Taneja, K. and Singer, R. H.,** Use of oligodeoxynucleotide probes for quantitative *in situ* hybridization to actin mRNA, *Anal. Biochem.,* 166, 389, 1987.
39. **Lathe, R.,** Synthetic oligonucleotide probes deduced from amino acid sequence data: theoretical and practical considerations, *J. Mol. Biol.,* 183, 1, 1985.
40. **Ikuta, S., Takagi, K., Wallace, R. B., and Itakura, K.,** Dissociation kinetics of 19 base paired oligonucleotide-DNA duplexes containing different single mismatched base pairs, *Nucleic Acids Res.,* 15, 797, 1987.
41. **Denny, P., Hamid, Q., Krause, J. E., Polak, J. M., and Legon, S.,** Oligoriboprobes: tools for *in situ* hybridization, *Histochemistry,* 89, 481, 1988.
42. **Bicknell, E. J. and Perrett, C. W.,** unpublished data, 1992.
43. **Ludwig, J. and Eckstein, F.,** Rapid and efficient synthesis of nucleoside 5′-O-(1-thiotriphosphates), 5′-triphosphates and 2′,3′-cyclophosphorothioates using 2-chloro-4H-1,3,2-benzodioxaphosphorin-4-one, *J. Org. Chem.,* 54, 631, 1989.
44. **Heiniger, H. J., Chen, H. W., and Commerford, S. L.,** Iodination of ribosomal RNA *in vitro, Int. J. Appl. Rad. Isotopes,* 24, 425, 1973.
45. **Höfler, H.,** Principles of *in situ* hybridization, in *In Situ Hybridization: Principles and Practice,* Polak, J. M. and McGee, J. O'D., Eds., Oxford University Press, Oxford, 1990, 15.
46. **Forster, A. C., McInnes, J., Skingle, D. C., and Symons, R. H.,** Non-radioactive hybridization probes prepared by chemical labelling of DNA and RNA with a modified novel reagent, photobiotin, *Nucleic Acids Res.,* 13, 745, 1985.
47. **Bolton, E. T. and McCarthy, B. J.,** A general method for the isolation of RNA complementary to DNA, *Proc. Natl. Acad. Sci. U.S.A.,* 48, 1390, 1962.
48. **Anderson, M. L. M. and Young, B. D.,** Quantitative filter hybridisation, in *Nucleic Acid Hybridisation: A Practical Approach,* Hames, B. D. and Higgins, S. J., Eds., IRL Press, Oxford, 1985, 73.

49. **Bonner, T. I., Brenner, D. J., Neufeld, B. R., and Britten, R. J.,** Reduction in the rate of DNA reassociation by sequence divergence, *J. Mol. Biol.,* 81, 123, 1973.
50. **Bodkin, D. K. and Knudson, D. L.,** Assessment of sequence relatedness of double-stranded RNA genes by RNA—RNA blot hybridization, *J. Virol. Methods,* 10, 45, 1985.
51. **Casey, J. and Davidson, N.,** Rates of formation and thermal stabilities of RNA:DNA and DNA:DNA duplexes at high concentrations of formamide, *Nucleic Acids Res.,* 4, 1539, 1977.
52. **Itakura, K., Rossi, J. J., and Wallace, R. B.,** Synthesis and use of synthetic oligonucleotides, *Annu. Rev. Biochem.,* 53, 323, 1984.
53. **Wood, W. I., Gitschier, J., Lasky, L. A., and Lawn, R. M.,** Base composition — independent-hybridization in tetramethylammonium chloride: A method for oligonucleotide screening of highly complex gene libraries, *Proc. Natl. Acad. Sci. U.S.A.,* 82, 1585, 1985.
54. **Young, B. D. and Anderson, M. L. M.,** Quantitative analysis of solution hybridisation, in *Nucleic Acid Hybridisation: A Practical Approach,* Hames, B. D. and Higgins, S. J., Eds., IRL Press, Oxford, 1985, 47.
55. **Amasino, R. M.,** Acceleration of nucleic acid hybridisation rate by polyethylene glycol, *Anal. Biochem.,* 152, 304, 1986.
56. **Studier, F. W.,** Effects of the conformation of single-stranded DNA on renaturation and aggregation, *J. Mol. Biol.,* 41, 199, 1969.
57. **Chan, V. T.-W., Herrington, C. S., and McGee, J. O'D.,** Basic background of molecular biology, in *In Situ Hybridization: Principles and Practice,* Polak, J. M. and McGee, J. O'D., Eds., Oxford University Press, Oxford, 1990, 1.

Chapter 2

PRINCIPLES OF PROBE PREPARATION FOR *IN SITU* HYBRIDIZATION

Pierre Couble

TABLE OF CONTENTS

0-8493-4414-X/93/$0.00 + $.50

I. INTRODUCTION

The purpose of this chapter is to introduce the principles of probe preparation by the most usually employed methods. The reader is referred to the appropriate chapters for detailed descriptions of the protocols and for the illustrations of the topographical resolution achieved with the various types of probes.

There are basically two split strategies for preparing labeled nucleic acids to be used as probes for detecting sequences of interest. One consists of enzymatically neosynthesizing DNA or RNA in the presence of labeled precursors (labeled ribo- or deoxyribonucleotides triphosphate) from templates made of either cloned DNA, *in vitro* amplified DNA, or chemically synthesized oligonucleotides.

The other way aims to graft a molecule or a residue onto purified DNA or RNA by chemical or physical means. For example, nonradioactive probes can be generated by the addition of biotin (a member of the vitamin B complex) or of digoxigenin (a plant steroid hormone) on nucleic acids, by using photoacetate derivatives. These contain an azide aryl group allowing covalent linking of the organic molecule on DNA or RNA in a reaction mediated by visible light.[6] The sulfonation of cytosines[16] obtained by incubating DNA in concentrated sodium bisulfite is also an alternative to the enzymatic preparation of nonradioactive probes for *in situ* hybridization.[4] Many other low-molecular weight reporter molecules (di- or trinitrophenol, N2 acetylaminofluoren, or iodofluoren, etc.) can be possibly linked to DNA or RNA, but, to my knowledge, have not led to applications to *in situ* cytodetection of hybridized sequences.

As a whole, the enzymatic neosynthesis of nucleic acids is preferable (and indeed most widely used) because it offers a larger choice of the nature of label — either radioactive or nonradioactive — and, most important, provides probes with the highest specific activities. To illustrate the versatility of this approach, I will review successively the various types of radioactive and nonradioactive available precursors, the properties of the enzymes selected for DNA and RNA polymerization, and then present the principles of the now routine, or to be routine methods of enzymatic preparation of probes to hybridize *in situ*.

II. LABELED NUCLEOTIDE PROBE PRECURSORS

There are two distinct classes of probe precursors: those radioactive, carrying one or more radioisotopes, and those nonradioactive — or "cold" — on which bases are grafted diverse structural motifs.

The radiolabeled ribo- or deoxyribonucleotides resemble closely the natural substrates of the enzymes used for preparing probes. They are available at different specific activities and, according to the nature of the radioelement and the procedure of autoradiography, allow various degrees of resolution and sensitivity.

The introduction of cold precursors for hybridization *in situ* has permitted freedom of the constraints inherent to handling radioactive molecules. Cold probes are therefore widely used now; however, at optimal specific activity, their sensitivity is significantly less than that of radioactive probes.

A. RADIOACTIVE PRECURSORS

The most common labeled nucleotides for *in situ* hybridization carry either atoms of phosphorus-32 (^{32}P), of sulfur-35 (^{35}S), or of tritium (^{3}H). No application to *in situ* hybridization has yet been reported with newly available precursors labeled with the isotope 33 of phosphorus (^{33}P), which may also be adequate.

The specific activity of radiolabeled precursors is expressed in becquerel per millimole (Bq/mmole) or in curie per millimole (Ci/mmole). One becquerel — recently referred to as the official radioactivity unit — corresponds to one desintegration (dpm) per second whereas one curie is 2.22×10^{12} disintegrations per minute. Thus, one curie is equivalent to 37 gigabecquerels.

1. Phosphorus-32 Labeled Nucleotides

The isotope 32 of phosphorus emits a β-radiation (an electron with a negative charge) of high energy of emission (1.71 megaelectronvolts). Probes labeled with this radioactive element confer the highest sensitivity of detection but a rather poor autoradiographic resolution. The half-life of phosphorus-32 is 14.3 d (the half-life is the time in which the amount of a radioactive nuclide decays to half of its initial value). The most common specific activity of ^{32}P nucleotides commercially available is 100 terabecquerels/mmol (3000 Ci/mmole). Given the maximal specific activity (carrier-free preparation) at 338 TBq/milliatom, the commercialized solution of phosphorus-32 labeled nucleotides contains 1 radioactive residue out of 3.

The high energy of emission and specific activity of phosphorus-32 pose problems of shielding (a 8-mm thick transparent plexiglass shield efficiently protects the manipulator) and of waste handling and storage, which are submitted to drastic regulations.

a. Alpha ^{32}P-Labeled Nucleotides

The three phosphate groups of NTPs are referred to as alpha, beta, and gamma from the carbon 5′ of the sugar (Figure 1). Such probes prepared by "nick translation" or "random priming" utilize nucleotides labeled on the alpha position. This phosphate group is indeed involved in the phosphodiester bonding of two nucleotides, during the process of polymerization.

The specific activity of probes made with one ^{32}P-labeled NTP (alpha ^{32}P-dCTP for example) at 110 TBq/mmol can reach more than 10^{9} dpm/μg

Alpha-^{32}P deoxythymidine 5' triphosphate

[1',2',5-^{3}H] deoxycytidine 5' triphosphate

Alpha-^{35}S deoxyadenosine 5' triphosphate

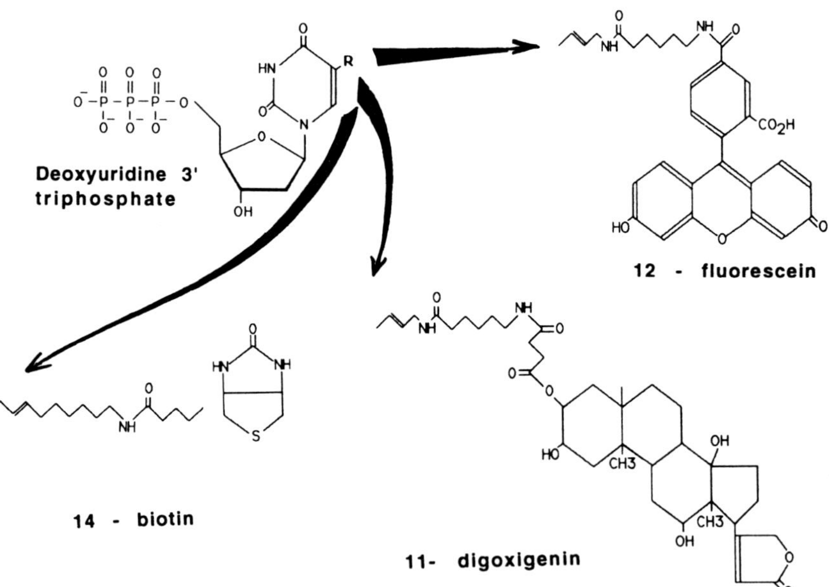

FIGURE 1. Chemical structure of selected radioactive and nonradioactive deoxynucleosides 5′ triphosphate used as precursors of probes for *in situ* hybridization. The biotin, digoxigenin, and fluorescein derivatives are shown branched on a dUTP nucleotide, via spacers of, respectively, 14, 11, and 12 atoms of carbon and nitrogen.

(disintegrations per minute and per microgram of polymerized nucleic acid). It should be noted that every β-transformation of the incorporated isotope transforms the alpha phosphorus-32 in sulfur:

$$^{32}_{15}P \rightarrow {}^{32}_{16}S + \beta^{-}$$

This will lead to breaking of the phosphodiester covalent bond established between adjacent nucleotides. As a consequence and given the short half-life of the isotope, the probe will reduce rapidly in size, which may be a source of artifacts. It is thus recommended that ^{32}P-labeled probes are utilized at the earliest after they are synthesized and should not be stored.

b. Gamma ^{32}P-Adenosine Triphosphate

This ribonucleotide is the substrate of the polynucleotidyl kinase that is used for preparing 5′-labeled oligonucleotides probes. In the reaction, the enzyme catalyzes the transfer of the gamma phosphate of ATP to the phosphate-free end of the DNA fragment. The specific activity of gamma ^{32}P-ATP for routine end-labeling is 110 TBq/mmole, but the same molecule is available at 250 TBq/mmole.

2. Sulfur-35 Labeled Nucleotides

Sulfur-35 is also a β-emitting radioisotope whose radiation develops an energy of 0.167 MeV, ten times less than the radioisotope 32 of phosphorus. The sensitivity of detection reached with ^{35}S-labeled probes is thus less than with ^{32}P-labeled probes, but the resolution achieved is better. The half life of ^{35}S is 87.4 d and therefore probes carrying this isotope are rather stable with time. Also, the need for speed in probe preparation is not as pressing as for the phosphorus-32 radioisotope.

In ^{35}S-labeled precursors, the radioactive element is substituted to an atom of oxygen of the alpha phosphate group (Figure 1). In the process of β transformation, ^{35}S gets transformed in chlorine, which does not induce breakage of the internucleotide covalent link.

The specific activity of ^{35}S-labeled ribo- or deoxyribo-nucleotides varies from 15 to 37 Tbq/mmol (400 to 1000 Ci/mmol). This corresponds to one and two radioactive nucleotides out of three since the specific activity at the highest isotopic abundance (carrier-free preparation) is 55 TBq/milliatom.

3. Tritiated Nucleotides

Tritium, the hydrogen-3 radioisotope, confers the best topological resolution of radioactive probes for *in situ* hybridization. This isotope is very stable (its half-life is 12.3 years) but is of weak energy of radiation (18.6

KeV). To enhance the specific activity of tritiated nucleotides, precursors are synthesized with more than one tritium substituted at different hydrogen positions of the sugar or the base (up to 4 radioisotopes per molecule) (Figure 1). In this way, the specific activity can reach 3.7 TBq/mmol (100 Ci/mmol). Increasing the specific activity of the probe can also be reached by using more than one tritiated nucleotide precursor.

B. NONRADIOACTIVE NUCLEOTIDES

The advantage of using cold probes is the fact that they do not impose special laboratory settings for the manipulation of radioactive probes. Once prepared, cold probes are stable for months and, moreover, they permit, after hybridization and washing, immediate revelation, observation, and interpretation.

The precursors for nonradioactive probes are triphosphate ribo- or deoxyribo-nucleotides carrying an additional covalently bound marker molecule. This molecule that will serve to reveal the hybridized probe is grafted via a spacer to a chemical group of the base. Spacers are made of 7 to 21 carbon and nitrogen chains. They limit the steric hydrance at the step of polymerization and enhance the accessibility of the marker molecule during the step reactions at revelation. dUTP, dATP, and UTP are usually available as cold-labeled precursors.

1. Biotin-Labeled Nucleotides

Biotin-coupled nucleotides were among the first cold precursors applied to *in situ* hybridization[9] (Figure 1). Revelation of biotinylated probes after hybridization makes use of the high affinity binding of the vitamin to its natural ligands: avidin or, more commonly used, streptavidin. Ligands are utilized as conjugated molecules with diverse enzymes (alcalin phosphatase, peroxidase . . .), chemical complexes (gold particles . . .), or hapten (fluorescein, rhodamin).[13] Alternatively, anti-biotin immunoglobulins coupled with signaling molecules have also been used for detection of hybrids.

2. Digoxigenin NTP

Digoxigenin, a plant steroid hormone, has been introduced as a label for hybridization by the company Boehringer. Its detection after hybridization utilizes an anti-digoxigenin Fab fragment coupled with either reporter enzymes or fluorescent compounds (fluorescein, rhodamin . . .).[2]

3. Fluorescein-Labeled Precursors

dUTP- or UTP-fluorescein are also interesting substrates for cold probe labeling. Depending on the abundance of the sequence to be hybridized in the tissue, the revelation of the hybrids may either be simply achieved by observation of the treated sections under UV light or after incubation with an antifluorescein antibody conjugated with a reporter enzyme.

III. DNA AND RNA POLYMERASES

The neosynthesis of DNA or RNA is the most current mode of making probes. The choice of the polymerases used to ensure proper incorporation of the labeled precursors depends on the type of probe that one desires (DNA or RNA probes), the achieved specific activity, and the type of template (cloned DNA, PCR amplified DNA, oligonucleotides) one starts with.

Table 1 summarizes the properties of the different DNA and RNA polymerases involved in probe preparation via neosynthesis of labeled nucleic acids. Together with the polymerase activity per se that proceeds from 5′ to 3′, DNA polymerases may or may not carry 3′ to 5′ and 5′ to 3′ exonuclease activities. Some of these properties are exploited for the preparation of probes.

The DNA polymerase of *E. coli* is the most common source of native enzyme with 5′ to 3′ polymerase, 3′ to 5′, and 5′ to 3′ exonuclease activities. The enzyme of Klenow is derived from the bacterial holoenzyme by proteolytic cleavage with subtilisin. It lacks the 5′ to 3′ exonuclease activity. These two enzymes initiate DNA synthesis on double-stranded DNA template altered by nicks (local breakage of the phosphodiester bond between two adjacent nucleotides) or copy single-stranded DNA after initiating neosynthesis on local double-stranded regions. Their 3′ to 5′ exonuclease activity is inhibited by the presence of the free dNTPs.

In contrast with these DNA polymerases that function with a model template, the enzyme deoxynucleotidyl terminal transferase polymerizes available nucleotides without template, starting from a 3′ end of DNA. These properties are utilized for oligonucleotide labeling by tailing their 3′ extremity.

Taq DNA polymerase and vent DNA polymerase are thermostable enzymes coded by thermophile bacteria (respectively, *Thermus aquaticus* and *Thermococcus litoralis*). This property is exploited for bicatenary amplification of DNA (PCR) that involves repeated denaturation and neosynthesis of DNA strands (see below). The vent DNA polymerase has a 3′ to 5′ exonuclease activity that allows a proofreading function during the process of amplification.

DNA dependent-RNA polymerases are used for preparing single-stranded RNA probes from cloned DNA. These enzymes are encoded by *E. coli* (T3 and T7) or salmonella (SP6) bacteriophages and the corresponding genes have been eventually cloned for their biotechnological production. Each initiates on a specific promotor (short DNA sequence motif) which has been inserted at one or the other boundary of the polylinker sequence of various plasmid vectors. These RNA polymerases are very stable and efficient *in vitro* and can be performed on several rounds of initiation on the same template, leading to substantial amounts of neosynthesized RNA.

TABLE 1
Properties of the DNA and RNA Polymerases Used for Generating DNA and RNA Probes

	$5' \rightarrow 3'$ Polymerase activity	$5' \rightarrow 3'$ Exonuclease activity	$3' \rightarrow 5'$ Exonuclease activity	Applications	Ref.
DNA Polymerases					
Native DNA polymerase of *E. coli*	On single-stranded DNA from double-stranded primer	DNA or RNA on DNA-DNA or DNA-RNA duplex	More intense on s.s. DNA inhibited by polymerase activity	"Nick-translation"	Kelley & Stump, 1979
Klenow fragment	On single-stranded DNA from double-stranded primer	None	More intense on s.s. DNA inhibited by polymerase activity	"Random priming"	Klenow et al., 1971
Taq DNA polymerase	On single-stranded DNA from double-stranded primer	None	None	"PCR" probes	Chien et al., 1976
Vent DNA polymerase	On single-stranded DNA from double-stranded primer	None	3′–5′ Proofreading activity	"PCR" probes	Mattila et al., 1991
Terminal transferase	From single-stranded or 3′ protruding double-stranded DNA	None	None	3′ Labeling of oligonucleotides	Bollum (1974)
RNA Polymerases					
From SP6 bacteriophage of *S. typhimurium*	From double-stranded DNA with a sequence-specific promoter	None	None	Single-stranded RNA probes	Melton et al., 1984
From T3 bacteriophage of *E. coli*	From double-stranded DNA with a sequence-specific promoter	None	None	Single-stranded RNA probes	Morris et al., 1986
From T7 bacteriophage of *E. coli*	From double-stranded DNA with a sequence-specific promoter	None	None	Single-stranded RNA probes	Tabor & Richardson, 1985

IV. PRINCIPLES OF PROBE PREPARATION

The principles of most common and established methods for preparing probes for *in situ* hybridization are described in this chapter. It should be noted that the usage of commercial labeling kits by these different methods is now extremely popular. Most of them proved to be very efficient and reliable.

A. DOUBLE-STRANDED DNA PROBES

For double-stranded DNA probes, the techniques of ''nick translation'' and ''random priming'' differ essentially in the procedure employed to promote DNA polymerase initiation sites. They end up, however, to produce probes with different specific activities. For the same amount of label incorporated, the former procedure leads to probes with 5 to 10 times lower concentration of label (generally expressed in quantity of incorporated labeled precursor per μg of total DNA) than the latter. However, the average length of the neosynthesized DNA fragments may be larger in the protocol of ''nick translation''.

1. Method of ''Nick Translation''

The principle of this method consists of initiating polymerization of labeled precursors on nicks provoked by limited DNase activity[14] (Figure 2). The template-dependent polymerization of DNA then proceeds by the degradation of one strand of the model DNA through 5′ to 3′ exonuclease activity and by its 5′ to 3′ neosynthesis by the enzyme. The 3′ to 5′ exonuclease activity of the native *E. coli* is inhibited in the reaction by the presence of the free nucleosides triphosphate. In practice, DNase nicking and polymerization occur simultaneously; in a typical reaction, incubation is only 2 to 3 h.

The yield of the reaction depends in particular on the amplor of the DNase activity, which is predosed in commercial kits of preparation of nick-translated probes. Also, most of the now available kits propose independent mixtures of the four cold nucleotides, allowing labeling of DNA with either one or the other, or eventually more than one, of the four possible labeled nucleotides.

All radioactive nucleotides 5′-triphosphate are properly incorporated by the bacterial polymerase. Cold precursors labeled with biotin, digoxigenin, or fluorescein are efficiently incorporated also, as long as the ratio of the labeled and nonlabeled versions in the pool of nucleotides (for example the ratio of dATP and dATP-biotin or the ratio of dTTP and dUTP-digoxigenin) is at its optimum. Again one would recommend using already prepared mixtures of nucleotides, like those available in labeling kits.

The starting of double-stranded DNA can either consist of amplified and purified recombinant plasmid DNA (plasmid DNA plus insert DNA), gel purified insert DNA, or PCR-amplified genomic DNA fragments. The starting

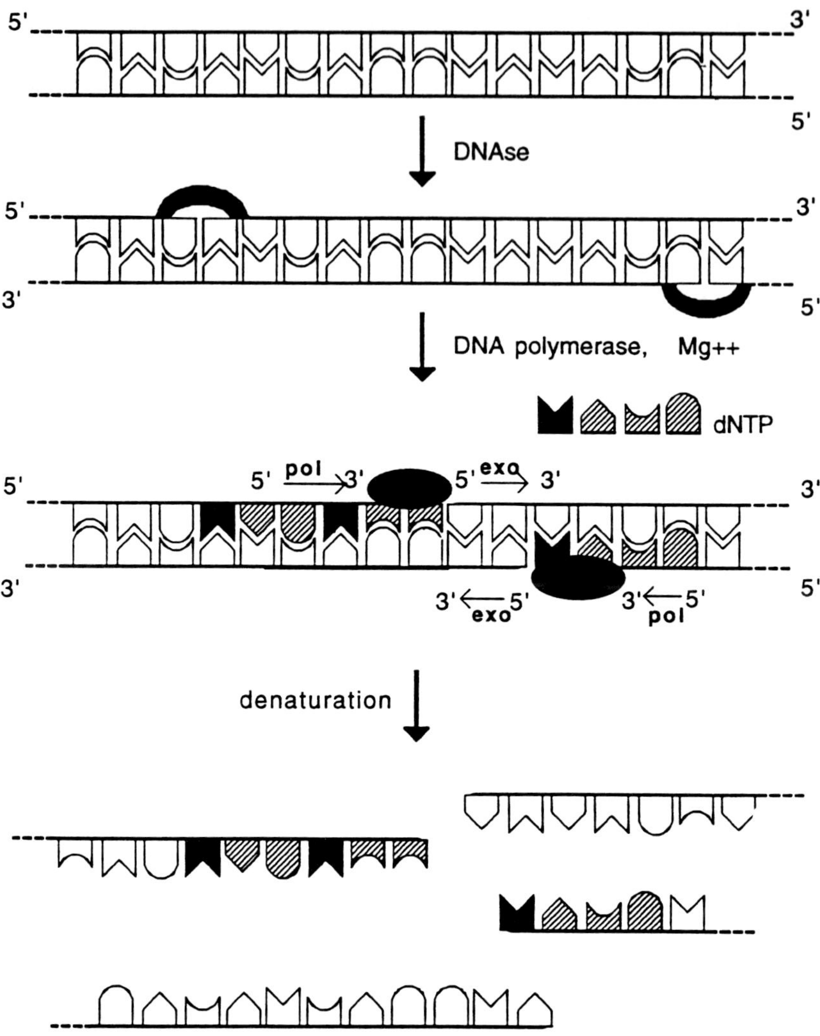

FIGURE 2. Procedure of labeling DNA by nick translation. The nicking and polymerization steps are shown successively. "Poly" and "exo" illustrate the 5′ to 3′ polymerization and 5′ to 3′ exonuclease activities of the *E. coli* native DNA polymerase. The four deoxyribonucleosides 5′ triphosphate are represented as hatched boxes, the darker one acting as the labeled precursor.

amount of double-stranded DNA in a typical reaction is 200 ng. The average length of the labeled fragment at completion of reaction is 200 to 400 nucleotides.

2. Method of "Random Priming"

The major constraint inherent to this method compared to that of "nick translation" is to start with linearized DNA, since it has to be denatured to

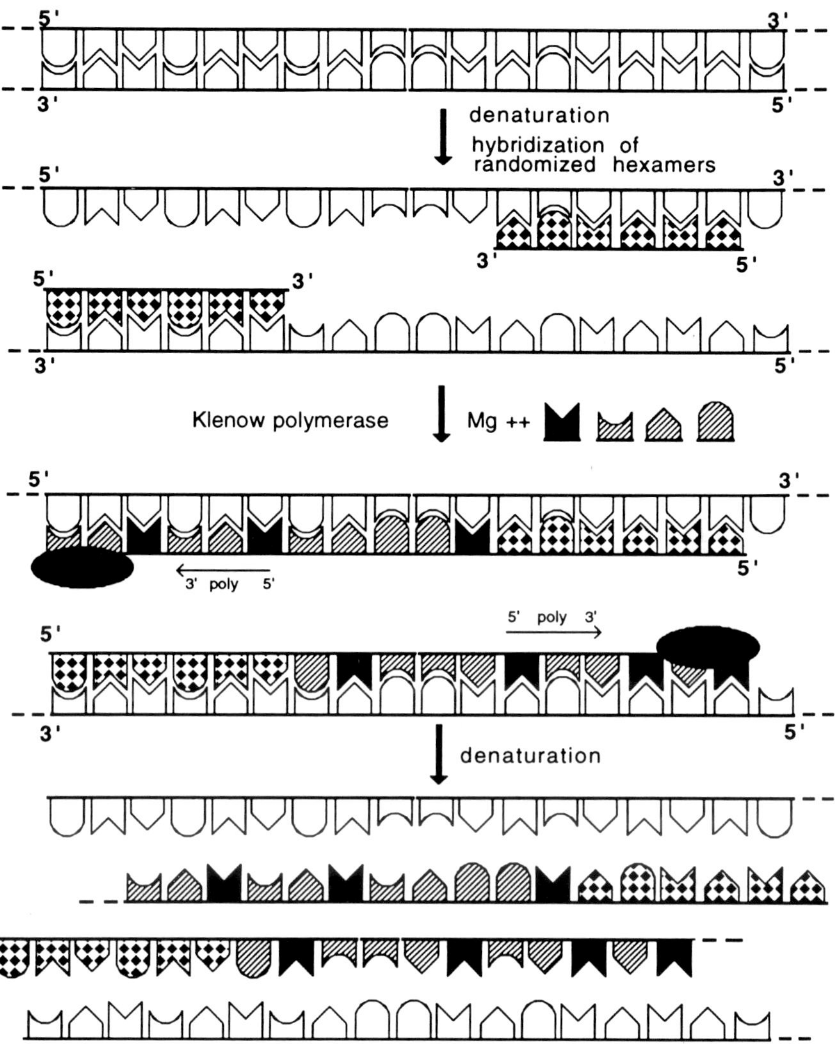

FIGURE 3. Principles of labeling of DNA probes by the method of "random priming". Starting DNA should be linearized to facilitate annealing of the hexamer primers of random sequence. Local double-stranded stretches allow initiation of DNA neosynthesis by the Klenow fragment of polymerase 1 (devoid of the 5′ to 3′ exonuclease activity). The free deoxyribonucleotide precursors are shown as hatched boxes, the darker acting as the labeled one.

permit annealing of the hexanucleotide primers. This concerns, in particular, recombinant supertwisted plasmid DNA which should be cut with a restriction enzyme prior to serving as a source of DNA for "random priming".

The annealing of the hexamers of random sequence (therefore complementary to any sequence of the probe DNA) provides as many primers for the initiation of polymerization by the Klenow fragment of the *E. coli* DNA polymerase (this lacks the here useless 5′ to 3′ exonuclease activity of the holoenzyme).

A typical reaction utilizes around 20 ng of starting double-stranded DNA and yields as much incorporated precursor as does a reaction of "nick translation" using 200 ng of the same DNA. The average length of random primed DNA is 200 to 300 nucleotides.

3. Bicatenary DNA Amplification (PCR Probes)

DNA probes can also be prepared by amplifying bicatenary DNA *in vitro* with the method referred to as the "polymerase chain reaction". This reaction drives exponential amplification of theoretically any DNA stretch situated in between two regions of known nucleotide sequence. This method, when applied to the generation of probes, allows a large versatility in the selection of the length and the type of DNA to hybridize, starting either from purified genomic DNA (which permits bypassing subsequent steps of cloning) or from already cloned materials. More, it is possible to envisage the production of probes of unknown sequence, as long as, again, the region amplified is bounded by known sequences.

The principle of bicatenary DNA amplification, which makes use of two synthetic primers, is illustrated in Figure 4. The temperature of incubation at the step of primer annealing is critical and depends on the length and G + C content of the primers. Designed at its origin to utilize the polymerize activity of the fragment of Klenow,[16] the protocol of amplification has been greatly ameliorated by using thermostable polymerases of extreme thermophile bacteria. Taq polymerase, the enzyme from *Thermophilus aquaticus,*[1] has been extensively employed for this purpose, whereas enzymes from other sources like *Thermococcus litoralis,*[10] which eventually copies DNA with reduced error incidence, are also developed.

Three methods involving an *in vitro* DNA amplification step can be used for preparing labeled DNA probes:

- The DNA sequence of interest is amplified by PCR with cold dNTPs (from genomic DNA for example), and a fraction of the resulting gel-purified fragment is subsequently labeled by one of the above described methods ("nick translation" or "random priming").
- The selected DNA sequence is amplified and labeled simultaneously by incorporation of a labeled precursor. This can be applied to generate nonradioactive probes labeled, for example, with digoxigenin-dUTP. In such a reaction, the size of the pool of the labeled precursor (actually a mixture of dig-dUTP and dTTP which, in a typical PCR reaction, are adjusted to, respectively, 0.13 and 0.07 mmol) can sustain numerous cycles of exponential amplification. The incorporation is however twice less as for unlabeled DNA. Radioactive probes can be also prepared

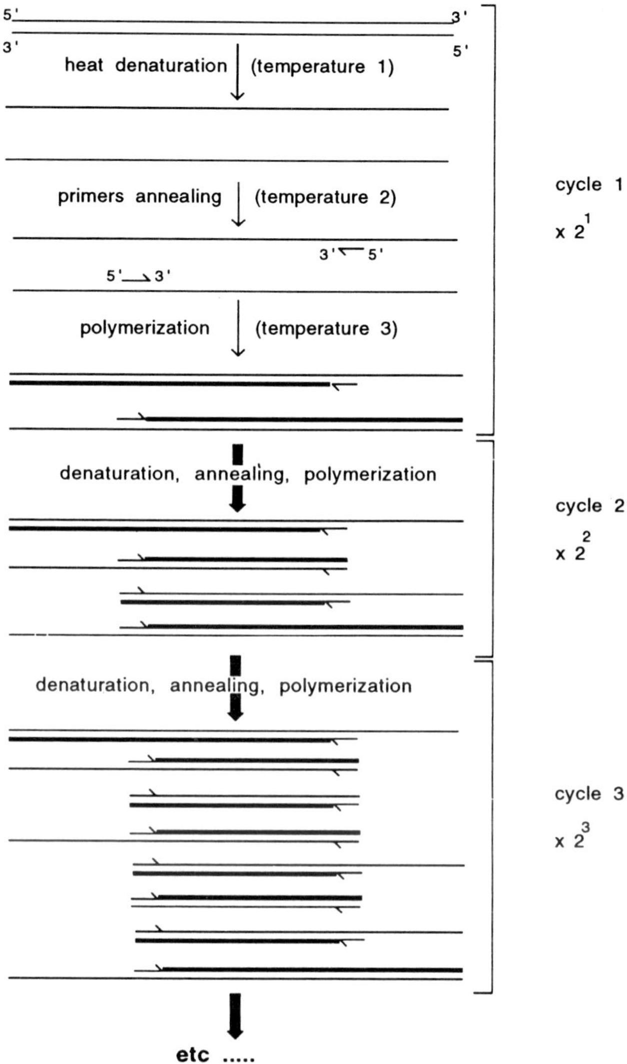

FIGURE 4. *In vitro* bicatenary amplification of DNA by PCR. This figure schematizes the principles of exponential amplification of a DNA segment comprised in between the two primers of known sequence. The original template DNA (top) is denatured at temperature 1 to allow the annealing of the two primers at temperature 2. This temperature is critical for proper amplification; it is driven by the G + C content of the primers and by the salinity of the reaction mixture. Polymerization can proceed at rather high temperature (around 76°C) with thermostable enzymes like the Taq polymerase. Thick links represent the neosynthesized DNA strands at one given cycle, illustrating the exponential accumulation of the segment to amplify. Amplification of DNA from eucaryote genomic DNA requires 20 to 30 cycles.

this way by selecting a sequence included into a recombinant plasmid DNA, for example. Care should be taken to adjust the degree of amplification (i.e., the number of cycles) to the size of the pool of the labeled precursor.

- The sequence is amplified by PCR as cold DNA then gel-purified to eliminate unincorporated primers and cold nucleotides. A fraction is then submitted to 1 to 10 (or more) PCR cycles in presence of one internal primer and of labeled nucleotides. In this step, the amplification is proportionate to the number of cycle and leads to labeling of only one of the two strands initially amplified (Figure 5).

B. SINGLE-STRANDED RNA PROBES

When incubating sections with double-stranded DNA as a probe, the hybridization of the strand complementary to the mRNA or (DNA) of interest is competed away by the autoreassociation of the complementary strands of the probes. This limitation can be circumvented by using *in vitro* transcribed RNA probes which are single-stranded. Such RNA probes can be of long size (more than 1000 bases) and of high specific activity. Their preparation requires work under RNase-free conditions, however. Also, when hybridizing mRNA with RNA probes, one should be aware that RNA-RNA duplexes are more stable than DNA-RNA ones, and that, therefore, physicochemical (temperature and salinity) parameters of the reaction must be adjusted accordingly.[5]

The neosynthesis of labeled RNA is realized by *in vitro* transcription of genomic DNA or cDNA inserted into one of these plasmids equipped with promoters for bacteriophage RNA polymerases, either T3, T7, or SP6 (see Table 1). As shown in Figure 6, two of these promoters (specific DNA sequence motifs recognized by the enzymes) are placed at each boundary of the plasmid and of the inserted DNA.

The transcription from one or the other promotor generates either an ''antisense'' probe (i.e., complementary to the mRNA encoded by the inserted gene fragment) or a probe with the same sequence as that of the mRNA of interest (this is often used as a negative control of hybridization). The knowledge of the transcriptional orientation of the inserted gene fragment drives the choice of the adequate enzyme to use (see Figure 6).

Calibration of the transcripts can be easily obtained by cutting the recombinant plasmid DNA with a restriction enzyme (it has been reported, however, that the restriction enzymes which generate 3′ protruding ends may not be appropriate). The removal of the double stranded DNA template is generally carried out by digestion with a RNase-free DNase.

Radioactive (more often ^{35}S-labeled NTPs) or nonradioactive (biotin and digoxigenin) precursors are used for RNA probes.

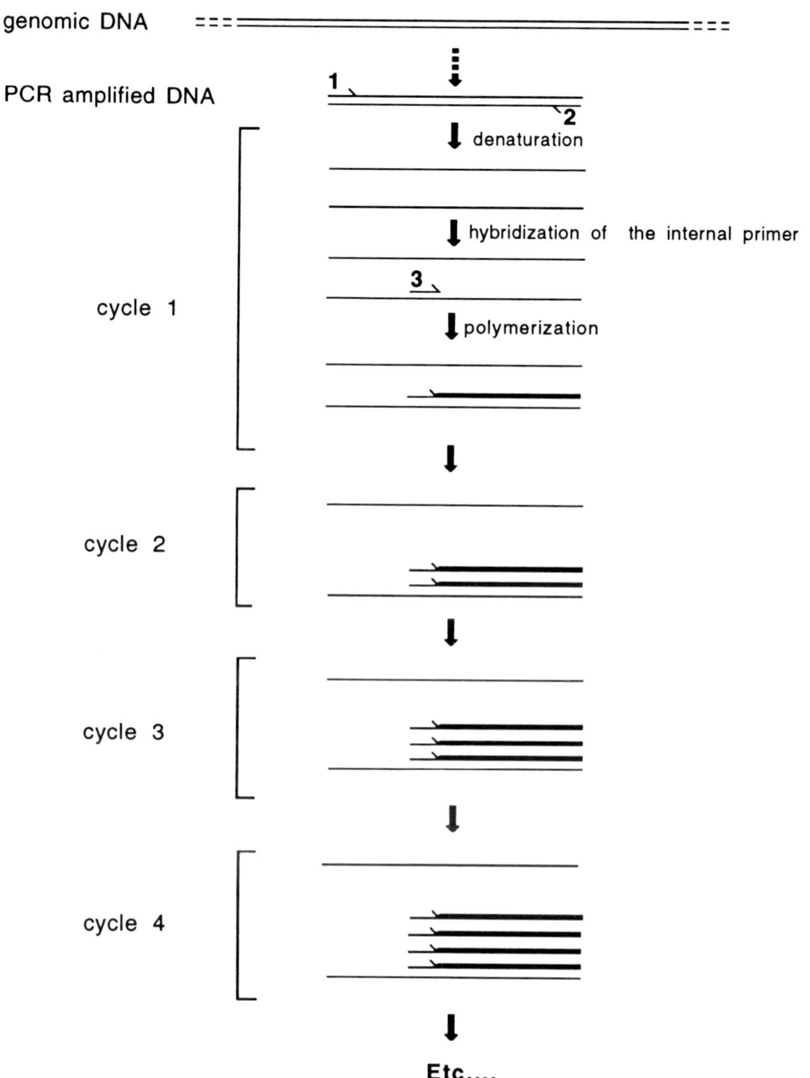

FIGURE 5. Single-strand amplification of an already PCR-amplified DNA fragment. The procedure is similar to the one described in Figure 4 except that only one primer (internal to the already amplified DNA) is annealed at each cycle, leading to the accumulation of one strand in proportion of the number of cycles realized. This method can be applied to asymmetric labeling of DNA for preparing hybridization probes by using radioactive or nonradioactive precursors.

C. OLIGONUCLEOTIDES PROBES

1. 5′ Terminus Labeling

Oligonucleotides can be labeled with ^{32}P in a reaction mediated by the polynucleotidyl kinase, an enzyme encoded by the bacteriophage T4 (Figure 7a). This enzyme catalyzes the transfer of the gamma phosphate residue of

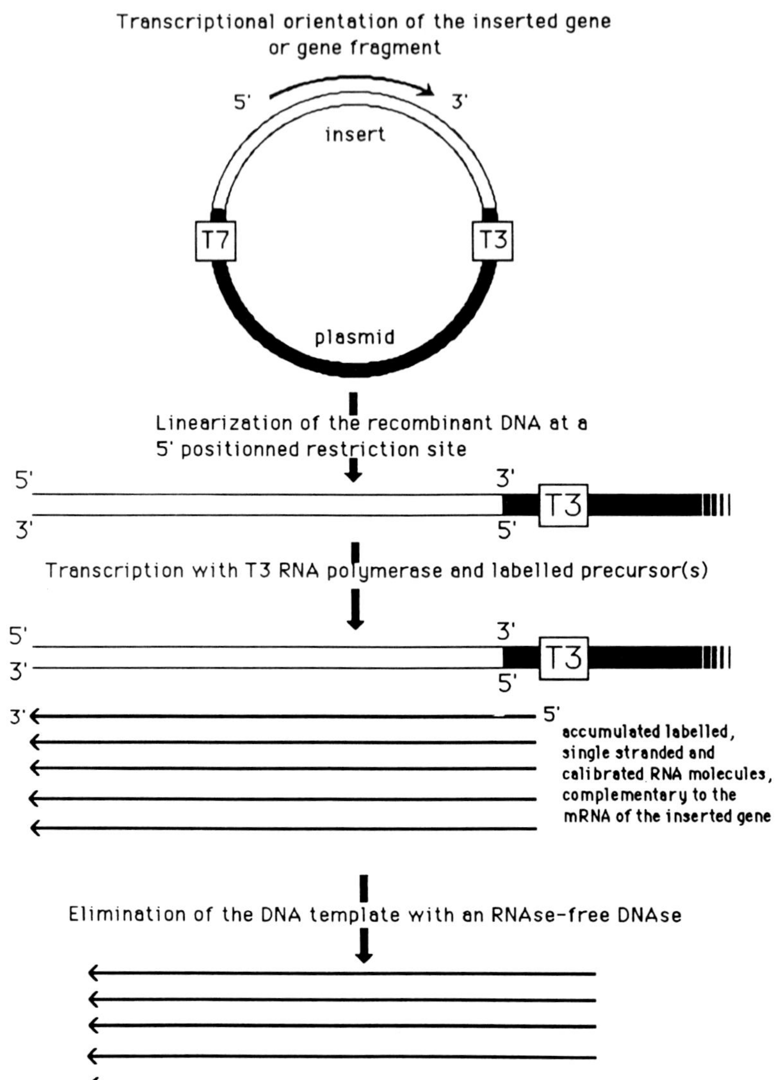

FIGURE 6. Preparation of single-stranded RNA probes by DNA-dependent RNA polymerases. RNA probes are generated by RNA transcription of recombinant DNA equipped with specific bacteriophage RNA polymerase promotors (T3 and T7 on the example). Linearization of the recombinant DNA at the 5′ extremity of the insert leads to single-stranded RNA molecules complementary to the mRNA of the gene of interest. Linearization at the opposite side of the insert and transcription using the other polymerase yield labeled RNA of same sequence as the mRNA to hybridize, which may constitute a negative control probe.

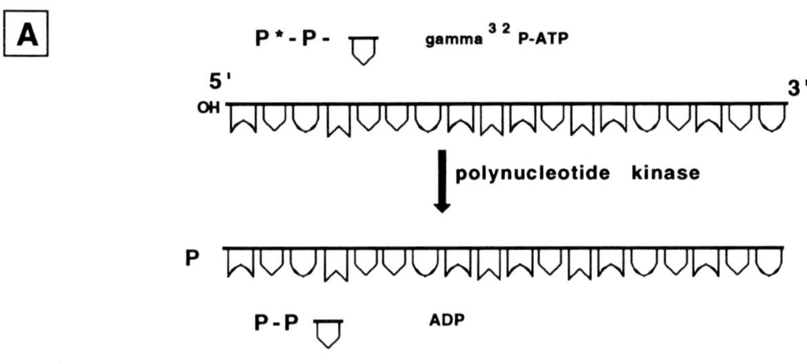

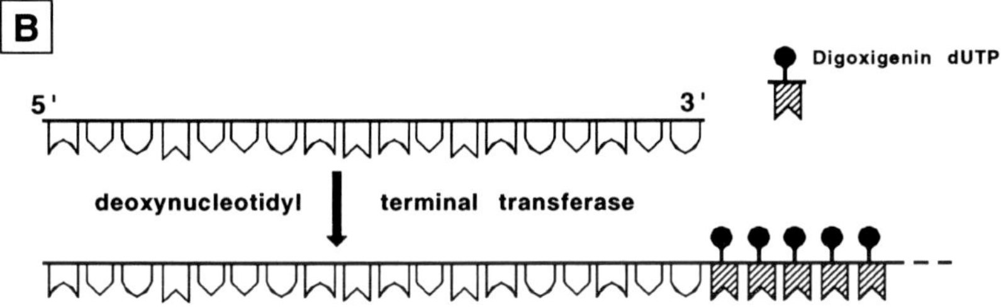

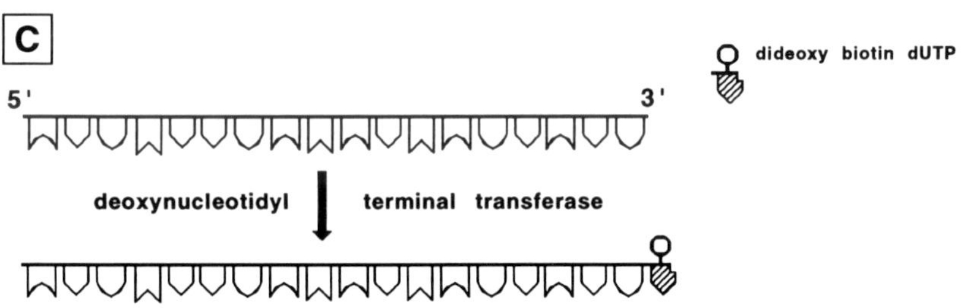

FIGURE 7. End labeling of oligonucleotides: (A) illustrates the transfer of the gamma phosphorus-32 labeled phosphate group of ATP to the 5′ hydroxyl residue of the oligonucleotide, mediated by the polynucleotidyl kinase; (B) and (C) show labeling at the 3′ end of the oligonucleotide using the deoxynucleotidyl transferase, with either a biotin dUTP (or any labeled nucleotides) or a dideoxy digoxigenin dUTP (or any labeled dideoxynucleotides).

ATP (donor molecule) to the 5′ terminus of DNA (or RNA). The absence of phosphate group at the 5′ terminus of chemically synthesized oligonucleotides renders the reaction of transfer very efficient.

Alternatively, oligonucleotides terminus labeling with ^{35}S can be achieved with ^{35}S-gamma ATP whose gamma phosphate residue carries an atom of sulfur-35.

2. Labeling on the 3′ Side

Adding labeled nucleotides on the 3′ end of a synthetic oligonucleotide is the method to use for preparing nonradioactive oligoprobes. This reaction is carried out with the enzyme deoxynucleotidyl terminal transferase.[15] This enzyme initiated 5′ to 3′ polymerization on oligonucleotides and incorporates available triphosphated nucleosides (Figure 7b). This reaction leads to tailing the oligonucleotide which has the advantage of increasing substantially the specific activity of the probe. It should be noted, however, that the addition of an homopolymer of varying sizes may alter the specificity of hybridization. Also, the hybridization of the prolonged oligonucleotide may need readjustment of the physicochemical parameters of the annealing reaction.

Labeling on the 3′ end can be constrained to the addition of only one nucleotide by using a chain terminator precursor, such as a nucleotide devoid of hydroxyl residue on the carbon 3′ of the pentose, the one involved in the establishment of the phosphodiester bonding during polymerization (such precursors are referred to as di-deoxyribonucleotides) (Figure 7c). Chain terminators are available labeled with ^{32}P, ^{35}S, biotin, digoxigenin, or fluorescein.

V. CONCLUSION

The plurality of the now established protocols for labeling DNA and RNA, together with the availability of a large variety of precursors (and others are to be developed), either radioactive or nonradioactive has led to significant improvement of the sensitivity of the method of *in situ* hybridization, i.e., detecting DNA or RNA sequences with very low abundance in the cell, and of the resolution of the topological detection of hybrids, at the tissular, cellular, and subcellular levels, either at the light or the electron microscope. By these advancements, preparing one's own probe (synthetic oligonucleotide or PCR-amplified DNA) provides also a certain degree of autonomy in experimentation.

The task the ''*in situ* hybridizer'' has to face is to select, among many, the proper protocol that will adequate the methodological strategy to the nature of the question to solve, and to the properties and parameters of the biological materials. The purpose of the following presentation is to illustrate this, by showing the manner by which the principles exposed in this chapter are converted into reality.

REFERENCES

1. **Chien, A., Edgar, D. B., and Trela, J. M.,** Deoxyribonucleic acid polymerase from the extreme thermophile *Thermus aquaticus.*, *J. Bacteriol.*, 127, 1550, 1976.
2. **Baldino, F. and Lewis, M. E.,** Non-radioactive *in situ* hybridization histochemistry with digoxigenin-dUTP labeled oligonucleotides, in *Methods in Neuroscience,* Conn, P. M., Ed., Academic Press, New York, 1989, 282.
3. **Bollum, F. J.,** *The Enzymes,* Vol. 10, Boyer, P. D., Ed., Academic Press, New York, 1974.
4. **Clavel, C., Doco-Fenzy, M., Lallemand, A., Binninger, I., and Birembaut, P.,** Detection par hybridation *in situ* des papillomavirus avec des sondes marquées par sulfonation, *Ann. Pathol.*, 2, 137, 1989.
5. **Cox, H. K., Deleon, D. Y., Angerer, L. M., and Angerer, R. C.,** Detection of mRNA in sea urchin ezmbryos by *in situ* hybridization using assymetric RNA probes, *Dev. Biol.*, 101, 485, 1984.
6. **Forster, A. C., Mc Innes, J. L., Skingle, D. C., and Symons, D. C.,** Nonradioactive hybridization probes prepared by the chemical reaction labeling of DNA and RNA with a novel reagent: photobiotin, *Nucleic Acids Res.*, 13, 745, 1985.
7. **Kelley, W. S. and Stump, K. H.,** A rapid procedure for isolation of large quantities of *Escherichia coli* DNA polymerase 1 utilizing a polA tranducing phage, *J. Biol. Chem.*, 254, 3206, 1979.
8. **Klenow, W. S., Overgaard-Hansen, K., and Pathar, S. A.,** Proteolytic cleavage of native DNA polymerase into two different catalytic fragments. Influence of assay conditions on the change of exonuclease activity accompanying cleavage, *Eur. J. Biochem.*, 22, 371, 1971.
9. **Langer, P. R., Waldrop, A. A., and Ward, D. C.,** Enzymatic synthesis of biotine labeled polynucleotides novel nucleic acid affinity probes, *Proc. Natl. Acad. Sci. U.S.A.*, 78, 6633, 1981.
10. **Mattila, P., Ronka, J., Tankanen, T., and Patkanen, K.,** Fidelity of DNA synthesis by the Thermococcus litoralis DNA polymerase — an extremely heat stable enzyme with proofreading activity, *Nucleic Acids Res.*, 19, 4967, 1991.
11. **Melton, D. A., Krieg, P. A., Rebagliati, T., Maniatis, T., Zinn, K., and Green, M. R.,** Efficient *in vitro* synthesis of biologically active RNA and RNA hybridization probes from plasmids containing a bacteriophage SP6 promotor, *Nucleic Acids Res.*, 12, 7035, 1984.
12. **Morris, C. E., Klement, J. F., and McAllister, W. T.,** Cloning and expression of the bacteriophage T3 RNA polymerase gene, *Gene,* 41, 193, 1986.
13. **Niedobitek, G., Finn, T., Herbst, H., and Stein, H.,** *In situ* hybridization using biotinylated probes. An evaluation of different detection systems, *Pathol. Res. Pract.*, 184, 343, 1989.
14. **Rigby, P. W. J., Dieckmann, M., Rhodes, C., and Berg, P.,** Labeling deoxyribonucleic acid to high specific activity *in vitro* by nick translation with DNA polymerase 1, *J. Mol. Biol.*, 113, 237, 1977.
15. **Riley, L. K., Marshall, M. E., and Coleman, M. S.,** A method for biotinylating oligonucleotide probe for use in molecular hybridizations, *DNA,* 5, 333, 1986.
16. **Saki, R. K., Sharf, S., Faloona, F., Mullis, K. B., Horn, G. T., Erlich, H. A., and Arnheim, N.,** Enzymatic amplification of β-globin genomic sequence and restriction site analysis for diagnosis of sickle cell anemia, *Science,* 230, 1350, 1985.
17. **Sverdlov, E. D., Monastyrskaya, G. S., Guskova, L. I., Levitan, T. L., Sheisenko, V. J., and Budowski, E. J.,** Modification of the cytidine residues with a bisulfite-*o*-methylhydroxyamine mixture, *Biochim. Biophys. Acta,* 340, 153, 1974.
18. **Tabor, S. and Richardson, C. C.,** A bacteriophage T7 RNA polymerase promotor system for controlled exclusive expression of specific genes, *Proc. Natl. Acad. Sci. U.S.A.*, 82, 1074, 1985.

Chapter 3

DNA *IN SITU* HYBRIDIZATION AND CONFOCAL MICROSCOPY

H. van Dekken, G. J. Brakenhoff, and J. G. J. Bauman

TABLE OF CONTENTS

0-8493-4414-X/93/$0.00 + $.50

I. GENERAL INTRODUCTION

Nonisotopic *in situ* hybridization of DNA has made tremendous progress in the last decade. Introduced as a new, but not very sensitive, tool for the localization of specific nucleic acids in microscopic preparations, it developed rapidly into a fast, high-sensitivity, high-resolution detection procedure. It allows for the cytogenetic analysis of specific molecular targets in cells and cell nuclei for clinical and cell biological purposes. The introduction of new and improved detection systems, such as confocal microscopes and CCD cameras, has greatly facilitated the analysis of the hybridization products. In this chapter an overview of present applications and approaches, combining DNA *in situ* hybridization and confocal microscopy, is described and shown.

II. NONISOTOPIC *IN SITU* HYBRIDIZATION OF DNA

A. METHODS FOR PROBE LABELING

The technique of *in situ* hybridization (ISH) was introduced by Gall and Pardue[1] and John et al.[2] in 1969, using radioactively labeled probes. At that time it was called ISH to emphasize the difference from the biochemical hybridization method, introduced in 1961 by Hall and Spiegelman.[3] Using radioactive isotopes, the insulin gene was mapped to chromosome 11 by Harper et al. in 1981.[4] The first reports on nonisotopic ISH came from Rudkin and Stollar in 1977.[5] After hybridization of the 5S ribosomal RNA genes on *Drosophila* polytene chromosomes with an RNA probe, they used a rabbit-derived antibody against DNA-RNA hybrids for detection. The antibody was visualized by a second fluorochrome-labeled antibody against rabbit immunoglobulins. This technique was further explored by Van Prooijen-Knegt et al.,[6] but the difficulty in obtaining good anti-DNA-RNA antibodies impaired further development. A direct approach was used by Bauman et al.,[7] who applied fluorochrome-labeled RNA as a probe for the detection of kinetoplast DNA in the insect Crithidiae Luciliae and of 5S ribosomal RNA genes in *Drosophila melanogaster.*

The most important labeling techniques for ISH came with the introduction of enzymatically or chemically modified nucleic acid probes. In 1981 Langer et al.[8] reported the first enzymatic synthesis of biotin-labeled polynucleotides. In 1982 the first applications were reported by using this technique for mapping genes on *Drosophila* chromosomes[9] and centromeric DNA on mammalian chromosomes.[10] The detection of the biotinylated hybrids is accomplished by anti-biotin antibodies or avidin, coupled to fluorochromes or enzymes. The first chemical introduction of a molecule, that could serve as a hapten for immunological detection, into the DNA was described by Tchen et al.[11] and Landegent et al.[12] (both in 1984). They used 2-acetylaminofluorene (AAF) to modify the guanine residues in the nucleic acids. This technique was further refined by Landegent et al.,[13] who reported the first detection of a single copy sequence by nonisotopic ISH. Using AAF-modified DNA probes and anti-AAF antibodies, these authors were able to localize the human thyroglobulin gene. Hopman et al.[14] modified the cytosine bases in a genomic human DNA probe with mercuric acetate to achieve the simultaneous detection of mercurated human DNA and biotinylated mouse satellite DNA in a hybrid cell line, using different colors for labeling the probes. The first triple color ISH was reported by Nederlof et al.,[15] using a combination of biotin-labeled, mercurated, and AAF-modified probes.

More recently reported modification procedures for ISH were the use of sulfonated probes[16] and the incorporation of digoxigenin into the cytosine residues of the DNA (e.g., Reference 17). In both cases, the hybrids are detected by specific monoclonal or polyclonal antibodies. The polymerase chain reaction (PCR) principle has been employed for a new and promising probe labeling method. Koch et al.[18] and Weiser et al.[19] used primers for chromosome-specific alpha-satellite repeats, followed by a PCR step to produce alfoid DNA probes for ISH. By including biotin-labeled nucleotides in this reaction the probes were labeled immediately. The same principle was applied for the synthesis of human chromosome(region)-specific DNA probes, derived from human DNA containing hybrid cell lines, by using alu or L1 repeats as a primer.[20]

At present the most frequently used labels for ISH are biotin and digoxigenin mediated systems, which are easy to use, commercially available, and of high sensitivity.

B. THE IMPROVEMENT OF SENSITIVITY

When nonisotopic *in situ* hybridization was introduced, only large targets could be detected. This was in contrast with the radioactively labeled probes, which were able to detect smaller targets. But the higher speed of analysis and the more precise localization properties made nonisotopic ISH very attractive for cytogeneticists. Due to the absence of radioactive waste, laboratory space and equipment could be reduced. Two detection schemes are generally employed. The first uses a fluorochrome, such as fluorescein (FITC), rhodamin (TRITC), or TexasRed. The second uses enzymes, such as peroxidase or alkaline phosphatase. The first might yield a better resolution, the second more permanent preparations. A rapid improvement came as a result of new and more sensitive labeling procedures. At first only repetitive DNA sequences[21-24] were used with targets ranging from 100 kb up to a few megabases (Mb) (megabase = 1000 kb). Among those targets were the ribosomal RNA genes on the acrocentric chromosomes, the alpha-satellite DNAs on the centromeres, and the repetitive sequences on the long arm of the Y chromosome.[10,25,26]

Nonisotopic *in situ* hybridization with chromosome-specific repetitive DNA probes, mostly alpha-satellite DNA, appeared to be very useful in the assessment of the copy number of a specific chromosome in interphase nuclei of isolated cells,[26-29] as well as in nuclei within deparaffinized tissue sections.[30-32] Hopman et al.[33] used centromeric probes for chromosomes 1 and 18 to detect aberrations in interphase cells of bladder cancer specimens. Van Dekken et al.[34-36] employed this technique to detect numerical chromosome changes in interphase nuclei of hematological, urological, and gastric cancers. In a recent study a set of 12 repetitive DNA probes was applied for the cytogenetic analysis of solid tumors of various histologies.[37] From comparisons of results obtained from metaphase and interphase cells it became clear that *in situ* hybridization with chromosome-specific repetitive DNA probes could be used reliably to investigate the presence of numerical chromosome aberrations in the interphase situation.[27-29,34,38-40] The investigation of chromosome aberrations in interphase nuclei is of great importance to cancer research, since most tumors have low mitotic indexes.[41]

As a consequence of these results, more scientists started to use ISH as an instrument to label specific DNA sequences. The refinement of the procedure ultimately led to the detection of single copy sequences. In 1985 Landegent et al.[13] were able to detect a 22-kb unique sequence using an AAF-modified probe. A few years later, several investigators reported the detection of single copy genes, a few kb in size.[42-44] Lawrence et al.[44] achieved to detect a 5-kb unique sequence in interphase nuclei. This made nonisotopic ISH suitable for gene mapping purposes and nearly as sensitive as radioactive ISH, but preferable to the latter with respect to speed and resolution.

An important development came with the availability of whole chromosome-specific DNA libraries as probes for the detection of numerical and structural abnormalities of a specific chromosome in metaphase cells and, with less accuracy, within interphase cell nuclei.[45-47] The utility of ISH with chromosome-specific DNA libraries for the visualization of numerical changes in interphase cells was demonstrated by Kuo et al.[48] In this study numerical aberrations were detected accurately in both metaphase and interphase cell preparations. Regional chromosome-specific probes were produced by using DNA from hybrid cell lines containing human DNA either directly as a probe[49,50] or indirectly through the application of PCR technology.[20] Also microdissection libraries were reported for this type of ISH.[51] New interesting probes for (interphase) ISH are provided by yeast artifical chromosomes, since they can contain several hundred kilobase inserts.[52,53] Since all these probes are genomic DNAs, a prehybridization of probe DNA and genomic human DNA is included in the procedure in order to suppress binding of the labeled repetitive sequences within the probe DNA to repetitive sequences in the cellular target. This competitive hybridization method has also been described for the *in situ* mapping of cosmid cloned genomic sequences, which also contain repetitive elements.[17,54-56] Cosmid clones that contain insert (probe) DNAs up to approximately 40 kb can provide a bigger target for analysis than the much smaller inserts in plasmid (or phage) vectors. Especially promising is their increasing availability,[57] which has made cosmid gene mapping by fluorescent ISH at present an important instrument in genetics.[17] Clinical applications of cosmid ISH were shown in a growing number of papers.[53,58,59] A useful addition to this is the simultaneous ISH and R-banding procedure that was developed by Cherif et al.[60] The visualization of the bcr/abl breakpoint in Ph+ CML blood and marrow interphase cells has been reported using cosmid and phage cloned DNA sequences.[61,62] It was shown by ISH to metaphase and interphase cells that, due to chromatin compaction, closely related unique sequences could be resolved more accurately in the interphase state.[63,64] Recently, this interphase approach to gene localization was demonstrated by using differently colored cosmid probes within the chromosome Xq28 region.[65]

At present, repetitive, unique, and library DNA probes for the majority of human chromosomes are available.[48,57,66] It is clear that with such small targets, analysis of hybridization results can be facilitated by digital fluorescence microscopy.

C. *IN SITU* HYBRIDIZATION AND CONFOCAL MICROSCOPY

With the advent of confocal microscopes, high-resolution analysis of fluorescently labeled biological structures became possible without the necessity of performing time-consuming computer deconvolution procedures to remove the out-of-focus flare.[67,68] In such a set-up the digitized images are readily accessible for measurement and analytical purposes.[69,70] At present confocal imaging in nonisotopic *in situ* hybridization is used mainly for two purposes: DNA probe (gene) mapping and three-dimensional analysis of specific genetic targets within cells or tissues.

The interest of the biomedical community in the use of ISH for probe or gene mapping grew quickly after the publication of a paper by the group of Ward in *Science*.[17] In this report 13 known genes and 36 random clones, all in cosmid vectors, were mapped to bands on prometaphase chromosomes.

The locations were confirmed by molecular techniques. Further, by hybridizing more cosmids simultaneously, using different colors, probe order could be established. Of crucial importance to this investigation was the application of confocal microscopy, rendering (1) high-resolution images and (2) the possibility of computerized measurements on the digitized pictures. The combination of fluorescent ISH and confocal imaging appeared fast and accurate, especially when compared with the traditional radioactive methods. This type of approach was followed rapidly in clinical applications (e.g., by Selleri et al.[59]). A comparison of confocal microscopy of single-copy hybridization signals with conventional fluorescence microscopy and high-sensitivity video cameras showed little difference in sensitivity.[71] The resolution of the images in the confocal microscope was, however, much better, facilitating DNA probe mapping.

The potential to label specific regions on chromosomes by the use of nonisotopic hybridocytochemistry prompted several scientists to use this technique for the study of chromosome organization in cells. The organization of chromosomes within cell nuclei has been investigated for over a century. The great pioneer in this field was Rabl,[72] who in 1885 proposed a model of organization, which is still under study. This model implies that centromeres and telomeres of chromosomes maintain the metaphase alignment in the interphase nucleus with centromeres occupying one pole and telomeres the other. In Rabl's view, this would make sense for reasons of cellular organization during the cell cycle.

An important technique with respect to the study of the spatial organization of chromosomes within nuclei was introduced by Agard and Sedat.[73] They applied optical sectioning of intact cells by sampling images of successive planes, while moving the microscope focus stepwise through the DAPI-stained nucleus. After using deconvolution algorithms to eliminate the out-of-focus fluorescence, three-dimensional reconstructions were made to analyze the DAPI banding patterns of polytene chromosomes in intact *Drosophilia* nuclei. It was found that the polytene chromosomes were in Rabl alignment and that each occupied a confined domain.[74] Using anticentromere antibody serum, Moroi et al.[75] found the centromeres in mammalian nuclei at the periphery, while Hadlaczky et al.[76] observed nonrandom chromosome patterns, e.g., somatic pairing of homologs.

From *in situ* hybridizations on slides it was known that human centromeres, hybridized with chromosome-specific centromeric DNA probes, occupied distinct regions within interphase nuclei.[26,27,77] It was also noted that human chromosomes in hybrid cell line nuclei were visible as confined domains.[78,79] Rappold et al.[80] hybridized X and Y specific probes to interphase nuclei of two cell lines, one of the two carrying a X-Y translocation. By measuring the distances between the hybridization spots, they found that in the cell line, carrying the translocation, the distance between the two chromosomes was significantly smaller. The first nonisotopic ISH to nuclei in suspension was reported by Trask et al.[81] They hybridized mouse thymocyte nuclei with mouse satellite DNA and measured the probe-related fluorescence on a flow cytometer. In this way the quantitation of target sequences by flow cytometry became possible. The first ISH to metaphase chromosomes in suspension was described by Dudin et al.[82] They were able to visualize the presence of human chromosomes in a human × hamster hybrid cell line. Next, the flow cytometric detection of specific chromosomes within human cell line nuclei was described.[83,84] Since the procedure was performed in suspension, the nuclei could also be used to determine the three-dimensional (3-D) localizations of the resulting specific hybridization signals by means of computer-aided fluorescence microscopy.[85-88] The first ISH data, derived from 3-D sources, were reported by Manuelidis.[89] Using physical sectioning and 3-D image reconstruction, she found a nonrandom distribution of mouse satellite DNA in different cell types of nervous tissues. Pinkel et al.[90] applied optical sectioning after ISH to hybrid nuclei in suspension, thereby preserving the 3-D morphology, to demonstrate human chromosomal domains in 3-D reconstructions of human × hamster hybrid cells. Manuelidis and Borden[91] used optical sectioning to reconstruct centromeric and telomeric sequences in human CNS cells after ISH to brain tissue sections. They observed that the chromosome-specific regions occupied specific compartments in these nuclei. This compartmentalization appeared to be dynamic. A shift of the X chromosome was visualized in epileptic cortex nuclei.[92] In 1989 Van Dekken et al.[85] used a similar procedure to reconstruct the positions of human genomic, centromeric, and telomeric regions in hybrid and human cell nuclei. This investigation was extended by means of confocal microscopy to statistically quantify the x-y-z localization of the pericentromeric region of chromosome 1 in different blood cell nuclei.[86,87] These DNA sequences were found to be present predominantly in the nuclear periphery. A nonrandom positioning of centromeric and ribosomal repetitive DNA sequences was also demonstrated by Popp et al.[93] and Rawlins and Shaw[94] after optical sectioning of hybridized cells.

III. CONFOCAL MICROSCOPY

A. INTRODUCTION

The aspect that has turned confocal microscopy into such a powerful technique is its ability to make selective images of just one plane inside a specimen. This property called optical sectioning is based on the virtual exclusion of information from out-of-focus regions located above and below the selected plane and is accompanied by excellent imaging transverse to the optical axis. These two optical aspects of confocal microscopy make such an instrument, when coupled to a computer system, ideal for direct mapping of 3-D data from extended specimens into computer memory. A confocal microscope proves to be particularly effective when operating in fluorescence. The resolution capabilities are indicated by the fact that under optimal circumstances the volume element contributing to each data point has dimensions of 0.2 μm laterally and 0.8 μm axially. The particular optical conditions in the microscope determine the actual performance of the system and how the optical sampling volume may be optimally fitted to the 3-D data collection grid.

Confocal scanning microscopy is beginning to take its place among the recognized techniques for the study of the spatial organization of biological structures. As far as resolution is concerned it forms a bridge between the limited resolving capability of conventional light microscopy and the superior potential of electron microscopy. However, in comparison with the latter technique, confocal microscopy has two great advantages. First, biological objects can be studied in their natural watery environment, thus avoiding the negative influence on morphology of the preparation techniques necessary for electron microscopy, e.g., chemical fixation, dehydration, embedding, and sectioning. Volume shrinkages of up to 50% have, for instance, been observed in bacteria as a consequence of such preparation steps.[95] Second, whereas in electron microscopy ultramicrotomy techniques have to be employed, in the confocal technique the spatial information can — due to its optical sectioning capability — be obtained directly, leaving the specimen basically intact.

The idea of applying the confocal concept to light microscopy dates from the late 1970s. It was shown by Sheppard and Choudhury[96] that in confocal scanning microscopy a fundamental improvement in imaging, as compared to normal microscopy, could be expected. Brakenhoff et al.[97,98] showed that this expectation could be realized in practice at high numerical apertures (NA), where the gain in resolution is of real practical value, together with the optical sectioning effect in the situation of diffraction-limited imaging. Earlier, optical sectioning was demonstrated in the range governed by geometrical optics by Egger and Petran.[99]

B. THE CONFOCAL MICROSCOPE TECHNIQUE

The basic principle of confocal microscopy is explained in Figure 1. The arrangement shown is for operation in fluorescence. The principle can also be applied in transmission. For the theory of image formation we refer to Brakenhoff et al.[97] and Wilson and Sheppard.[100] Sufficient to say here that the imaging can be considered to be determined by the interaction of the object with the product of the illumination distribution and the detection sensitivity distribution. If these 3-D distributions are diffraction limited and overlap optimally, it can be shown that improved imaging results, when compared with conventional microscopy.[96,100] For point objects, for instance, lateral and axial resolution improve by a factor of 1.38. In order to obtain an absolute gain with respect to nonconfocal microscopy, optics have to be used with maximum NA, i.e., with NA = 1.3 to 1.4. With good optics, the expected improved confocal imaging can be shown to be realizable in practice,[97,98] resulting in observed point resolutions down to 130 to 140 nm.

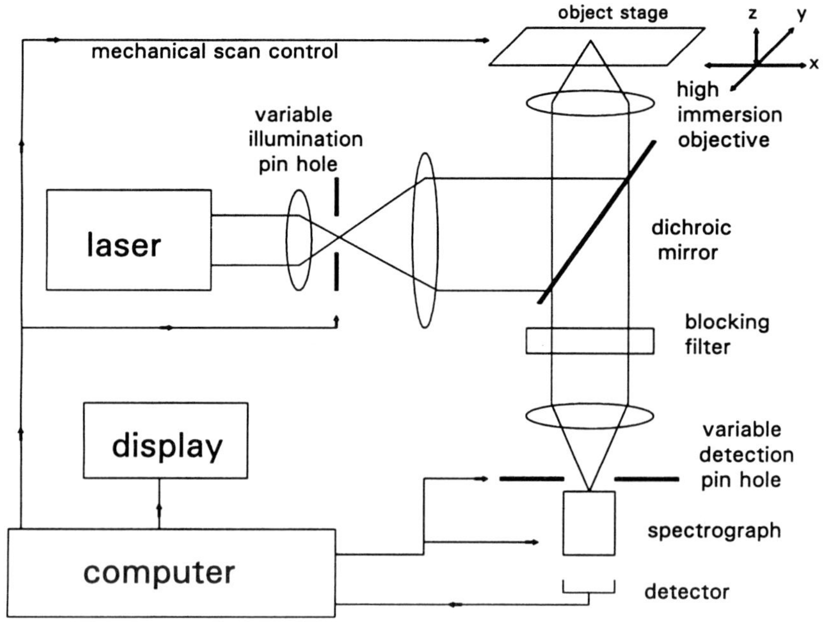

FIGURE 1. Principle of the "Amsterdam" confocal scanning fluorescence microscope. Light originating from a laser-illuminated pinhole is focused on a certain point of the object while the same point is also precisely imaged on a detector pinhole. The term confocal relates to the fact that the image of the illumination pinhole and the back projection of the detection pinhole have a common focus in the object. The specimen in the arrangement shown is scanned mechanically through this confocal point and the image data, in addition to being displayed directly, are digitized and stored in a computer system. The optical arrangement, the dichroic beam splitter/ blocking filter, is as commonly used in fluorescence microscopy. The spectrograph enables one to select a certain band from the fluorescence radiation for image formation. Both the band width and the wavelength on the spectrograph can be set by the computer system. For confocal reflection microscopy, the blocking filter is omitted and the dichroic mirror is replaced by a beamsplitter.

The sectioning property in confocal microscopy deserves special attention. In normal microscopy all the radiation generated at various levels in the specimen will reach the image plane. This often causes a strong reduction in the contrast of the image of the in-focus level of the specimen. However, in confocal microscopy the optical arrangement results in almost complete suppression of the contributions from the out-of-focus levels in the specimen. The sectioning power of confocal microscopy is perhaps best illustrated by the fact that in normal fluorescence microscopy one cannot find the vertical position of a uniformly fluorescing layer in a specimen, while in confocal microscopy its height can be determined with the precision as defined by the improved axial response function.

In confocal microscopy one can either scan the beam[101,102] or scan the specimen[67,97] to obtain the specimen data. Other approaches are also possible.[103] The instrument shown in Figure 1 is of the specimen-scanning type.

It is completely controlled by a computer system which takes care both of the instrument operation and the 3-D data collection, representation, and processing functions. The 3-D image acquisition is done with the help of a software routine, automatically collecting the data from a series of 2-D images at various heights and storing these in memory. Such a 3-D data set is called a 3-D image. We typically collect data from 16 sections on a 256 × 256 pixel grid for each section. Also larger formats are possible. The collection of such a 3-D image takes about 1 min.

C. CONFOCAL IMAGING AND REPRESENTATION

As indicated above the main property of confocal microscopy is that one is able to sample just one volume element in a specimen, independently of its surroundings. For basic processing of these data the usual filter routines, adapted for 3-D image data sets, can be applied. The data collected and processed by these methods can be represented in various ways. A straightforward and very informative representation mode of 3-D data sets is of a series of sections acquired at increasing depth. The depth information in the specimen can be effectively presented as a number of x to z cross sections, where the z-axis is along the optical axis, and x and y are the lateral coordinates. Series of cross sections like these can provide information on the exact spatial positioning of specific structures. In addition some more visually appealing methods are available. Pairs of stereoscopic images can, for instance, be constructed by the following method. One image of such a pair is generated in the computer by projecting the 3-D data set on the image plane along a certain projection direction. Each pixel in the image-to-be is generated by means of a projection line through the data set. To determine the value to be assigned to that pixel from the voxel values encountered along the projection line, one can use various algorithms such as additions, retaining the maximum, or retaining the minimum. Two images with different projection directions form a stereo pair. In our group, images are also created by the so-called ''simulated fluorescence process'', which gives a direct impression of the 3-D structure of the biological object without recourse to stereoscopic methods. The idea behind this method is that in the computer the excitation and emission steps of the fluorescence process are simulated, operating directly on the values of the collected 3-D image. This is done by illuminating the 3-D data set from a certain angle, using a computer program. The 3-D image elements absorb this (artificial) radiation in proportion to their value, and subsequently emit radiation in proportion to the absorbed quantity. While passing a 3-D image element, the radiation is absorbed in proportion to the element value, both in the absorption and reemission phase. Details of the system and the algorithms were described by Van der Voort et al.[70,104] This process, in which the illumination and emission directions can be different, also allows adjustment of the absorption and emission parameters used. A background is added to provide better depth perception. The result is a very realistic representation of the object, due to the natural way structural elements are highlighted and shadowed.

D. BILATERAL CONFOCAL SCANNING WITH CCD DETECTORS

While excellent 3-D confocal imaging has been produced by the approaches indicated above, some limitations of these systems are presently surfacing, especially regarding high speed applications. These are partly associated with the poor photon economy of such systems. For instance, the photomultipliers used for signal detection have quantum efficiencies of maximally 20 to 25% in the blue, falling off strongly towards the red where they are in the order of 5 to 10%. This latter range is especially important for detection in fluorescence. High speed image acquisition with photo-multipliers necessitates a wide frequency bandwidth in the detection chain. This results in increasingly nonoptimal signal-to-noise conditions during image acquisition at higher speeds.[105] In a bilateral scanning approach a two-sided scanning mirror element is utilized to scan both the confocal illumination spot and direct the photons produced to the detection system.[106]

One of the main reasons the bilateral scanning concept is so useful is that it permits the use of CCDs (charge coupled device) as detectors. The main reasons why this device is particularly suitable for this purpose are the following:

1. High quantum efficiencies (up to 80 to 100%) for specialized devices, with 50% attainable in the regular commercial types.
2. The high dynamic range of CCDs matches the image contrast range of confocal images, the high linearity creating optimal conditions for subsequent image processing.
3. The integration capability on the photon/electron level enables decoupling of scan-speed and read-out speed, thus improving basic signal-to-noise detection conditions. This aspect is also instrumental for the creating of real-time (stereoscopic) 3-D images of spatial objects.
4. In this set-up the use of a CCD assures imaging with high geometric fidelity.

In the above we have described the present confocal approaches. These can be tailored towards specific imaging applications such as high or low speed, or the desirability of having a specific optical probe available for imaging. While the older systems have proved their value, it may well be that the novel bilateral scan approach combined with CCDs can provide the optimal confocal imaging solution for weakly fluorescing or bleach-sensitive biological specimens.

IV. MATERIALS AND METHODS

A. LIST OF MATERIALS

Material	Manufacturer	Code
	(addresses below)	
Chromosome preparation		
Chromosome medium	Boehringer	807311
Colcemid	Boehringer	295892
Potassium chloride	Sigma	P4504
Methanol p.a.	Merck	6009
Acetic acid 100%	Merck	63
Pretreatments		
3-amino-propyltriethoxysilane	Serva	13165
Acetone p.a.	Merck	14
Xylene	Merck	8687
Ethanol p.a.	Merck	983
Hydrogen peroxide	Merck	7209
Pepsin	Sigma	P 7000
Tween-20	Merck	822184
Hydrochloric acid	Merck	317
Formalin solution p.a.	Merck	4003
RNase A	Boehringer	109169
Hybridization components		
Formamide	Merck	9684
Mixed bed resin (Dowex AG501-X8)	Bio-Rad	142–6424
Dextran sulfate	Pharmacia	17–0340–02
Herring sperm DNA	Boehringer	223646
Human Cot-1 DNA	Gibco BRL	5279SA
Biotin labeling of DNA		
Nick translation kit	Gibco BRL	8247SA
• 10× dNTP mix (including BIO-14-dATP)		
• 10× enzyme mix		
• Autoclaved H_2O		
• STOP buffer		
Fluorescent detection reagents		
Nonfat dry milk	Carnation, or any other brand	
Avidin-FITC	Vector Lab.	A2011
Biotin-labeled goat-α-avidin	Vector Lab.	BA0300
Propidium iodide	Sigma	P4170
PPD (*p*-phenylenediamine dihydrochloride)	Sigma	P1519
Glycerol	Sigma	G5516
Buffers		
PBS (pH 7.4)		
Sodium chloride	Merck	6404
Potassium dihydrogenphosphate	Merck	4873
Disodium hydrogenphosphate	Merck	6580

IV. MATERIALS AND METHODS

A. LIST OF MATERIALS

Material	Manufacturer	Code
2× SSC (pH 7)		
Sodium chloride (0.3*M*)	Merck	6404
Sodium citrate (0.03 *M*)	Merck	6448
4× SSCT (pH 7.5)		
Sodium chloride (0.6 *M*)	Merck	6404
Sodium citrate (0.06 *M*)	Merck	6448
Tween-20 (0.1%)	Merck	822184
4× SSCTM (pH 7.5)		
Sodium chloride (0.6 *M*)	Merck	6404
Sodium citrate (0.06 *M*)	Merck	6448
Tween-20 (0.1%)	Merck	822184
Nonfat dry milk (5% w/v)	Carnation, or any other brand	

Manufacturers' addresses	
Bio-Rad	Richmond, CA, U.S.
Boehringer	Mannheim, Germany
Gibco BRL	Gaithersburg, MD, U.S.
Merck	Darmstadt, Germany
Pharmacia	Uppsala, Sweden
Serva	Heidelberg, Germany
Sigma	St. Louis, MO, U.S.
Vector Lab.	Burlingame, CA, U.S.

B. FUNDAMENTAL CONCEPTS OF *IN SITU* HYBRIDIZATION

1. Introduction

The *in situ* hybridization procedure contains aspects related to both the maintenance of the morphology of the material under investigation and the achievement of specificity in the hybridization reaction. The final morphology of the tissue or cells will be determined by the fixation, pretreatment, and hybridization conditions. The specificity of the hybridization reaction will be determined by the properties of the probe and by the hybridization conditions. Additionally, obtaining high efficiency in both hybridization and reaction steps imposes separate requirements to the entire procedure. The following steps can be distinguished:

1. Preparation of the labeled probe
2. Preparation of the slides, fixation, and pretreatment of the material on the slide
3. Denaturation of the target DNA

4. The hybridization reaction, and the removal of aspecific and nonbound probe
5. Detection of bound probe

Each of these steps will be discussed in the next section. Probe properties determine the specificity and sensitivity of the hybridization reaction. The tissue and slide fixation and pretreatments are important to preserve the topological position of the target DNA, as well as to allow the penetration of the probe and detection reagents. The specificity of the hybridization is governed by precise conditions of hybridization mixture composition and temperature. The final washing and detection steps are of great importance in visualizing the position and quantities of hybridized probe with a minimum of background. The type of microscopy, fluorescence or transmittance, is related to the type of detection reagents used.

2. Probe Preparation and Labeling

Probes used for *in situ* hybridization are generally prepared using standard molecular techniques and we refer to the available manuals, e.g., Davis et al.[107] The most widely used probes are double-stranded DNA probes consisting of the DNA of interest cloned into a plasmid, cosmid, or bacteriophage DNA vector. Further, synthetic oligonucleotide probes are being increasingly utilized.

Labeling of DNA probes can be performed either enzymatically or chemically. Enzymatically, modified nucleotides are incorporated using the techniques generally adopted for radioactive labeling. Biotin- or digoxigenin-labeled nucleoside triphosphates are most commonly applied in procedures, such as nick translation or random primer extension. Chemically, a variety of methods, such as photobiotinylation, sulfonation, mercuration, and AAF-modification, are available. A more detailed overview can be found in Bauman et al.[108] Before, during, or after modification, the probe size should be controlled in order to obtain maximum penetration of the labeled probe fragments into the target cells. Optimum probe fragment size ranges from 50 to 500 nucleotides.

Probe and detection reagent accessibility can be improved by pepsin or proteinase treatment. For most tissues, pepsin digestion proved most satisfactory with respect to preservation of morphology.

The kinetics of *in situ* hybridization depend on probe-to-tissue penetration and the hybridization reaction itself. The latter follows the laws as determined in biochemical assay conditions, i.e., filter hybridization (see Hames and Higgins).[109] Tissue dependent fixation parameters, proteolysis, and lipid extractions on submicroscopic network spacing are poorly understood, so that optimal conditions are best set empirically.

3. Hybridization Reaction Conditions

The functional properties of nucleic acids are encoded in their nucleotide base sequence. The nucleotides in the two opposite strands within the double-stranded DNA (ds-DNA) are pairwise complementary. Adenine (A) on one strand is always paired with thymine (T) on the other strand, likewise guanine (G) always pairs with cytosine (C). In RNA the thymine is replaced by uridine (U). The complementary nucleotide strands are held together by hydrogen bonds. The two complementary strands can be separated by agents that disrupt these hydrogen bonds, such as heat, extreme acid, or alkaline pH, and hydrogen bond-breaking chaotropic agents like formamide. The energy needed to break the bonds between the two base pairs is different. It is lower for A.T pairs than for G.C pairs, since A.T is bound by only two hydrogen bonds, whereas a G.C pair is formed by three. This implies that relatively G.C-rich (unique) DNA denatures at a higher temperature than relatively A.T-rich (repetitive) DNA. The separated strands remain intact by the sugar-phosphate moieties forming the backbone structure of nucleic acids. The backbone phosphate groups are negatively charged, and this electrostatic repulsion assists in the separation of the two strands once the hydrogen bonds are broken. Strand separation or melting occurs at specific temperatures ruled by the base composition of the nucleic acid, the length of the molecules, and the chemical composition, e.g., salt and formamide concentrations, of the buffer used. Empirical formulas can be used to calculate this so called melting temperature Tm in °C:

$$\mathrm{Tm} = 81.5 + 0.41\,(\%\mathrm{GC}) + 16.6 \log\,[\mathrm{Na}^+] - 0.72\,(\%\ \mathrm{formamide})$$

where Tm is the melting temperature in °C; (%GC) the percentage G + C of the DNA; $[Na^+]$ the molar concentration of Na^+ in the buffer solution; and (% formamide) the concentration of formamide, the most widely used hydrogen bond-disrupting agent.

The melting temperature of ds-DNA is lower than that of DNA.RNA hybrids, which in turn is lower than that of ds-RNA. Melting temperature properties govern the reaction conditions for optimal hybridization. Single-stranded DNA, denatured by heat, forms double-stranded hybrids, when the temperature is lowered again. The optimal hybrid formation is about 25 °C below the Tm of ds-DNA. The perfection of base pairing, also called the stringency required for hybrid formation, is thus a function of the difference between the Tm and the hybridization temperature. The reaction rate is a function of the concentrations of probe and target sequences. The effective concentration of probe DNA in the hybridization solution can be increased by the inclusion of water capturing polymers, e.g., dextran sulfate. For more theoretical information, see Cantor and Schimmel,[110] and Hames and Higgins.[109]

4. Hybridization, Washing, and Staining Conditions

Sensitive detection of small DNA targets is maximal, when high signals are obtained with minimal background reactions. Background signals can be derived from nonspecific probe binding and nonspecific binding of reporter molecules. Probe binding specificity is determined by the *stringency* of the hybridization and washing conditions, while *specificity of detection* is established by the type of reporter molecules and incubation conditions.

Stringency — Probe molecules binding to target sequences that do not have a full sequence homology will form hybrids with a high percentage of bases forming mismatches (%Mm). The formation of such hybrids can be prevented by modifying the stringency of the hybridization conditions, i.e., temperature, salt concentration (SSC), or % formamide. Likewise the washing conditions after hybridization can be adjusted. Stringent hybridization conditions, hybridizing less than 25°C below the Tm of the desired hybrids, are in general more effective than increasing the stringency of washing.

Specificity of detection — Background reduction can be achieved by staining in high salt conditions like 4× SSC,[44] which prevents ionogenic binding of detector molecules. This is most clear for highly charged molecules, such as avidin conjugates. Also the addition of nonfat dry milk, instead of serum, in the staining solution is effective in background reduction. The sensitivity can be increased by amplification of detector signals by immunocytochemical methods, directed against the reporter molecule in use. Antibodies, specific for FITC, avidin, horseradish peroxidase, or alkaline phosphatase have been employed for this purpose.

C. DNA *IN SITU* HYBRIDIZATION TO CELLS ON SLIDES

DNA-ISH to metaphase or interphase cells, fixed on glass microscope slides, is the most widely used form of the technique. It is illustrated in Figure 3.

1. Probe and Probe Labeling

The human chromosome 6p21 genomic DNA probe, in a yeast artificial chromosome (YAC) vector, was obtained from Dr. A. Geurts van Kessel (KUN, Nijmegen, The Netherlands). For *in situ* hybridization, the complete vector plus insert DNA (this accounts for all types of vectors and inserts!) was labeled with biotin-14-dATP by nick translation according to the manufacturers' directions. The DNA probe was stored at −20°C.

2. Metaphase Preparation

Ten ml of heparinized whole peripheral blood was cultured in 100 ml complete chromosome medium. This chromosome medium is based on medium F12 with fetal calf serum, $NaHCO_3$, glutamine, streptomycin, penicillin, and PHA-M. After culturing the blood cells for 72 h under standard conditions, Colcemid (0.2 μg/ml final concentration) was added for another 90 min to collect mitotic cells. Then the cells were centrifuged (10 min, 200 *g*), and washed once in PBS at room temperature (RT). The cells were spun down, resuspended, and incubated in an excess of hypotonic solution (75 m*M* KCl) for 15 min at 37°C. After centrifugation the supernatant was removed carefully, and methanol/acetic acid (3:1, v/v) fixative was added dropwise under continuous gentle mixing. Fixation was allowed for 10 min at RT, followed by centrifugation. This fixation procedure was repeated (3 to 5 times), until the pellet turned from brown to gray-white. The latter indicates the complete removal of erythrocytes. The cells, stored in the fixative at −20°C, could be used for experiments during several months.

3. Pretreatment

The cells, in fixative, were dropped onto ethanol-cleaned microscope glass slides, air-dried for 2 to 4 hours, and used for ISH. Before denaturation and ISH the cells on the slides were incubated with RNase A (200 μg/ml in 2× SSC, pH 7) for 30 min at 37°C to remove the cellular RNA. This RNA can lead to background fluorescence due to both specific (unique transcribed DNA sequences) and nonspecific binding of the probe. After rinsing twice in 2× SSC at RT, the specimens were pretreated in order to improve target accessibility, remove cellular debris, and preserve the nuclear material on the slides throughout the procedure (see Protocol 1):

PROTOCOL 1 — PRETREATMENT FOR CELLS ON SLIDES

- Perform protein extraction by immersing the slides in 0.1 *N* HCl + 0.05% Tween-20 for 15 min at RT.
- Rinse once in 2× SSC and then in PBS, both at RT.
- Fix the cells in 1% formalin in PBS for 5 min at RT.
- Rinse once in PBS and immerse the slides in 2× SSC until denaturation.

4. *In Situ* Hybridization

Production of ss-DNA for ISH was accomplished with heat denaturation. Slides were immersed for 2 min at 70°C in 70% deionized formamide, 2× SSC at pH 7. Formamide was deionized by treatment with a mixed-bed ion-exchange resin.[109] Then the slides were dehydrated in an ethanol series (70% [cold], 83%, 96%: 1 min cach) and air-dried at RT.

The DNA probes were denatured for 5 min at 70°C in the hybridization mixture.

For competitive ISH with unique genomic DNA probes this mixture contained (final concentrations):

- 5 to 10 μg/ml Probe DNA
- 500 μg/ml Sonicated genomic human DNA
- 250 μg/ml Human Cot-1 DNA
- 125 μg/ml Sonicated herring sperm DNA
- 0.1% Tween-20

- 10% Dextran sulfate
- 50% Deionized formamide
- In 2× SSC at pH 7

After a "pre-annealing" period of 30 min at 37°C (to block the repetitive sequences within this genomic probe fragment), the hybridization mixture (12.5 μl) was put on the heat-denatured slides under a 18 × 18-mm glass coverslip.

For ISH with repetitive (peri)centromeric DNA probes the hybridization mixture contained (final concentrations):

- 1 to 2 μg/ml Probe DNA
- 500 μg/ml Sonicated herring sperm DNA
- 0.1% Tween-20
- 10% Dextran sulfate
- 60% Deionized formamide
- In 2× SSC at pH 7

After denaturation the mixture was quickly cooled on ice and subsequently placed on the heat-denatured slides under a 18 × 18-mm glass coverslip.

The differences between ISH with genomic and repetitive DNA probes are outlined in Protocol 2.

PROTOCOL 2 — GENOMIC VS. REPETITIVE DNA PROBE SCHEME

	Unique genomic (e.g., YACs, cosmids	**Repetitive (e.g., centromeric)**
RNase treatment:	Yes	Optional
Pre-annealing:	Yes	No
Competition:	Cot-1 and/or genomic human DNA	No
% Formamide in hybmix and wash:	50	60

The ISH was performed overnight at 37°C in a humidified chamber. Then the coverslips were removed by immersing the slides briefly in 2× SSC. Next, the slides were washed 3 times for 3 min each in 50% formamide (60% for repetitive DNA probes; see Protocol 2) in 2× SSC, pH 7 at 42°C and subsequently 2 times for 5 min each in 2× SSC at 42°C. The slides were then immersed in 4× SSCT.

5. Cytochemistry

Detection of the biotin-labeled probes was accomplished with FITC conjugated avidin (5 μg/ml in 4× SSCTM) for 20 min at 37°C. The nonfat dry milk was added to the 4× SSCT to prevent nonspecific binding of the avidin. The slides were then washed 2 times in 4× SSCT for 5 min at 42°C. The probe-linked fluorescence was amplified, if necessary, by incubation with biotin-labeled goat-α-avidin (5 μg/ml in 4× SSCTM; 20 min at 37°C), washed in 4× SSCT, and incubated again for 20 min with the avidin-FITC (Protocol 3).

PROTOCOL 3 — ISH SIGNAL AMPLIFICATION PRINCIPLE

- Visualize biotin-labeled hybridized DNA probe with avidin conjugate (e.g., avidin-FITC).
- Wash the slides (in 4× SSCT).
- Tag bound avidin conjugate with biotin-labeled goat-α-avidin.
- Wash.
- Amplify ISH signal by second incubation with the avidin conjugate.
- Wash.

The DNA of the cells was then counterstained with propidium iodide (2.5 μg/ml) to allow simultaneous observation of total DNA and hybridized probe. The DNA stain was carried in an antifade solution (1 mg/ml PPD in 90% glycerol, 10% PBS, pH 8 [v/v]) to preserve the fluorescein fluorescence during prolonged microscopy.[111]

D. DNA *IN SITU* HYBRIDIZATION TO CELLS IN SUSPENSION

An example of this hybridization protocol that can be used for cell biological studies is shown in Figure 2.

1. Probe and Probe Labeling

The human chromosome 1q12 specific repetitive satellite DNA probe pUC1.77, 1770 bp in pUC18, was kindly supplied by Dr. H. Cook (MRC, Edinburgh, U.K.). For *in situ* hybridization, complete plasmid DNA was labeled with biotin-14-dATP by nick translation according to the manufacturers' directions. The DNA probe was stored at −20°C.

2. Cell Preparation

The promyelocytic cell line HL60 was cultured in RPMI medium, containing 10% fetal calf serum. In culture, this cell line displayed a tendency to become tetraploid, resulting in a mixture of diploid and tetraploid cells.

Whole cells, used for *in situ* hybridization, were washed once in PBS and fixed for later use (Protocol 4).

PROTOCOL 4 — FIXATION OF CELLS IN SUSPENSION

- Spin the cells (each centrifugation: 10 min, 200 *g*).
- Resuspend the cell pellet in 70% ethanol and fix for 10 min at 4°C.
- Spin the cells down and resuspend in 100% ethanol for 10 min at 4°C.

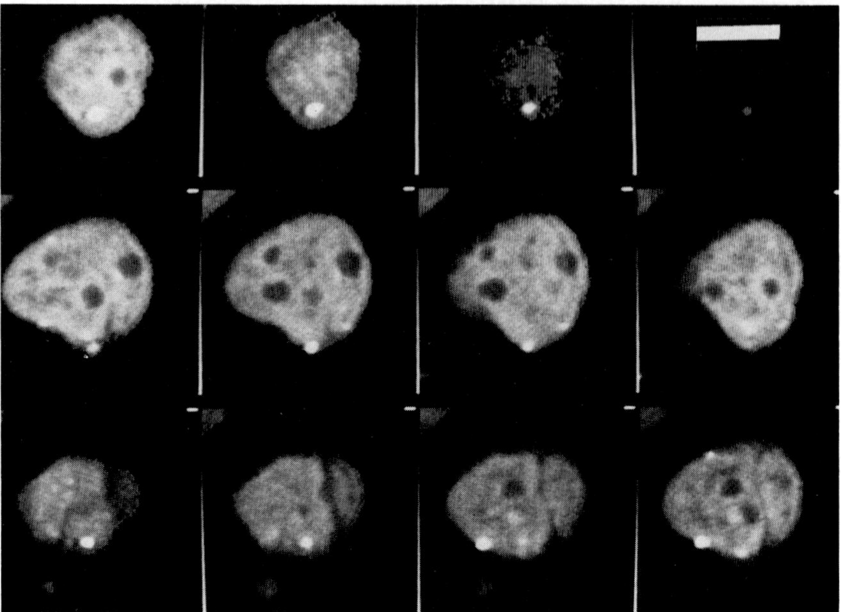

FIGURE 2. Photomicrographs demonstrating 12 optical sections (starting with the upper right image), obtained by the confocal microscope, described in Figure 1. A tetraploid HL60 cell nucleus was scanned after fluorescent *in situ* hybridization (ISH) with biotin-labeled 1q12 repetitive DNA to suspended HL60 cells. The hybridized probe was detected with fluoresceinated avidin, appearing white in the photomicrograph. The nuclear DNA was counterstained with propidium iodide (gray). Sections were typically obtained at sampling intervals of 1 μm (bar is 10 μm).

- Spin again, resuspend, and fix in 0.5% formalin in PBS for 5 min at 4°C.
- Wash once in PBS, and once in PBS + 0.1% Tween-20.
- Resuspend in 70% ethanol, and store the cells at −20°C at a concentration of 10^7/ml.

3. Pretreatment

The cells, when used for ISH, were washed once in PBS, and protein extraction was carried out (see Protocol 5).

PROTOCOL 5 — PRETREATMENT FOR FIXED CELLS IN SUSPENSION

- Resuspend the cell pellet in 0.1 *N* HCl + 0.05% Tween-20 at 4°C for 15 min.
- Spin the cells (200 *g*) for 10 min.

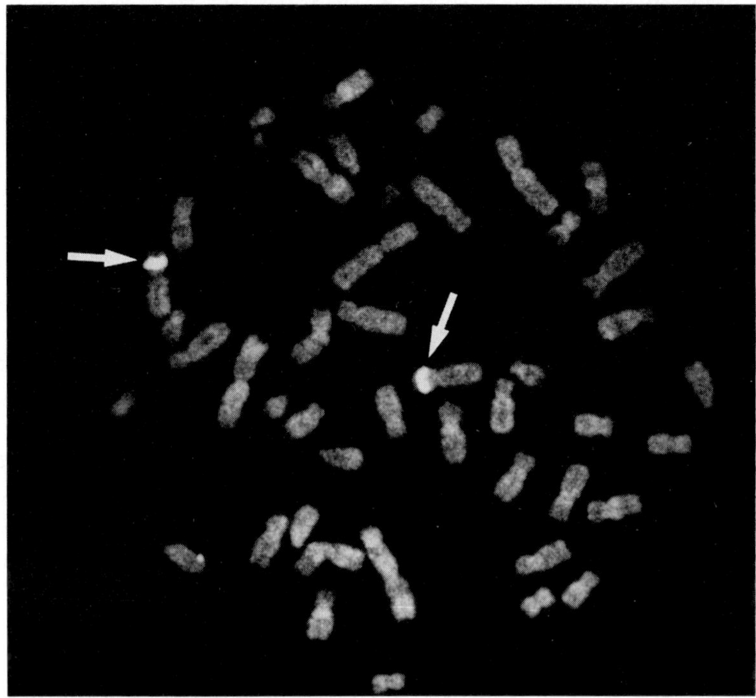

FIGURE 3. Confocal digitized image of biotin-labeled DNA sequences hybridized to band 2.1 of the short arm of human chromosome 6 (6p21) after fluorescent ISH to a human lymphocyte metaphase plate. The hybridized probe (white) is easily localized to the appropriate chromosome band (arrows). All of the DNA was counterstained with propidium iodide. This image was produced on a Bio-Rad MRC 600 confocal setup (KUN, Nijmegen, The Netherlands).

- Wash once in PBS at 4°C for 5 min.
- Resuspend pellet in PBS at 4°C at a concentration of 5 × 10^7/ml.

4. ISH to DNA in Suspended Cells

Isolated cells and probe were separately denatured by heating: 2-μl cell suspension was added to 8 μl of hybridization mix (final concentrations):

- 0.1% Tween-20
- 10% dextran sulfate
- 60% deionized formamide
- in 2× SSC at pH 7

This cell mixture was denatured for 10 min at 70°C. Probe DNA (1 to 2 μg/ml) and sonicated herring sperm carrier DNA (500 μg/ml) were also denatured for 5 min at 70°C in hybridization mix.

Then both mixtures were quickly cooled on ice and 20 μl of probe mix was added to the 10-μl cell mixture in an Eppendorf tube. The sample was mixed and incubated for 14 to 18 h at 37°C in a ventilating oven. Then cells were washed in 1 ml 60% formamide, 2× SSC, pH 7 for 10 min at 42°C. After centrifugation (each centrifugation: 10 min, 200 *g*), the pellet was resuspended in 1 ml 2× SSC for 10 min at RT and washed again in 1 ml 4× SSCT for 10 min at RT.

5. Cytochemistry

After resuspending the pellet in 100 μl 4× SSCTM for 5 min at RT, the biotin-labeled probe was stained by adding an equal volume of staining buffer (10 μg/ml avidin-FITC in 4× SSCTM). The sample was mixed and incubated for 30 min at 37°C in a ventilating oven. After centrifugation the pellet was washed in 1 ml 4× SSCT for 10 min at RT. The probe-linked fluorescence intensity was amplified by successive treatments with biotin-labeled goat-α-avidin (10 μg/ml in 4× SSCTM) and avidin-FITC (see principle in Protocol 3), and finally resuspended in 50 μl 4× SSCT.

Then 50 μl of the PPD antifade solution was added, containing 1 μg/ml propidium iodide. Due to the presence of glycerol the solution is very viscous, which prevents the cells from moving during microscopy. Now 5 μl of the mixture was put on a glass microscope slide under an 18 × 18-mm glass coverslip. Special care was taken to avoid pressure on the coverslip, which could have affected the morphology of the cells and nuclei.

E. DNA *IN SITU* HYBRIDIZATION TO CELLS WITHIN SECTIONS

This type of DNA-ISH is illustrated in Figure 4. It is of importance in pathology (and cytogenetics), since it allows histological and genetic evaluation on the same slide.

1. Probe and Probe Labeling

The human chromosome 7 specific centromeric repetitive satellite DNA probe was kindly supplied by Dr. H. Willard. For *in situ* hybridization, complete plasmid DNA was labeled with biotin-14-dATP by nick translation according to the manufacturers' directions. The DNA probe was stored at −20°C.

2. Tissue Preparation

The 4-μm paraffin sections were floated in warm (40°C) demineralized water and mounted on aminoalkylsilane-coated glass slides (see Protocol 6). Coating of the slides was performed according to the procedure of Rentrop et al.[112]

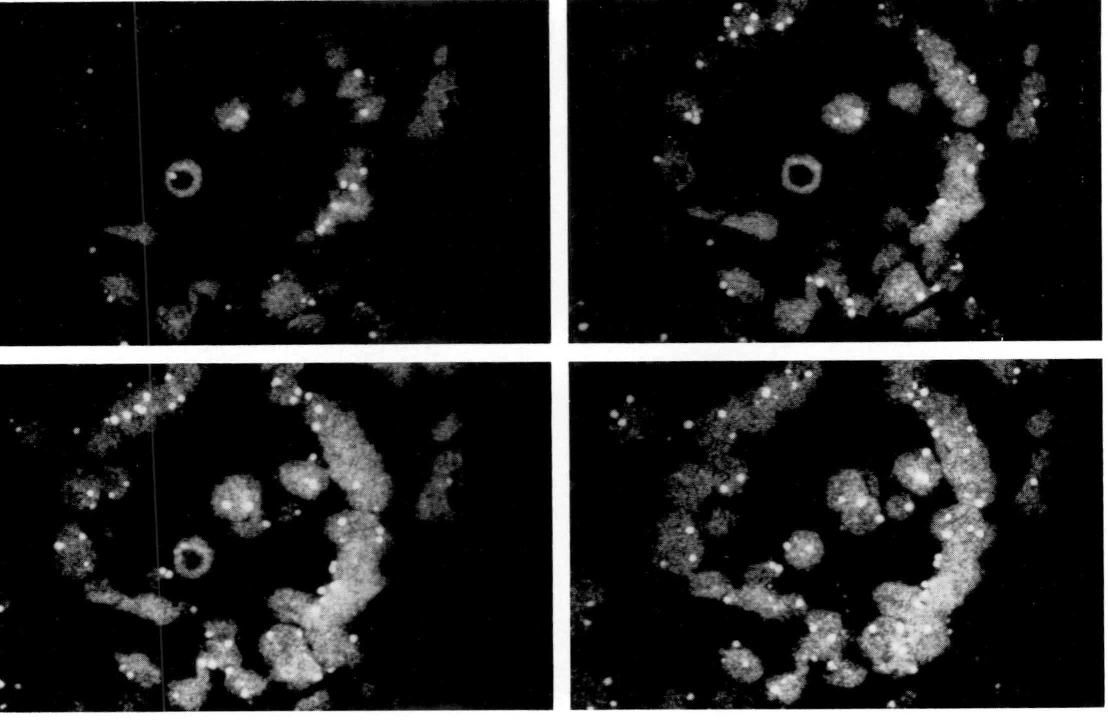

FIGURE 4. Four consecutive confocal images of fluorescent *in situ* hybridization to a 4-μm tissue section of a human endocrine tumor with a chromosome 7 specific centrometric DNA probe. Multiple hybridization spots (white), indicating aneuploidy of chromosome 7, can be seen in the nuclear sections within the tissue fragments. The sections were taken at 1 μm intervals, starting at the position shown at the upper left. These images were made on a Bio-Rad MRC 600 machine (ITRI-TNO, Rijswijk, The Netherlands).

PROTOCOL 6 — AMINOALKYLSILANE COATING OF GLASS SLIDES

- Put dry-cleaned glass slides (maximum 500) in glass racks; do not use metal racks and jars.
- Rinse slides in acetone at RT.
- Immerse racks in 2% aminoalkylsilane in acetone for 60 s at RT.
- Rinse in acetone.
- Dry the slides overnight at 60°C in a ventilating oven.
- Store coated slides in refrigerator.

The mounted sections were air-dried for several hours at 56°C in a ventilating oven.

3. Pretreatment

The sections were deparaffinized at RT as shown in Protocol 7.

PROTOCOL 7 — DEWAXING OF PARAFFIN-EMBEDDED SECTIONS

- Take the slides through 3 changes of xylene, 10 min each.
- Go through 4 changes of ethanol, 2 min each.
- Immerse in PBS.
- Block endogenous peroxidase activity in 0.3% H_2O_2 in PBS for 20 min.
- Rinse the slides twice in PBS.

Now an incubation followed in 2× SSC in order to "open-up" the tissue for the subsequent proteolytic treatment. It results in shorter pepsin digestion times and increased preservation of morphology (Protocol 8).

PROTOCOL 8 — PRETREATMENT OF TISSUE SECTIONS

- Incubate sections in 2× SSC at 70°C for 30 min.
- Digest the specimens with pepsin (4 mg/ml in 0.2 *N* HCl) for 7.5 min at 37°C.
- Rinse twice in distilled water at RT.
- Immerse the slides in 2× SSC until denaturation.

The pepsin digestion time had been evaluated in a previous experiment by a time series (5/10/15/20 min). The optimal pepsin time was judged by the following two criteria: ISH signal intensity and tissue morphology.

4. *In Situ* Hybridization

Production of ss-DNA for ISH was accomplished with heat denaturation. Slides were immersed for 2 min at 70°C in 70% deionized formamide, 2×

SSC at pH 7. Then the slides were dehydrated in an ethanol series (70% [cold], 83%, 96%: 1 min each) and air-dried at RT.

The chromosome 7 specific repetitive DNA probe was denatured for 5 min at 70°C in the hybridization mixture, containing (final concentrations):

- 1 to 2 μg/ml Probe DNA
- 500 μg/ml sonicated herring sperm DNA
- 0.1% Tween-20
- 10% Dextran sulfate
- 60% Deionized formamide
- In 2× SSC at pH 7

After denaturation the probe mixture was cooled on ice and put immediately under a glass coverslip covering the section. The ISH was performed overnight at 37°C in a humidified chamber. Then the coverslips were removed by immersing the slides briefly in 2× SSC. Next, the slides were washed 3 times for 3 min each in 60% formamide in 2× SSC, pH 7 at 42°C, and subsequently 2 times for 5 min each in 2× SSC at 42°C. The slides were then immersed in 4× SSCT.

5. Histochemistry

For fluorescent staining of the hybridized probe (see Figure 4), the procedure was used, as described in the Cytochemistry section of the DNA-ISH protocol to cells on slides.

F. CONFOCAL SCANNING AND IMAGE ANALYSIS

The confocal scanning laser microscope[67,69] was used to perform 3-D analysis on the hybridized cells. A krypton laser (Spectra Physics Series 2000, Mountain View, CA), tuned at 482.5 nm (10 mW), was used to excite both the probe-linked FITC and the total nuclear DNA stain propidium iodide. The resulting green and red fluorescence, respectively, was measured through a 500-nm long pass filter (Corion Corp., Holliston, MA). The laser output was kept as low as possible to minimize bleaching of the fluorochromes. The specimen was scanned mechanically along three axes through the confocal point under microprocessor control to produce a 3-D image. Typically, the sampling interval in this image is 1 μm along the optical axis and 90 nm in the lateral plane. The fluorescence intensity data of the objects were stored in a 256 × 256 × 16 8-bit memory array as a function of specimen position.

Image processing was performed as described by Van der Voort et al.[70] Preprocessing of the images was done by means of a median filtering procedure with a 7-voxel window. Briefly, this implicates that the 8-bit gray level value of each voxel (volume element) becomes the median of itself and the six perpendicularly adjacent voxels. This noise-reducing filter results in images with more defined contours, surrounding areas of different fluorescence intensities.

V. RESULTS AND DISCUSSION

In this section, three examples of DNA-ISH, evaluated by confocal microscopy, will be shown and discussed. In Figure 2 a cell biological application, i.e., the position of specific structures (chromosomal centromere regions) inside three-dimensionally preserved cell nuclei, is demonstrated. Cells of the promyelocytic leukemia cell line HL60 were labeled in suspension by ISH with a chromosome 1 specific pericentromeric (1q12) probe. Hybridization was visualized using fluoresceinated avidin; the nucleus was counterstained with propidium iodide. The labeled nuclei were analyzed by optical sectioning, as provided by a scanning confocal microscope (Figure 1). The 1q12 regions of chromosome 1 could be seen as compact domains in the cell nuclei (Figure 2). Four pericentromeric 1q12 regions can be distinguished in this tetraploid polymorphic nucleus. Also regions without DNA stain can be seen, most likely nucleoli. In order to investigate the 3-D topography of the centromeric ISH regions, software was made to determine the position of the 1q12 regions within the nucleus. This was feasible, since all of the x-y-z information was stored in computer memory. In a series of nuclei it was found that these sequences were located in the periphery of the HL60 cell nuclei.

The confocal microscope is becoming increasingly important as an apparatus to analyze the 3-D topography of cellular objects. The main reasons for this are its high resolution optical sectioning capacity, its noninvasiveness, which leaves the objects intact, and its imaging capabilities. Combining such a machine with a flow cytometer creates a powerful tool for the study of spatial and quantitative aspects of fluorescent objects. The quantitation by flow cytometry of hybridized ribosomal RNA sequences[113] and repetitive DNA probes[83,84] in suspended cells has been described. An application, which is increasingly being used at (cyto)genetic laboratories, is the mapping of DNA sequences onto metaphases. This is illustrated in Figure 3. In this image the localization of a 100-kb unique genomic DNA segment in a YAC vector is visualized by fluorescent ISH and confocal imaging. The DNA sequence is localized on chromosome band 21 on the short arm of chromosome 6 (band 6p21). The location on both #6 homologs can easily be distinguished. For real mapping purposes, the approach of Cherif et al.[60] is convenient. It combines ISH with chromosome banding, thus facilitating probe mapping.

As mentioned before, this type of analysis gained enormous attention after a publication in *Science*,[17] in which cosmid cloned unique DNA sequences were located on chromosome 11, showing the usefulness of fluorescent ISH and confocal microscopy in cytogenetics. It can be used for the mapping of unknown probes and for the analysis of the presence of a known DNA sequence. The increasing availability of cosmid and YAC cloned sequences is important in this respect.

A new area for confocal ISH imaging might be the visualization of target DNA in interphase nuclei in tissue sections. This approach combines a cytogenetic analysis with histological evaluation, when conventional nonfluorescent staining is used. This is especially promising in cancer research, since the tumor cells within the tissue can easily be recognized histologically. Simultaneously, ISH with appropriate DNA probes can be utilized for genetic analysis of the tumor cells. In a pilot experiment, shown in Figure 4, a 4-μm tissue section of an endocrine tumor was hybridized with a centromeric chromosome 7 specific repetitive DNA probe, and fluorescently labeled. The presence of multiple copies of centromere (and chromosome) 7 in this tumor is obvious. By combining the somewhat higher sensitivity of the confocal setup with its digital imaging capacities, it should be possible to visualize unique cosmid and YAC probes, containing cancer-related genes. This is presently at the borderline of detection in sections using immunoperoxidase staining of the probes. Confocal fluorescent ISH imaging could be sensitive enough for this application.

An important problem in analyzing fluorescently labeled biological structures by confocal microscopy is photobleaching. This phenomenon can be minimized by embedding the cells in an antifade-containing medium (e.g., PPD).[111] Also the intensity of the laser light can be reduced without having an impact on the quality of the produced images. This is further optimized by the use of more sensitive detection systems, such as CCD cameras. Difficulties associated with light absorption, e.g., scanning inside thick specimens, can be corrected with computer algorithms.

REFERENCES

1. **Gall, J. G. and Pardue, M. L.,** Molecular hybridisation of radioactive DNA to the DNA of cytological preparations, *Proc. Natl. Acad. Sci. U.S.A.*, 64, 600, 1969.
2. **John, H., Birnsteil, M. L., and Jones, K. W.,** RNA-DNA hybrids at cytological levels, *Nature*, 223, 582, 1969.
3. **Hall, B. D. and Spiegelman, S.,** Sequence complementarity of T2-DNA and T2-specific RNA, *Proc. Natl. Acad. Sci. U.S.A.*, 47, 137, 1961.
4. **Harper, M. E., Ullrich, A., and Saunders, G. F.,** Localization of the human insulin gene to the distal end of the short arm of chromosome 11, *Proc. Natl. Acad. Sci. U.S.A.*, 78, 4458, 1981.
5. **Rudkin, G. T. and Stollar, B. D.,** High resolution detection of DNA-RNA hybrids in situ by indirect immunofluorescence, *Nature*, 265, 472, 1977.
6. **Van Prooijen-Knegt, A. C., van Hoek, J. F. M., Bauman, J. G. J., van Duijn, P., Wool, I. G., and van der Ploeg, M.,** In situ hybridization of DNA sequences in human metaphase chromosomes visualized by an indirect fluorescent immunocytochemical procedure, *Exp. Cell Res.*, 141, 397, 1982.
7. **Bauman, J. G. J., Wiegant, J., Borst, P., and van Duijn, P.,** A new method for fluorescence microscopical localization of specific DNA sequences by in situ hybridization of fluorochrome labeled RNA, *Exp. Cell Res.*, 138, 485, 1980.
8. **Langer, P. R., Waldrop, A. A., and Ward, D. C.,** Enzymatic synthesis of biotin-labeled polynucleotides: Novel nucleic acid affinity probes, *Proc. Natl. Acad. Sci. U.S.A.*, 78, 6633, 1981.

9. **Langer-Safer, P. R., Levine, M., and Ward, D. C.,** Immunological method for mapping genes on Drosophila polytene chromosomes, *Proc. Natl. Acad. Sci. U.S.A.*, 79, 4381, 1982.
10. **Manuelidis, L., Langer-Safer, P. R., and Ward, D. C.,** High-resolution mapping of satellite DNA using biotin-labeled DNA probes, *J. Cell Biol.*, 95, 619, 1982.
11. **Tchen, P., Fuchs, R. P. P., Sage, E., and Leng, M.,** Chemically modified nucleic acids as immunodetectable probes in hybridization experiments, *Proc. Natl. Acad. Sci. U.S.A.*, 81, 3466, 1984.
12. **Landegent, J. E., Jansen in de Wal, N., Baan, R. A., Hoeijmakers, J. H. J., and van der Ploeg, M.,** 2-Acetylaminofluorene-modified probes for the indirect hybridocytochemical detection of specific nucleic acid sequences, *Exp. Cell Res.*, 153, 61, 1984.
13. **Landegent, J. E., Jansen in de Wal, N., van Ommen, G. J. B., Baas, F., de Vijlder, J. J. M., van Duijn, P., and van der Ploeg, M.,** Chromosomal localization of a unique gene by non-autoradiographic in situ hybridization, *Nature*, 317, 175, 1985.
14. **Hopman, A. H. N., Wiegant, J., Raap, A. K., Landegent, J. E., van der Ploeg, M., and van Duijn, P.,** Bi-color detection of two target DNAs by non-radioactive in situ hybridization, *Histochemistry*, 85, 1, 1986.
15. **Nederlof, P. M., Robinson, D., Abuknesha, R., Wiegant, J., Hopman, A. H. N., Tanke, H. J., and Raap, A. K.,** Three-color fluorescence *in situ* hybridization for the simultaneous detection of multiple nucleic acid sequences, *Cytometry*, 10, 20, 1989.
16. **Morimoto, H., Monden, T., Shimano, T., Higashiyama, M., Tomita, N., Murotani, M., Matsuura, N., Okuda, H., and Mori, T.,** Use of sulfonated probes *in situ* detection of amylase mRNA in formalin-fixed paraffin sections of human pancreas and submaxillary gland, *Lab. Invest.*, 57, 737, 1987.
17. **Lichter, P., Chang Tang, C., Call, K., Hermanson, G., Evans, G. A., Housman, D., and Ward, D. C.,** High-resolution mapping of human chromosome 11 by *in situ* hybridization with cosmid clones, *Science*, 247, 64, 1990.
18. **Koch, J. E., Kolvraa, S., Petersen, K. B., Gregersen, N., and Bolund, L.,** Oligonucleotide-priming methods for the chromosome-specific labelling of alpha satellite DNA *in situ*, *Chromosoma*, 98, 259, 1989.
19. **Weier, H.-U. G., Segraves, R., Pinkel, D., and Gray, J. W.,** Synthesis of Y chromosome-specific labeled DNA probes by *in vitro* DNA amplification, *J. Histochem. Cytochem.*, 38, 421, 1990.
20. **Lichter, P., Ledbetter, S. A., Ledbetter, D. H., and Ward, D. C.,** Fluorescence *in situ* hybridization with alu and L1 polymerase chain reaction probes for rapid characterization of human chromosomes in hybrid cell lines, *Proc. Natl. Acad. Sci. U.S.A.*, 87, 6634, 1990.
21. **Cook, H. J. and Hindley, J.,** Cloning of human satellite III DNA: different components are on different chromosomes, *Nucleic Acids Res.*, 6, 3177, 1979.
22. **Burk, R. D., Szabo, P., O'Brien, S., Nash, W. G., Yu, L., and Smith, K. D.,** Organization and chromosomal specificity of autosomal homologs of human Y chromosome repeated DNA, *Chromosoma*, 92, 225, 1985.
23. **Willard, H.,** Chromosome-specific organization of the human alpha satellite DNA, *Am. J. Hum. Genet.*, 37, 524, 1985.
24. **Buroker, N., Bestwick, R., Haight, G., Magenis, R. E., and Litt, M.,** A hypervariable repeated sequence on human chromosome 1p36, *Hum. Genet.*, 77, 175, 1987.
25. **Bauman, J. G. J., Wiegant, J., van Duijn, P., Lubsen, N. H., Sondermeijer, P. J. A., Hennig, W., and Kubli, W.,** Rapid and high resolution detection of *in situ* hybridisation to polytene chromosomes using fluorochrome-labeled RNA, *Chromosoma*, 84, 1, 1981.
26. **Pinkel, D., Straume, T., and Gray, J. W.,** Cytogenetic analysis using quantitative, high-sensitivity, fluorescence hybridization, *Proc. Natl. Acad. Sci. U.S.A.*, 83, 2934, 1986.

27. **Moyzis, R. K., Albright, K. L., Bartholdi, M. F., Cram, L. S., Deaven, L. L., Hildebrand, C. E., Joste, N. E., Longmire, J. L., Meyne, J., and Schwarzacher-Robinson, T.,** Human chromosome-specific repetitive DNA sequences: novel markers for genetic analysis, *Chromosoma,* 95, 375, 1987.
28. **Devilee, P., Thierry, R. F., Kievits, T., Kolluri, R., Hopman, A. H. N., Willard, H. F., Pearson, P. L., and Cornelisse, C. J.,** Detection of chromosome aneuploidy in interphase nuclei from human primary breast tumors using chromosome-specific repetitive DNA probes, *Cancer Res.,* 48, 5825, 1988.
29. **Cremer, T., Tesin, D., Hopman, A. H. N., and Manuelidis, L.,** Rapid interphase and metaphase assessment of specific chromosomal changes in neuroectodermal tumor cells by in situ hybridization with chemically modified DNA probes, *Exp. Cell Res.,* 176, 199, 1988.
30. **Burns, J., Redfern, D. R. M., Esiri, M. M., and McGee, J. O'D.,** Human and viral gene detection in routine paraffin embedded tissue by *in situ* hybridisation with biotinylated probes: viral localisation in herpes encephalitis, *J. Clin. Pathol.,* 39, 1066, 1986.
31. **Pringle, J. H., Homer, C. E., Warford, A., Kendall, C. H., and Lauder, I.,** *In situ* hybridization: alkaline phosphatase visualization of biotinylated probes in cryostat and paraffin sections, *Histochem. J.,* 19, 488, 1987.
32. **Emmerich, P., Jauch, A., Hofmann, M.-C., Cremer, T., and Walt, H.,** Interphase cytogenetics in paraffin embedded sections from human testicular germ cell tumor xenografts and in corresponding cultured cells, *Lab. Invest.,* 61, 235, 1989.
33. **Hopman, A. H. N., Ramaekers, F. C. S., Raap, A. K., Beck, J. L. M., Devilee, P., van der Ploeg, M., and Vooijs, G. P.,** *In situ* hybridization as a tool to study numerical chromosome aberrations in solid bladder tumors, *Histochemistry,* 89, 307, 1988.
34. **Van Dekken, H., Hagenbeek, A., and Bauman, J. G. J.,** Detection of host cells following sex-mismatched bone marrow transplantation by fluorescent *in situ* hybridization with a Y-chromosome specific probe, *Leukemia,* 3, 724, 1989.
35. **Van Dekken, H., Pizzolo, J. G., Kelsen, D. P., and Melamed, M. R.,** Targeted cytogenetic analysis of gastric tumors by in situ hybridization with a set of chromosome specific DNA probes, *Cancer,* 66, 491, 1990.
36. **Van Dekken, H., Schervish, E., Pizzolo, J. G., Fair, W. R., and Melamed, M. R.,** Simultaneous detection of fluorescent in situ hybridization and *in vivo* incorporated BrdU in a human bladder tumor, *J. Pathol.,* 164, 17, 1991.
37. **Van Dekken, H., Pizzolo, J. G., Reuter, V. E., and Melamed, M. R.,** Cytogenetic analysis of human solid tumors by *in situ* hybridization with a set of 12 chromosome specific DNA probes, *Cytogenet. Cell. Genet.,* 54, 103, 1990.
38. **Van Dekken, H. and Bauman, J. G. J.,** A new application of *in situ* hybridization: detection of numerical and structural chromosome aberrations with a combination centromeric-telomeric DNA probe, *Cytogenet. Cell. Genet.,* 48, 188, 1988.
39. **Anastasi, J., Le Beau, M. M., Vardiman, J. W., and Westbrook, C. A.,** Detection of numerical chromosomal abnormalities in neoplastic hematopoietic cells by *in situ* hybridization with a chromosome-specific probe, *Am. J. Pathol.,* 136, 131, 1990.
40. **Hopman, A. H. N., Moesker, O., Smeets, A. W. G. B., Pauwels, R. P. E., Vooijs, G. P., and Ramaekers, F. C. S.,** Numerical chromosome 1, 7, 9, and 11 aberrations in bladder cancer detected by *in situ* hybridization, *Cancer Res.,* 51, 644, 1991.
41. **Sandberg, A. A., Turc-Carel, C., and Gemmill, R. M.,** Chromosomes in solid tumors and beyond, *Cancer Res.,* 48, 1049, 1988.
42. **Garson, J. A., van den Berghe, J. A., and Kemshead, J. T.,** Novel non-isotopic *in situ* hybridization technique detects small (1 Kb) unique sequences in routinely G-banded human chromosomes: fine mapping of N-myc and beta-NGF genes, *Nucleic Acids Res.,* 15, 4761, 1987.

43. **Ambros, P. F. and Karlic, H. I.,** Chromosomal insertion of human papillomavirus 18 sequences in HeLa cells detected by non-isotopic *in situ* hybridization and reflection contrast microscopy, *Hum. Genet.,* 77, 251, 1987.
44. **Lawrence, J. B., Vilnave, C. A., and Singer, R. H.,** Sensitive, high-resolution chromatin and chromosome mapping *in situ:* presence and orientation of two closely integrated copies of EBV in a lymphoma line, *Cell,* 52, 51, 1988.
45. **Cremer, T., Lichter, P., Borden, J., Ward, D. C., and Manuelidis, L.,** Detection of chromosomes aberrations in metaphase and interphase tumor cells by *in situ* hybridization using chromosome-specific library probes, *Hum. Genet.,* 80, 235, 1988.
46. **Pinkel, D., Landegent, J., Collins, C., Fuscoe, J., Segraves, R., Lucas, J., and Gray, J. W.,** Fluorescence *in situ* hybridization with human chromosome specific libraries: detection of trisomy 21 and translocations of chromosome 4, *Proc. Natl. Acad. Sci. U.S.A.,* 85, 9138, 1988.
47. **Lichter, P., Cremer, T., Chang Tang, C., Watkins, P. C., Manuelidis, M., and Ward, D. C.,** Rapid detection of human chromosome 21 aberrations by *in situ* hybridization, *Proc. Natl. Acad. Sci. U.S.A.,* 85, 9664, 1988.
48. **Kuo, W.-L., Tenjin, H., Segraves, R., Pinkel, D., Golbus, M. S., and Gray, J.,** Detection of aneuploidy involving chromosomes 13, 18, or 21, by fluorescence *in situ* hybridization (FISH) to interphase and metaphase amniocytes, *Am. J. Hum. Genet.,* 49, 112, 1991.
49. **Kievits, T., Devilee, P., Wiegant, J., Wapenaar, M. C., Cornelisse, C. J., van Ommen, G. J. B., and Pearson, P. L.,** Direct nonradioactive *in situ* hybridization of somatic cell hybrid DNA to human lymphocyte chromosomes, *Cytometry (Special issue: Analytical Cytogenetics),* 11, 105, 1990.
50. **Suijkerbuijk, R. F., van de Veen, A. Y., van Echten, J., Buys, C. H. C. M., de Jong, B., Oosterhuis, J. W., Warburton, D. A., Cassiman, J. J., Schonk, D., and Geurts van Kessel, A.,** Demonstration of the genuine iso-12p character of the standard marker chromosome of testicular germ cell tumors and identification of further chromosome 12 aberrations by competitive *in situ* hybridization, *Am. J. Hum. Genet.,* 48, 269, 1991.
51. **Trautman, U., Leuteritz, G., Senger, G., Claussen, U., and Ballhausen, W. G.,** Detection of APC region-specific signals by nonisotopic chromosomal *in situ* suppression (CISS)-hybridization using a microdissection library as a probe, *Hum. Genet.,* 87, 495, 1991.
52. **Riethman, H. C., Moyzis, R. K., Meyne, J., Burke, D. T., and Olson, M. V.,** Cloning human telomeric DNA fragments into *Saccharomyces cerevisiae* using a yeast-artificial-chromosome vector, *Proc. Natl. Acad. Sci. U.S.A.,* 86, 6240, 1989.
53. **Rowley, J. D., Diaz, M. O., Espinoza, III, R., Patel, Y. D., van Melle, E., Ziemin, S., Taillon-Miller, P., Lichter, P., Evans, G. A., Kersey, J. H., Ward, D. C., Domer, P. H., and Le Beau, M. M.,** Mapping chromosome band 11q23 in human acute leukemia with biotinylated probes: identification of 11q23 translocation breakpoints with a yeast artificial chromosome, *Proc. Natl. Acad. Sci. U.S.A.,* 87, 9358, 1990.
54. **Albertson, D. G.,** Mapping muscle protein genes by *in situ* hybridization using biotin-labeled probes, *EMBO J.,* 4, 2493, 1985.
55. **Landegent, J. E., Jansen in de Wal, N., Dirks, R. W., Baas, F., and van der Ploeg, M.,** Use of whole cosmid cloned genomic sequences for chromosomal localization by nonradioactive *in situ* hybridization, *Hum. Genet.,* 73, 366, 1987.
56. **Kievits, T., Dauwerse, J. G., Wiegant, J., Devilee, P., Breuning, M. H., Cornelisse, C. J., van Ommen, G.-J. B., and Pearson, P. L.,** Rapid subchromosomal localization of cosmids by nonradioactive *in situ* hybridization, *Cytogenet. Cell. Genet.,* 53, 134, 1990.
57. **Van Dilla, M. A., et al.,** Human chromosome-specific DNA libraries: construction and availability, *Bio/technology,* 4, 537, 1986.

58. **Ried, T., Mahler, V., Vogt, P., Blonden, L., van Ommen, G. J. B., Cremer, T., and Cremer, M.,** Direct carrier detection by in situ suppression hybridization with cosmid clones of the Duchenne/Becker muscular dystrophy locus, *Hum. Genet.*, 85, 581, 1990.
59. **Selleri, L., Hermanson, G. G., Eubanks, J. H., Lewis, K. A., and Evans, G. A.,** Molecular localization of the t(11;22) (q24;q12) translocation of Ewing sarcoma by chromosomal *in situ* suppression hybridization, *Proc. Natl. Acad. Sci. U.S.A.*, 88, 887, 1991.
60. **Cherif, D., Julier, C., Delattre, O., Derre, Lathrop, G. M., and Berger, R.,** Simultaneous localization of cosmids and chromosome R-banding by fluorescence microscopy: application to regional mapping of human chromosome 11, *Proc. Natl. Acad. Sci. U.S.A.*, 87, 6639, 1990.
61. **Arnoldus, E. P. J., Wiegant, J., Noordermeer, I. A., Wessels, J. W., Beverstock, G. C., Grosveld, G. C., van der Ploeg, M., and Raap, A. K.,** Detection of the Philadelphia chromosome in interphase nuclei, *Cytogenet. Cell. Genet.*, 54, 108, 1990.
62. **Tkachuk, D. C., Westbrook, C. A., Andreeff, M., Donlon, T. A., Cleary, M. L., Suryanaryan, K., Homge, M., Redner, A., Gray, J., and Pinkel, D.,** Detection of bcr-abl fusion in chronic myelogeneous leukemia by *in situ* hybridization, *Science*, 250, 559, 1990.
63. **Trask, B. J., Pinkel, D., and van den Engh, G.,** The proximity of DNA sequences in interphase cell nuclei is correlated to genomic distance and permits ordering of cosmids spanning 250 kilobase pairs, *Genomics*, 5, 710, 1989.
64. **Lawrence, J. B., Singer, R. H., and McNeil, J. A.,** Interphase and metaphase resolution in different distances within the human dystrophin gene, *Science*, 249, 928, 1990.
65. **Trask, B. J., Massa, H., Kenwrick, S., and Gitschier, J.,** Mapping of human chromosome Xq28 by two-color fluorescence in situ hybridization of DNA sequences to interphase cell nuclei, *Am. J. Hum. Genet.*, 48, 1, 1991.
66. **Willard, H. F. and Waye, J. S.,** Hierarchical order in chromosome-specific human alpha satellite DNA, *Trends Genet.*, 3, 192, 1987.
67. **Brakenhoff, G. J., van der Voort, H. T. M., van Spronsen, E. A., Linnemans, W. A. M., and Nanninga, N.,** Three-dimensional chromatin distribution in neuroblastoma nuclei shown by confocal scanning laser microscopy, *Nature*, 317, 748, 1985.
68. **White, J. G., Amos, W. B., and Fordham, M.,** An evaluation of confocal versus conventional imaging of biological structures by fluorescence light microscopy, *J. Cell Biol.*, 105, 41, 1987.
69. **Brakenhoff, G. J., van der Voort, H. T. M., van Spronsen, E. A., and Nanninga, N.,** Three-dimensional imaging in fluorescence by confocal scanning microscopy, *J. Microsc.*, 153, 151, 1989.
70. **Van der Voort, H. T. M., Brakenhoff, G. J., and Baarslag, M. W.,** Three-dimensional visualization methods for confocal microscopy, *J. Microsc.*, 153, 123, 1989.
71. **Albertson, D. G., Sherrington, P., and Vaudin, M.,** Mapping nonisotopically labeled DNA probes to human chromosome bands by confocal microscopy, *Genomics*, 10, 143, 1991.
72. **Rabl, C.,** in "Uber Zelltheilung. Morphologisches Jahrbuch." 10, 214, 1885.
73. **Agard, D. A. and Sedat, J. W.,** Three-dimensional architecture of a polytene nucleus, *Nature*, 302, 676, 1983.
74. **Mathog, D., Hochstrasser, M., Gruenbaum, Y., Saumweber, H., and Sedat, J.,** Characteristic folding pattern of polytene chromosomes in Dorsophila salivary gland nuclei, *Nature*, 308, 414, 1984.
75. **Moroi, Y., Hartman, A. L., Nakane, P. K., and Tan, E. M.,** Distribution of kinetochore (centromere) antigen in mammalian cell nuclei, *J. Cell Biol.*, 90, 254, 1981.
76. **Hadlaczky, G., Went, M., and Ringertz, N. R.,** Direct evidence for the non-random organization of mammalian chromosomes in the interphase nucleus, *Exp. Cell Res.*, 167, 1, 1986.

77. **Van Dekken, H., Pinkel, D., Mullikin, J., and Gray, J. W.,** Enzymatic production of single-stranded DNA as a target for fluorescence *in situ* hybridization, *Chromosoma,* 97, 1, 1988.
78. **Manuelidis, L.,** Individual interphase chromosome domains revealed by *in situ* hybridization, *Hum. Genet.,* 71, 288, 1985.
79. **Schardin, M., Cremer, T., Hager, H. D., and Lang, M.,** Specific staining of human chromosomes in Chinese hamster × man Hybrid cell lines demonstrates interphase chromosome territories, *Hum. Genet.,* 71, 281, 1985.
80. **Rappold, G. A., Cremer, T., Hager, H. D., Davies, K. E., Muller, C. R., and Yang, T.,** Sex chromosome positions in human interphase nuclei as studied by *in situ* hybridization with chromosome specific DNA probes, *Hum. Genet.,* 67, 317, 1984.
81. **Trask, B., van den Engh, G., Landegent, J., Jansen in de Wal, N., and van der Ploeg, M.,** Detection of DNA sequences in nuclei in suspension by *in situ* hybridization and dual beam flow cytometry, *Science,* 230, 1401, 1985.
82. **Dudin, G., Cremer, T., Schardin, M., Hausmann, M., Bier, F., and Cremer, C.,** A method for nucleic acid hybridization to isolated chromosomes in suspension, *Hum. Genet.,* 76, 290, 1987.
83. **Trask, B., van den Engh, G., Pinkel, D., Mullikin, J., Waldman, F., van Dekken, H., and Gray, J.,** Fluorescence *in situ* hybridization to interphase cell nuclei in suspension allows flow cytometric analysis of chromosome content and microscopic analysis of nuclear organization, *Hum. Genet.,* 78, 251, 1988.
84. **Van Dekken, H., Arkesteijn, G. J. A., Visser, J. W. M., and Bauman, J. G. J.,** Flow cytometric quantification of human chromosome specific repetitive DNA sequences by single and bicolor fluorescent *in situ* hybridization to lymphocyte interphase nuclei, *Cytometry (Special issue: analytical cytogenetics),* 11, 153, 1990.
85. **Van Dekken, H., Pinkel, D., Mullikin, J., Trask, B., van den Engh, G., and Gray, J.,** Three-dimensional analysis of the organization of human chromosome domains in human and human-hamster hybrid interphase nuclei, *J. Cell Sci.,* 94, 299, 1989.
86. **Van Dekken, H., van Rotterdam, A., Jonker, R., van der Voort, H. T. M., Brakenhoff, G. J., and Bauman, J. G. J.,** Confocal microscopy as a tool for the study of the intranuclear topography of chromosomes, *J. Microsc.,* 158, 207, 1990.
87. **Van Dekken, H., van Rotterdam, A., Jonker, R. R., van der Voort, H. T. M., Brakenhoff, G. J., and Bauman, J. G. J.,** Spatial topography of a pericentromeric region (1q12) in lymphoid and myeloid hemopoietic cells studied by *in situ* hybridization and confocal microscopy, *Cytometry,* 11, 570, 1990.
88. **Van Dekken, H., van der Voort, H. T. M., Brakenhoff, G. J., and Bauman, J. G. J.,** Three-dimensional reconstruction of pericentromeric (1q12) DNA and ribosomal RNA sequences in HL60 cells after double-target *in situ* hybridization and confocal microscopy, *Cytometry,* 11, 579, 1990.
89. **Manuelidis, L.,** Different central nervous system cell types display distinct and nonrandom arrangements of satellite DNA sequences, *Proc. Natl. Acad. Sci. U.S.A.,* 81, 3123, 1984.
90. **Pinkel, D., Gray, J. W., Trask, B., van den Engh, G., Fuscoe, J., and van Dekken, H.,** Cytogenetic analysis by *in situ* hybridization with fluorescently labeled nucleic acid probes, *Cold Spring Harbor Symp. Quant. Biol.,* 51, 151, 1986.
91. **Manuelidis, L. and Borden, J.,** Reproducible compartmentalization of individual chromosome domains in human CNS cells revealed by in situ hybridization and three-dimensional reconstruction, *Chromosoma,* 96, 397, 1988.
92. **Borden, J. and Manuelidis, L.,** Movement of the X chromosome in epilepsy, *Science,* 242, 1687, 1988.
93. **Popp, S., Scholl, H. P., Loos, P., Jauch, A., Stelzer, E., Cremer, C., and Cremer, T.,** Distribution of chromosome 18 and X centric heterochromatin in the interphase nucleus of cultured human cells, *Exp. Cell Res.,* 189, 1, 1990.

94. **Rawlins, D. J. and Shaw, P. J.,** Three-dimensional organization of ribosomal DNA in interphase nuclei of Pisum sativum by in situ hybridization and optical tomography, *Chromosoma,* 99, 143, 1990.
95. **Woldringh, G. L., de Jong, M., van den Berg, W., and Koppes, L.,** Morphological analysis of the division cycle of *Escherichia coli* substrains during slow growth, *J. Bacteriol.,* 131, 270, 1976.
96. **Sheppard, C. J. R. and Choudhury, A.,** Image formation in the scanning microscope, *Opt. Acta,* 24, 1051, 1977.
97. **Brakenhoff, G. J., Blom, P., and Barends, P.,** Confocal scanning light microscopy with high aperture immersion lenses, *J. Microsc.,* 117, 219, 1979.
98. **Brakenhoff, G. J., Blom, P., and Woldringh, C. L.,** Developments in high resolution confocal scanning light microscopy (CSLM), in *Scanned Image Microscopy,* Ash, E. A., Ed., Academic Press, New York, 1980.
99. **Egger, M. D. and Petran, M.,** New reflected light microscopy for viewing unstained brain and ganglion cells, *Science,* 157, 305, 1967.
100. **Wilson, T. and Sheppard, C. J. R.,** *Theory and Practice of Scanning Optical Microscopy,* Academic Press, New York, 1984.
101. **Wijnaendts van Resandt, R. W., Marsman, H. J. B., Kaplan, R., Davoust, J., Steltzer, E. H. K., and Sticker, R.,** Optical fluorescence microscopy in 3 dimensions: microtomoscopy, *J. Microsc.,* 138, 29, 1985.
102. **Carlsson, K., Danielson, P. E., Lenz, R., Liljeborg, A., Majlof, L., and Asland, N.,** Three dimensional microscopy using a confocal scanning laser microscope, *Opt. Lett.,* 10, 53, 1985.
103. **Petran, M. and Hadravsky, M.,** Tandem-scanning reflected light microscope, *J. Opt. Soc.,* 58, 661, 1968.
104. **Van der Voort, H. T. M., Brakenhoff, G. J., Valkenburg, J. A. C., and Nanninga, N.,** Design and use of a computer controlled confocal microscope, *Scanning,* 7, 66, 1985.
105. **Wells, K. S., Sandison, D. R., Strickler, J., and Webb, W. W.,** Quantitative fluorescence imaging with laser scanning confocal microscopy, in *Handbook of Biological Confocal Microscopy,* Pawley, F., Ed., Plenum Press, London, 1989, 27.
106. **Brakenhoff, G. J. and van der Voort, H. T. M.,** Illumination and detection strategies for confocal microscopy, *SPIE Proc.,* 1139, 117, 1989.
107. **Davis, L. G., Dibner, M. D., and Battey, J. F.,** in *Basic Methods in Molecular Biology,* 1986.
108. **Bauman, J. G. J., Pinkel, D., Trask, B. J., and van der Ploeg, M.,** Flow cytometric measurements of specific DNA and RNA sequences, in *Flow cytogenetics,* Gray, J. W., Ed., Academic Press, New York, 1989, 275.
109. **Hames, B. D. and Higgins, S. J.,** Eds., Nucleic acid hybridization, a practical approach, *IRL Press,* Oxford, 1985.
110. **Cantor, C. R. and Schimmel, P. R.,** in *Biophysical Chemistry Part III: The Behaviour of Biological Macromolecules,* W. H. Freeman, San Francisco, 1980.
111. **Johnson, G. D. and de C. Nogeira Araujo, G. M.,** A simple method of reducing the fading of immunofluorescence during microscopy, *J. Immunol. Methods,* 43, 349, 1981.
112. **Rentrop, M., Knapp, B., Winter, H., and Schweizer, J.,** Aminoalkylsilane-treated glass slides as support for *in situ* hybridization of keratin cDNA's to frozen tissue sections under varying fixation and pretreatment conditions, *Histochem. J.,* 18, 271, 1986.
113. **Bauman, J. G. J. and Bentvelzen, P.,** Flow cytometric detection of ribosomal RNA in suspended cells by fluorescent *in situ* hybridization, *Cytometry,* 9, 517, 1988.

Chapter 4

IN SITU HYBRIDIZATION IN SEMITHIN SECTIONS

Christine Pechoux and Gérard Morel

TABLE OF CONTENTS

0-8493-4414-X/93/$0.00 + $.50

I. INTRODUCTION

In situ hybridization applied to the detection of messenger RNA (mRNA) in tissue sections is now a widely used technique in light microscopy.[1] Most of the studies in this field have been carried out on frozen or paraffin sections. After the *in situ* hybridization stage, the localization of the RNA sequences was carried out by autoradiography methods, but this procedure does not give precise resolution. Large quantities of DNA hybridization probes are biotin-labeled or digoxigenin-labeled by *in vitro* transcription. The use of nonradioactive labeling gives these DNA probes a long storage life, along with the possibility of enzymatic detection and the use of ready-made detection systems. The end result is an inexpensive, rapid, easy-to-use hybridization protocol, which generates improved signal-to-noise ratios, and substantially better resolution in light microscopy studies.

In general, the amount of image information available from a tissue section depends on the thickness of the section. Semithin sections (0.5 to 1 μm) obtained from plastic-embedded tissue or frozen tissue contain more image information than paraffin or cryostat sections. Plastic-embedded material and frozen material can also be used to detect mRNA at the ultrastructural level.

We describe an *in situ* hybridization method, using radioactive-labeled probes and nonradioactive-labeled probes, applied to semithin plastic sections: epoxy resin-embedded, London Resin White (LR White), Lowicryl K4M, Spurr medium, and frozen semithin sections obtained by cryoultramicrotomy.

The different detection methods are autoradiography for the radioactive-labeled probes, enzymatic, and fluorescent or colloidal gold detection for the nonradioactive labeled probes.

II. TISSUE PREPARATION

In situ hybridization in semithin sections presented the same difficulties as *in situ* hybridization at the light and electron microscopic level, with regard to the optimal preservation of structures and the conservation of the nucleic acids studied. The first step was fixation tissue, followed by either embedding or freezing. Semithin sections were transferred to pretreated slides (see Protocol 1) prior to the application of the *in situ* hybridization technique.

Tissues, biopsies, cultured cells, and fractionated cells can all constitute sources of semithin sections (see Protocol 2) for use in the *in situ* hybridization technique. The same standard precautions regarding removal and delay must be performed for tissue samples as those applied to electron microscopy procedures.

PROTOCOL 1 — PREPARATION OF SLIDES FOR SEMITHIN SECTIONS

After the washing step (1), there are two possibilities: poly-L-lysine (2) or 3-aminopropyl thrietoxysilane (3).

1. Washing step:
   a. Dip histological slides overnight in ethanol 95% containing HCl 1%
   b. Wash in running water at least 2 h
   c. Wash in distilled water for 2 × 5 min
2. Poly-L-lysine treatment:
   a. Dip 5 min without drying in a solution of poly-L-lysine (mol wt 70 to 150,000) in 2.5 μg/ml distilled water
   b. Dry in an oven (50°C)
   c. Store at the same temperature with desiccant
3. 3-aminopropyl thrietoxysilane treatment:[2]
   a. Dip in 3-aminopropyl thietoxysilane 2% in acetone
   b. Dip in acetone for 2 × 2 min
   c. Dip in distilled water for 2 × 2 min
   d. Dry in an oven (40°C)
   e. Store at room temperature

PROTOCOL 2 — PROCEDURE FOR PREPARING SEMITHIN SECTIONS

1. Semithin frozen sections
   a. Sections of tissue at 0.5- to 1-μm thickness are obtained with a cryoultramicrotome (specimen temperature between −60 and −90°C; knife temperature between −40 and −60°C)
   b. Transfer the sections from the knife to a histological pretreated slide with a drop of saturated sucrose
   c. Dry the sections for 1 h at room temperature
2. Semithin plastic sections
   a. Sections of tissue at 1-μm thickness are obtained with a histological diamond knife, at a cutting speed of about 20 mm/s
   b. Transfer them to a histological pretreated slide (see Protocol 1) with a drop of distilled water
   c. Dry the sections for 30 min at 70°C

A. FIXATION (SEE PROTOCOL 3)

This step can be carried out by perfusion or immersion. Differences were observed in the morphological aspects but not in the *in situ* hybridization signal.[3]

The fixative most often used was freshly prepared paraformaldehyde at a concentration of between 2 and 8% in a phosphate buffer (mono-di-phosphate buffer (0.1 *M*) with or without 0.15 *M* NaCl, pH 7.4), in which the tissue was placed for about 2 h at 4°C before being washed in the same buffer for the same period. For better preservation of ultrastructural morphology, glutaraldehyde can be added at a concentration lower than 0.5%.[4] The use of these cross-linking fixatives, particularly glutaraldehyde, has the disadvantage of limiting the access of the probe to the nucleic acid. The signal decreased after the addition of glutaraldehyde and paraformaldehyde at higher levels of concentration. Prior fixation with osmium tetroxide (O_sO_4) compromises mRNA-cDNA hybridization,[5] but if carried out after the hybridization procedure as a secondary fixative for electron microscopy purposes, it does not seem to interfere with the labeling.

PROCOTOL 3 — TISSUE FIXATION

1. Fix fresh biopsies in 2 to 4% paraformaldehyde, in a 0.1 *M* mono-di-phosphate buffer (pH 7.4), with or without 0.2% glutaraldehyde, for 2 h at 4°C
2. Rinse them in the same buffer for 2 h at 4°C

B. TREATMENT OF TISSUE SAMPLES

Tissue was treated with proteases or detergents before embedding was carried out, in the case of pre-embedding method only, in order to increase the penetration of the probe.[6] Pretreatment of the tissue with proteinase K, or Triton X-100 detergent, has been reported, though they have deleterious effects on the ultrastructure. High concentrations (0.05 to 0.1%) of these agents have been used to enhance signals from long cDNA probes labeled with (α^{35}S)-dCTP in random primer and nick translation techniques[7] (Figure 1).

C. PREPARATION PROCEDURE

1. Frozen Tissue

The main advantage of freezing is to retain water in the tissue, since no dehydration or embedding procedures were carried out. The details of the optimal freezing step used for electron microscopy are described in Chapter 6. For semithin frozen sections, it was possible to limit both the difficulties presented by this procedure and the quantity of materials that had to be frozen. This simplified form of the procedure (see Protocol 4) did however involve a loss of structure in some tissue which was sensitive to osmolarity variations. The use of a cryoprotectant hyperosmolar, such as saccharose, 2.3 *M*,[8] limited the free water within the cells, but gave osmolarity artifacts.

PROTOCOL 4 — FREEZING OF SEMITHIN SECTIONS

1. Wash in mono-di-phosphate buffer 0.1 *M* (pH 7.4) for 15 min
2. Incubate in graded saccharose of 0.5, 1, 1.5, 2 *M* (1 h in each bath)

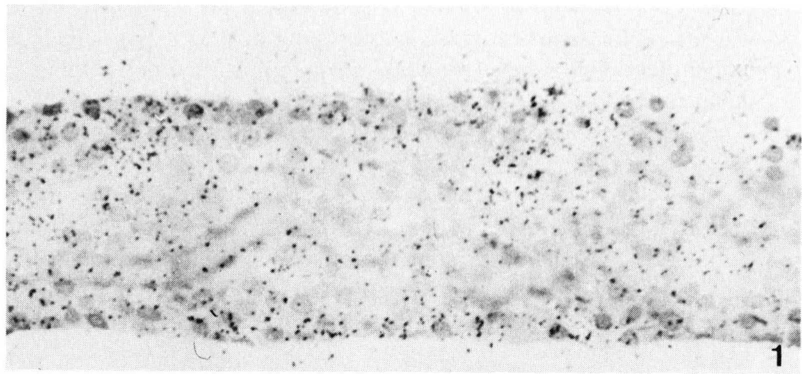

FIGURE 1. Checking cDNA penetration of vibratome sections 100-nm thick. The GH cDNA was labeled with ^{35}S by nick translation. *In situ* hybridization was done overnight by incubation of vibratome sections in a buffer, then washing and embedding in Araldite (for details see Chapter 5). Semithin sections of transversal vibratome section were dipped in nuclear emulsion. After one week of exposure, revelation was carried out using Dektol, followed by staining (see Protocol 24). All somatotrophs in the vibratome section were hybridized.

3. Incubate overnight in 2.3 *M* saccharose
4. Place the sample on a cryoultramicrotome support
5. Place in liquid nitrogen or solid nitrogen (see Protocol 5)
6. Store the sample in liquid nitrogen

PROTOCOL 5 — PREPARATION OF SOLID NITROGEN

1. Place the liquid nitrogen in a vacuum evaporator
2. Pump with a rotary pump until the nitrogen becomes solid
3. Quickly open the evaporator
4. Shoot the sample into the melting nitrogen

The advantage of this freezing procedure is the absence of the chelefactor effect.[9]

2. Epoxy Resin-Embedded Tissue[10,11]

No important difference is observed between the embedding method described for morphological studies and *in situ* hybridization studies (see Protocol 6).

PROTOCOL 6 — EPOXY RESIN-EMBEDDED TISSUE

1. Fixation (see Protocol 3)*
2. Inclusion
   a. Dehydration
      The tissue was dehydrated with graded ethanol (50, 70, 80, 85, 90, 95, 100%) for 10 min each
   b. Infiltration
      Tissues were infiltrated in a 1:1 mixture of the final embedding medium and propylene oxide for 4 h to overnight
   c. Embedding
      Tissue samples were immersed in Araldite 502 in inverted Beem capsules for 6 h at room temperature
   d. Polymerization
      Blocks were polymerizated at 56°C for 18 to 20 h
3. Semithin sections (See Protocol 2)

3. Hydrophilic Resin-Embedded Tissue

Water-soluble resin has the advantage that it can be used in the localization of antigens or nucleic acids in sections of fixed, embedded tissue showing good morphological preservation, and that it does not have to be removed from the resin, unlike the epoxy-resin used previously. A hydrophilic embedding agent allows the penetration by aqueous solutions, antibodies, and nucleic acid fragments. *In situ* hybridization with Lowicryl K4M embedded tissue (see Protocol 7) gave comparable results to those obtained with unembedded tissues.[12] Osmium tetroxide postfixation was generally omitted: in the case of LR White (see Protocol 8) this is because osmium tetroxide causes an exothermic reaction and prevents impregnation,[13] and in the case of Lowicryl K4M, because it prevents polymerization of the resin by UV light.[14] The disadvantages of Lowicryl K4M are its high toxicity and the necessity to work at low temperatures. LR White and Spurr (see Protocol 9) are less toxic than Lowicryl, and are easier to manipulate.

PROTOCOL 7 — LOWICRYL K4M RESIN-EMBEDDING METHOD[15]

1. Fixation (See Protocol 3)
2. Inclusion
   a. Dehydration
      30% ethanol for 30 min at 4°C
      50% ethanol for 1 h at −20°C
      70% ethanol for 1 h at −20°C

* Tissues were processed rapidly: prolonged storage in phosphate buffer and postfixation with osmium tetroxide were avoided where possible.

90% ethanol for 1 h at −20°C
100% ethanol for 2 × 1 h at −20°C with agitation

b. Infiltration
100% ethanol - Lowicryl 1-1 (v/v) for 4 h at −20°C
100% ethanol - Lowicryl 1-2 (v/v) for 4 h at −20°C
100% ethanol - Lowicryl 1-3 (v/v) for 4 h at −20°C
Lowicryl for 2 × 12 h at −20°C with agitation

c. Embedding
In gelatine capsule without air bubbles as far as possible

d. Polymerization
24 h at −20°C in UV light (360 nm)
48 h at +4°C in UV light
48 h at room temperature in UV light

PROTOCOL 8 — LONDON RESIN WHITE (LR WHITE) EMBEDDING METHOD[13]

1. Fixation (See Protocol 3)
2. Inclusion
   a. Dehydration
      Depending on the tissue sample size:
      - Partial dehydration
        - 70% ethanol for 2 × 20 min
      - Total dehydration
        - 70% ethanol for 2 × 20 min
        - 100% ethanol for 2 × 20 min

   b. Infiltration - embedding
      - After partial dehydration
        - 70% ethanol - LR White 2-1 (v/v) for 30 min
        - 70% ethanol - LR White 1-1 (v/v) for 30 min
        - 70% ethanol - LR White 1-2 (v/v) for 30 min
        - LR White for 1 h
        - LR White overnight
      - After total dehydration
        - 100% ethanol - LR White 2-1 (v/v) for 30 min
        - 100% ethanol - LR White 1-1 (v/v) for 30 min
        - 100% ethanol - LR White 1-2 (v/v) for 30 min
        - LR White 1 h
        - LR White overnight

   c. Polymerization
      The polymerization was carried out without an accelerator, at 60°C (± 2°C) for a period of 20 to 24 h in capsules, without air bubbles as far as possible.
3. Semithin sections (See Protocol 2)

The sections were dried in an oven at 50°C.

PROTOCOL 9 — SPURR-EMBEDDING METHOD[16]

1. Fixation (See Protocol 3)
2. Inclusion
   a. Dehydration (at room temperature)
      30% acetone for 30 min
      50% acetone for 30 min
      60% acetone for 30 min
      70% acetone overnight
      80% acetone for 30 min
      95% acetone for 30 min
      acetone for 1 h
      acetone for 2 × 30 min
   b. Infiltration
      acetone - Spurr 3-1 (v/v) for 1 h
      acetone - Spurr 1-1 (v/v) overnight
      acetone - Spurr 1-3 (v/v) for 1 h
      Spurr for 1 h
      Spurr for 2 h
   c. Embedding
      In gelatine capsules
   d. Polymerization
      polymerization at 60°C, overnight

III. PROBE PREPARATION

A. CHOICE OF PROBE

Recombinant DNA technology now provides the possibility of obtaining DNA and RNA probes of any desired sequence, and we can choose between single- and double-stranded probes. For hybridization to RNA, both double- and single-stranded probes have been used successfully.[17] DNA probes may form double strands in a hybridization buffer, and thus reduce the concentration of probe available for reaction. The second possibility is that of oligonucleotide probes obtained from known DNA sequences, with care being taken to ensure that a given sequence:

- Is in fact a complementary sequence, antisense to the nucleic acid studied
- Corresponds to the protein
- Contains between 20 and 30 nucleotides; short probes produce the highest rate of hybridization and do not affect specificity
- Includes sufficient G-C nucleotides; but the oligo-mRNA hybridization will be more stable if these nucleotides do not exceed 70%

TABLE 1
Characteristics of Isotopes Included in Nucleotides

	^{3}H	^{35}S	^{33}P	^{32}P
Emission	β	β	β	β
Half-life	12.3 years	87 d	25 d	14 d
E_{max}	0.018 MeV	0.17 MeV	0.223 MeV	1.6 MeV
Specific activity	1.07 TBq/mmol	55.5 TBq/mmol	111 TBq/mmol	222 TBq/mmol

The third possibility is that of RNA probes, which produce less background noise than DNA probes. But these probes are not easy to work with, and it is necessary to be careful about RNase action.

The stability of hybrids varies according to the nature of the nucleic acid studied. This stability follows the descending order DNA/DNA, DNA/RNA, RNA/RNA.

B. CHOICE OF LABEL

1. Radioactive Label Probes

The main radioisotopes used were tritium (^{3}H), sulfur-35 (^{35}S), phosphorus-33 (^{33} P), and sometimes phosphorus-32 (^{32}P) (see Chapter 2).

Each of these radioisotopes has advantages and disadvantages: the half-lives and energy permissions of these isotopes are presented in Table 1. Resolution is inversely proportional to emission energy, and exposure time is proportional to the half-life of the isotope. In this way, ^{32}P giving poor resolution can be used at short exposure times. Inversely ^{3}H gave good resolution at long exposure times. ^{35}S was the best compromise and the most easily used as a labeling isotope in nucleotides.[18,19]

2. Nonradioactive Label Probes

Nonisotopic labeled probes have been more and more frequently used in recent years to complement or even replace radioactive probes in techniques such as *in situ* hybridization.

Nonradioactive probes have many advantages over radioactive probes. The absence of radioactivity removes several disadvantages linked to the use of radioisotopes, for example, the necessary precautions and expenses linked to radioprotection, the short half-lives of radioisotopes, and the necessity for autoradiographic procedures, all of which are crucial in the practice *in situ* hybridization in nonspecialized laboratories, which are not likely to be equipped for use of radioisotope labels.[20]

The most successful of these nonradioactive labels uses either biotin-substituted nucleotides,[21] digoxigenin, fluorescent labeling, photobiotin, or chemical labeling such as sulfonation to label nucleic acids probe (see Chapter 2). After hybridization, the hybrids formed contained biotin, digoxigenin, or chemical labels which can be revealed using specific antibodies conjugated to enzymatic or fluorescent reagents, or colloidal gold particles.

The advantage of these nonradioactive probes is that they give high cellular and subcellular resolution, and that the reaction is confined to the cell compartment containing the nucleotide target. Another point is the rapidity of the revelation, compared to *in situ* hybridization with radioisotope probes.

A limitation of nonradioactive probes is that they are less sensitive than radioactive probes. They also hamper quantitative analysis because the procedure and intensity of the reaction are difficult to standardize. Revelation with colloidal gold particles followed or not by silver amplification can be used to produce a semiquantitative analysis.

C. cDNA LABELING (SEE ALSO CHAPTER 2)

1. Nick Translation (See Protocol 10)[22]

The nick translation system was designed for the efficient incorporation of labeled deoxynucleoside triphosphate into duplex DNA. The reaction involves the simultaneous action of two enzymes: pancreatic deoxyribonuclease I (DNase I) and *E. coli* DNA polymerase I (DNA pol I).

DNase I introduces nicks randomly in each strand of a DNA molecule, resulting in both free 3′ hydroxyl and free 5′ phosphate groups. DNA pol I then catalyzes the addition of a nucleotide residue to the 3′ hydroxyl terminus of the nick, using the complementary DNA strand as a template. At the same time the 5′ → 3′ exonuclease activity of this enzyme removes the nucleotide from the 5′ phosphoryl terminus of the nick. Thus, the initial nick is actually displaced along the DNA molecule in a 5′ → 3′ direction. This technique requires a large quantity of DNA (1 μg); the reaction time is short (60 min) but the temperature (15°C) is important.

PROTOCOL 10 — NICK TRANSLATION LABELING

1.	Mix	
	DNA	1 μg
	dNTP mixture	5 μl
	labeled nucleotide	× μl
	DNA pol I, 4U/ml - DNase I 40 pg/ml	5 μl
	H_2O	to 50 μl
2.	Incubate for 2 h at 15°C	

Labeled nucleotide:	0.3 m*M* biotin-dUTP or digoxigenin-dUTP 50 μCi (α^{35}S) dATP
dNTP mixture	0.2 m*M* of each dNTP in 500 m*M* Tris-HCl (pH 7.8), 50 m*M* $MgCl_2$, 100 m*M* 2-mercaptoethanol, 100 μg/ml nuclease-free BSA.

2. Random Priming Reaction (See Protocol 11)[23a,23b]

The random priming reaction uses a mixture of random hexanucleotides to prime DNA synthesis *in vitro* from a linear single-stranded DNA template, which is produced by brief heat denaturation of a double-stranded molecule. Primer extension method utilize the ability of DNA polymerases to synthesize a new DNA strand complementary to a template strand, starting from a free 3′ hydroxyl. The enzyme is the Klenow fragment of *E. coli* DNA pol I, which lacks 5′ to 3′ exonuclease activity and reverse transcriptase. Synthesis has been found to occur very efficiently with ng amounts of DNA (generally 20 to 30 ng).

PROTOCOL 11 — RANDOM PRIMED DNA LABELING

1.	Mix	
	DNA	25 ng
	Reaction mixture × 10	2 μl
	dNTP (0.5 mmol/l)	1 μl each
	labeled nucleotide	× μl
	Klenow enzyme (2 units)	1 μl
	H_2O	to 20 μl
2.	Incubate for 30 min at 37°C	

Reaction mixture × 10:	130 m*M* potassium phosphate, 6.5 m*M* $MgCl_2$, 33 μ*M* dTTP, 1 m*M* DTT, 0.032 mg/ml BSA (pH 7.4) containing hexanucleotide mixture
Labeled nucleotide:	0.3 m*M* biotin-dUTP or digoxigenin-dUTP 50 μCi (α^{35}S) dATP

D. OLIGONUCLEOTIDE LABELING[24]

1. Tailing 3′ End Labeling Method (See Protocol 12)

Terminal deoxynucleotidyl transferase (TdT) adds deoxyribonucleotide monophosphate onto the 3′ ends of oligonucleotides. This enzyme does not require a template, but simply adds a series of supplied deoxyribonucleotides onto the 3′ end of single-stranded DNA.

The precursor is usually a deoxynucleoside: biotin 11-dUTP, digoxigenin 11-dUTP, or 5′ triphosphate (α^{35}S) dATP.

PROTOCOL 12 — OLIGONUCLEOTIDE 3′ END LABELING

1.	Mix	
	Oligonucleotide	10 pmol
	$CoCl_2$ (1.5 m*M*)	2 μl
	Tailing buffer × 5	4 μl
	Labeled nucleotide	× μl
	TdT (25 units)	1 μl
	H_2O	to 20 μl
2.	Incubate for 30 min at 37°C	

Buffer × 5: 1 *M* potassium cacodylate, 125 m*M* Tris-HCl, 1.25 mg/ml BSA (pH 6.6 at 25°C)

E. PURIFICATION OF PROBE

After the labeling procedure the probes must be separated from unincorporated nucleotides by gel filtration on Sephadex G25 (see Protocol 13) or ethanol precipitation (see Protocol 14).

PROTOCOL 13 — PURIFICATION OF LABELED PROBES

Spun column procedure[25]

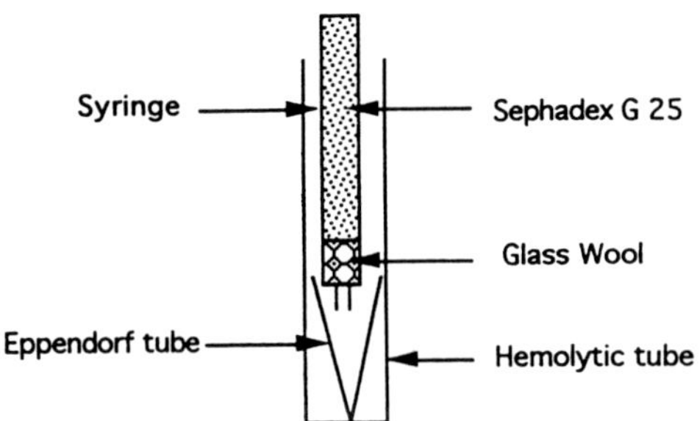

1. Preparation of spun column:
   a. Plug the bottom of a 1-ml disposable syringe with a small amount of sterile glass wool.
   b. In a syringe prepare a column (0.9 ml) of Sephadex G25, equilibrated in 10 m*M* Tris 100 m*M* EDTA (pH 8) containing 100 m*M* NaCl (STE).
   c. Insert the syringe into a glass centrifuge tube.
   d. Centrifuge 4 min at 4000 *g* at 4°C.
   e. Continue to add Sephadex until the packed column volume is 0.9 ml.
2. Add the DNA sample to the column into a total volume of 50 μl (use STE to make up the volume).
3. Centrifuge in the same conditions, collecting the 50 to 60 μl of effluent from the syringe in an uncapped Eppendorf tube. The unincorporated dNTP remains in the syringe, which would be carefully disposed of. The labeled cDNA or oligonucleotides is collected from the uncapped Eppendorf tube.
4. Use the DNA either in this form or concentrated by ethanol precipitation (see Protocol 14).

F. PRECIPITATION

DNA precipitation is necessary in two cases: after DNA purification and for eliminating the unincorporated nonradioactive nucleic acids and precipitating the labeled probe (see Protocol 14).

PROTOCOL 14 — CONCENTRATION OF NUCLEIC ACIDS

1. Estimate the volume of the DNA solution.
2. Add either 1/10 of final volume of sodium acetate 3 *M* pH 5.2 or ammonium acetate 7.5 *M*, 3 volumes of cold ethanol (−20°C), and 1 μl (10 mg/ml) of tRNA. Mix well. Cool to −20°C.
3. Store overnight at −20°C or for 1 h at −80°C to allow the nucleic acid precipitate to form.
4. Centrifuge 30 min at 14,000 *g* at +4°C.
5. Discard the supernatant.
6. Wash in ice-cold 70% ethanol.
7. Centrifuge 15 min at 14,000 *g* at +4°C.
8. Discard the supernatant; traces of supernatant may be removed by brief treatment (1 to 2 min) in a vacuum desiccator or lyophilizer.
9. Dissolve the probe pellet in the desired volume of buffer. (The labeled probe should be stored at −20°C until utilization.)

G. STORAGE

1. Radioactive Labeled Probes

The half-life of ^{35}S isotope is 87 d (see Table 1). It should be used within 1 month.

2. Nonradioactive Labeled Probes

It is possible to use labeled probes for up to 1 year after labeling.

H. LABELING CONTROLS

1. Radioactive Labeled Probes

It is necessary to determine the specific activity (SA) of the labeled DNA and the percentage of incorporation (see Chapter 2).

2. Nonradioactive Labeled Probes

It is possible to check incorporation of digoxigenin (see Protocol 15) or biotin (see Protocol 16).

After labeling with biotin 11-dUTP it is possible to check incorporation into the labeling procedure. The detectability of biotinylated probes is a function of the number of biotin molecules per kb. Under the standard nick translation conditions provided, ±50 biotin molecules are incorporated per kilobase of DNA. In this procedure, biotinylated DNA is spotted on nitrocellulose filters. The streptavidine-alkaline phosphatase conjugate binds strongly to the biotin and is visualized when the enzyme is provided with substrates that produce a colored precipitate. As little as 2 pg biotinylated DNA should be visible. This protocol could be used as a basis for detecting biotinylated DNA probes hybridized in Southern blots.

PROTOCOL 15 — CHECKING OF DIGOXIGENIN-LABELED DNA (FIGURE 2)[26]

1. Dilute linearized denatured pBR328 DNA from 10 pg to 0.01 pg with Tris-EDTA buffer. All dilution as well as the controls contained 50 ng of herring sperm DNA
2. Spot on a nitrocellulose or nylon membrane
3. Prehybridization for 1 h
4. Hybridization with a digoxigenin-labeled pBR328 probe for 16 h at 68°C
5. Wash filter in SSC × 2, 0.1% SDS for 2 × 5 min at room temperature
6. Wash filter in SSC × 0.1, 0.1% SDS for 2 × 15 min at 68°C
7. Color reaction:

a.	Incubate filter with 20-ml antibody anti-digoxigenin conjugate to alkaline phosphatase dilute in buffer 1	30 min
b.	Wash in 5-ml buffer 1	2 × 15 min
c.	Wash in 50-ml buffer 2	2 min
d.	Incubate in 7.5 ml buffer 2 added to 33 μl NBT + 25 μl BCIP overnight in the dark	
e.	Wash in TE	
f.	Dry in a vacuum oven for 2 min at 80°C	

Solutions:	Buffer 1:	100 m*M* Tris-HCl, 150 m*M* NaCl (pH 7.5)
	Buffer 2:	100 m*M* Tris-HCl, 100 m*M* NaCl, 50 m*M* $MgCl_2$ (pH 9.5)
	TE:	10 m*M* Tris, 1 m*M* EDTA
	NBT:	75 mg nitroblue tetratzolium/ml TE
	BCIP:	50 mg 5-bromo-4-chloro-3-indoyl phosphate/ml TE

PROTOCOL 16 — CHECKING OF BIOTINYLATED DNA (FIGURE 3)[27]

1. Dil ute biotinylated standard DNA and the biotinylated test DNA in concentrations of 0, 1, 5, 10, 20 pg/μl with Tris-EDTA buffer
2. Spot 3 μl of dilute biotinylated DNA and 1 μl of a spot of calf thymus DNA (1 μg/μl) on a small sheet of nitrocellulose (8 × 4 cm^2)
3. Spot 3 μl of diluted biotinylated control DNA dilution
4. Dry the filter in a vacuum oven for 1 to 2 h at 80°C
5. Revelation

a.	Block in 10 ml of buffer 1 containing 3% BSA	10 min
b.	Incubate in 10-ml buffer 1 containing 10 μl SA-AP conjugate	10 min

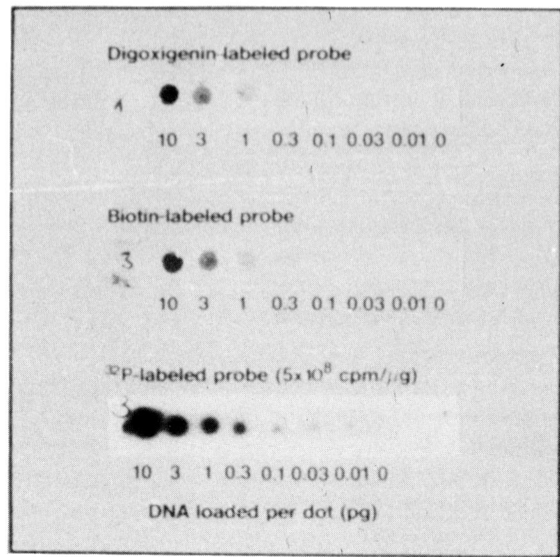

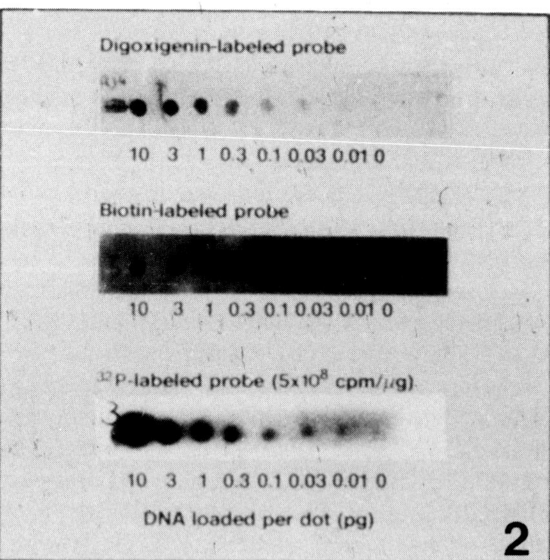

FIGURE 2. Comparison of DNA detection in dot-blot hybridization on nitrocellulose (a) or nylon membrane (b) with digoxigenin, biotin, and ^{32}P-labeled probes, using linearized pBR 328 DNA as target and probe for 16 h of detection. The photograph was a generous gift of Boehringer Mannheim.

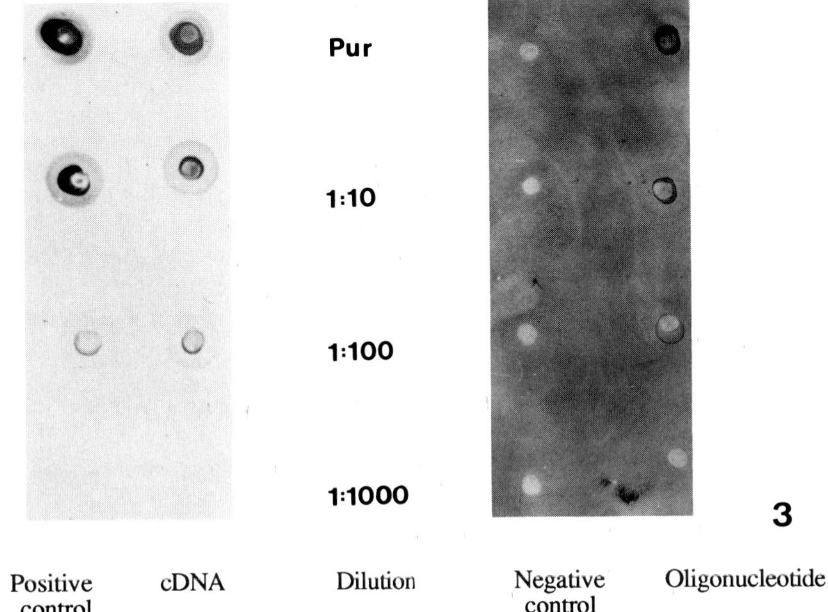

FIGURE 3. Checking of a labeling of the biotin-labeled DNA probe in dot-blot hybridization on nitrocellulose (a) and of digoxigenin-labeled oligonucleotide on nylon membrane (b) using revelation as described in Protocols 15 and 16, respectively.

c.	Wash in 50-ml buffer 1	2 × 15 min
d.	Wash in 50-ml buffer 2	10 min
e.	Incubate in 7.5-ml buffer 2 added to 33 μl NBT + 25 μl BCIP 30 min in the dark	
f.	Wash in TE	
g.	Dry in a vacuum oven for 2 min at 80°C	

Solutions:	Buffer 1:	100 m*M* Tris-HCl, 150 m*M* NaCl (pH 7.5)
	Buffer 2:	100 m*M* Tris-HCl, 100 m*M* NaCl, 50 m*M* $MgCl_2$ (pH 9.5)
	TE:	10 m*M* Tris, 1 m*M* EDTA
	SA-AP:	Streptavidine-alkaline phosphatase
	NBT:	75 mg nitroblue tetratzolium/ml TE
	BCIP:	50 mg 5-bromo-4-chloro-3-indoyl phosphate/ml TE

Check incorporation of biotinylated-dUTP by comparing the intensities of standard and test DNA. For the probe to be suitable for *in situ* hybridization, it should be at least half as intense as the standard DNA at the corresponding dilution.

IV. PRETREATMENTS

On semithin sections pretreatment depends on the embedding procedure.

A. PROTEOLYSE

After fixation of the tissue, it is necessary to pretreat tissue sections, since hybridization on sections of fixed material usually yields low signals. Most procedures therefore include treatment with proteases (proteinase K or pronase) (see Protocol 17) to partially remove cellular proteins. Treatment of sections with HCl (see Protocol 18) has also been reported to increase signals by several times. However, these treatments can cause substantial morphological deterioration.[28]

PROTOCOL 17 — PROTEOLYSE PRETREATMENTS

1. Proteinase K (storage in aliquot at a concentration of 1 mg/ml at −20° C
   a. Dip the sections in TE × 1 for 5 min at 37°C
   b. To each section activated (by heating for 30 min at 37°C) apply proteinase K (1 μg/ml) for 10 to 30 min at a temperature which will depend on level of proteolyse required.
   c. Wash in TE × 1 for 2 × 5 min
   d. Postfix in 4% paraformaldehyde buffered for 5 min at 4°C and wash in same buffer
   e. Dehydration in graded ethanol (95% − 2 × 100%) for 1 to 2 min
2. Pronase
   a. Apply the pronase (0.05 to 0.1 mg/ml) to each section for 1 to 2 min at room temperature
   b. Stop the reaction in 50 m*M* Tris-HCl, glycine
   c. Wash in 50 m*M* Tris-HCl or 100 m*M* phosphate buffer; add 15 m*M* NaCl with gentle agitation
   d. Postfix in 4% paraformaldehyde buffered for 5 min at 4°C, and wash in same buffer
   e. Dehydration in graded ethanol (95% − 2 × 100%) for 1 to 2 min

PROTOCOL 18 — ACID HYDROLYSIS

1. Dip the sections in 0.2 *N* HCl for 5 min at room temperature
2. Wash in 100 m*M* phosphate buffer for 2 × 5 min

B. OTHER

Another pretreatment often used for reducing background noise during *in situ* hybridization, by saturating nonspecific adsorption sites, involves acetylation (see Protocol 19). Another possibility is prehybridization, in which the slides bearing the sections are incubated in a hybridization buffer for 2 or 3 h.

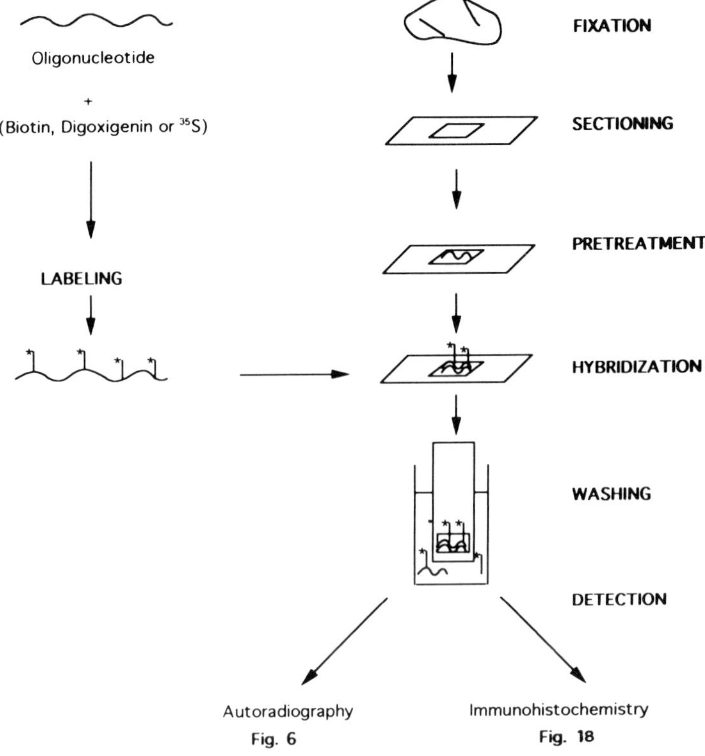

FIGURE 4. Schema of the principal steps of *in situ* hybridization.

PROTOCOL 19 — ACETYLATION[29]

1. Drain excess buffer
2. Dip the slides in 100 m*M* triethanolamine (Sigma) for 10 min, with agitation
3. Add 25 m*M* acetic anhydride for 10 min, with agitation
4. Wash in buffer for 5 min, with gentle agitation

It is possible to store the sections for some months at −20°C, after dehydration by immersion (1 to 2 min) in graded ethanol (95% − 2 × 100%), in containers containing a drying agent.

V. *IN SITU* HYBRIDIZATION (FIGURE 4)

A. GENERAL CONSIDERATION

The hybridization procedure consists of matching complementary strands of nucleic acid, DNA/DNA, DNA/RNA, or RNA/RNA. For *in situ* hybridization, the hybrids are formed between the nucleic acid present in the section and a labeled nucleic acid (the probe). Many factors influence the stability of hybrids. The most important of these are the sodium ion concentration (Na^+) of the hybridization buffer and the incubation temperature. These parameters determined the Tm (the temperature at which 50% of DNA is single-strand), depending on the sodium ion concentration, the length of the DNA sequence, and the percentage of G-C.[30]

$$Tm = 16.6 \log(Na^+) + 0.41\ (\%\ G + C) + 81.5$$

Addition of formamide reduces the Tm of 0.65°C/% formamide.[31]

B. HYBRIDIZATION TEMPERATURE

The hybridization temperature, which must be 10 to 15°C below the Tm, was determined by the sodium ion (SSC) and formamide concentrations.[32] To reduce nonspecific hybridization, Denhardt's solution and tRNA were added and it is also possible to add DNA of salmon sperm. Dextran sulfate increases the concentration of the probe.[33] If this probe is labeled with a radioactive nucleotide, 10 m*M* DTT are added to the buffer. These constituents were brought together as follows.

C. HYBRIDIZATION BUFFER

Components	Storage concentration	Final concentration
SSC	× 20	× 4
Deionized formamide	100%	50%
Denhardt's solution	× 50	× 1
Dextran sulfate	50%	10%
tRNA	10 mg/ml	250 μg/ml
DNA salmon sperm	10 mg/ml	400 μg/ml

Denhardt's solution × 50: 1% Polyvinylpyrrolidone, 1% Ficoll 400, 1% BSA (store in aliquot at −20°C)
SSC × 20: 3 *M* NaCl, 300 m*M* sodium citrate (pH 7, with NaOH 10 *N*)
Deionized formamide: Mix 50 ml of formamide with 5 g of mixed-bed ion exchange resin. Stir for 30 min at room temperature. Filter twice through Whatman number 1 filter paper. Dispense into 500 μl aliquots and store at −20°C
Dextran sulfate 50%: Dissolve 50 mg of dextran sulfate in 65 μl of distilled water. Stir gently with a Pasteur pipette. Centrifuge.
tRNA and DNA salmon sperm are stored in aliquot at −20°C
DTT (Dithiotreitol) (1 *M*): Dissolve 3.09 g of DTT in 20 ml of 0.01 *M* sodium acetate (pH 5.2). Sterilize by filtration. Dispense into 500-μl aliquot and store at −20°C

The dried pretreated sections were covered by in a volume of hybridization buffer which depended on the dimensions of the coverslide used (> 1 μl/10 mm²). The slides were incubated in a humid chamber containing SSC ×5 to avoid the necessity for sealing with rubber cement.

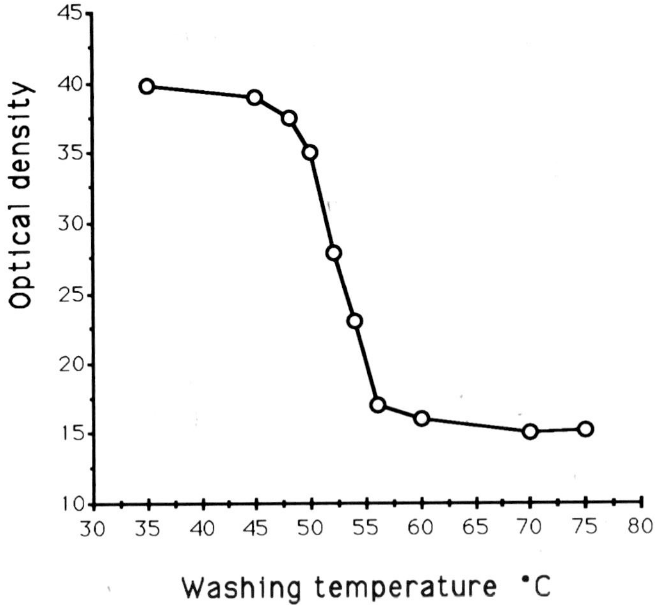

FIGURE 5. Washing temperature dissociation curve for *in situ* hybridization.

D. PROBE CONCENTRATION

The probe is added to the buffer in quantities which depend on the nature (double-stranded DNA or oligonucleotide) and the labeling of the probe (radioactive or nonradioactive), and on the type of preparation of the tissue (frozen or embedded sections), i.e., for semithin frozen sections radioactive probes are used at a concentration of 2 p*M*/ml of hybridization buffer, while nonradioactive probes are used at a concentration of 10 p*M*/ml of hybridization buffer; for semithin Lowicryl-embedded tissue, radioactive probes are used at 20 p*M*/ml of hybridization buffer and nonradioactive probes are used at a concentration of 100 p*M*/ml of hybridization buffer. *With double-stranded DNA probes it is necessary to apply heat denaturation for 5 min at 196°C, and then to rapidly chill the tube in an ice bath before use.*

E. WASHING

The stability of the hybrids obtained between the probe and the target nucleic acid is superior to that of nonhomologous hybrids and of nonspecific association. The aim of the washing steps (see Protocols 20 and 21) is to preserve specific hybrids and eliminate nonspecific binding by increasing stringency. Two parameters can be modified: sodium ion concentration and temperature (Figure 5).[34]

PROTOCOL 20 — WASHING AFTER *IN SITU* HYBRIDIZATION WITH RADIOACTIVE PROBE

1. Rinse in SSC × 4
2. Wash in:
   a. SSC ×2 for 30 min at a temperature depending upon the Tm
   b. SSC ×1 for 30 min at room temperature
   c. SSC ×0.5 for 30 min at room temperature
3. Dehydrate:
   a. 95% ethanol for 2 min
   b. 100% ethanol for 2 × 2 min

PROTOCOL 21 — WASHING AFTER *IN SITU* HYBRIDIZATION WITH NONRADIOACTIVE PROBE

1. Rinse in SSC ×4
2. Wash in:
   a. SSC ×2 for 30 min at room temperature
   b. SSC ×2 for 30 min at a temperature depending upon the Tm
   c. SSC ×1 for 1 h at a room temperature
   d. SSC ×0.5 for 1 h at a room temperature

VI. REVELATION

The aim of detection procedures is the visualization of hybrids. Two different forms exist: either the hybrids contain radioactive isotopes detected by autoradiography or they contain an antigenic component revealed by immunohistochemistry.

A. AUTORADIOGRAPHY

With radioactive probes, the hybrids formed are localized by autoradiography (see Protocol 22), with the detection of radiation from a radioactive source in a nuclear emulsion (see Protocol 23) covering a semithin section containing the radioactive isotope. The exposure time depends on the radioactive specificity of the probe, the quantity of hybrided probe, and the radioisotope used (^{3}H or ^{35}S).[18,19,35]

PROTOCOL 22 — PROCEDURE FOR LIGHT MICROSCOPIC AUTORADIOGRAPHY (FIGURE 6)

1. Check the darkroom conditions: safelight (compatible with nuclear emulsion), temperature (18 to 20°C), humidity (40 to 50%)
2. Apply nuclear emulsion (Amersham LM1, Ilford K5, Kodak NTB_2) by dipping, as described in Protocol 23

FIGURE 6. Different steps in the autoradiography procedure: (1) emulsion, (2) drying, (3) exposure.

3. Dry at room temperature on a horizontal support
4. Expose the coated slides at 4°C in light-tight containers containing drying agent for approximately 2 to 30 days
5. Develop the preparations (18°C in D19 or Dektol developer diluted v/v for 2 min)
6. Wash them briefly in water and then fix them in 30% sodium thiosulfate for at least 2 min
7. Rinse them carefully in water
8. Stain them for viewing by bright field or epifluorescence (see Protocol 24), or omit staining and view in dark field
9. Mount them on coverslips

PROTOCOL 23 — LIQUID EMULSION PREPARATION

1. Check the darkroom equipment:

   - A water bath at 45°C
   - A dipping jar containing distilled water to dilute nuclear emulsion (1:1 for Kodak NTB2; 2:1 for Amersham LM1; 3:1 for Ilford K5)
   - A horizontal support to dry the histological slides

2. Place the jar of emulsion in the water bath for 30 min. Place the dipping jar in the bath
3. Fill the dipping jar with emulsion to the required level
4. Dip a clean slide in the emulsion, shake it gently, and take it to the safelight to examine it for the presence of bubbles. If bubbles are present, gently tap the side of the jar and leave it for some minutes

5. Dry another clean slide, withdrawing it slowly and carefully
6. Drain the excess emulsion onto tissue paper
7. Leave the dipped slide on the horizontal support until it is dry in order to judge the thickness and regularity of the emulsion

PROTOCOL 24 — STAINING OF SEMITHIN SECTIONS

Autoradiographs produced by dipping in liquid emulsion can be easily stained with Toluidine Blue (or with one of several other dyes, according to the case) after development and fixation; staining before applying the emulsion is almost always unsuccessful due to positive chemographic effects.

1. Prepare 1% solution of Toluidine Blue in 1% sodium borate
2. Put the slide in a diluted solution of Toluidine Blue (1:5) for 5 to 10 min at room temperature
3. Differentiate the staining with 95% ethanol
4. Dehydrate quickly in 2 × 100% ethanol
5. Mount on coverslip.

Autoradiographic revelation can be used to visualize hybrids in frozen semithin sections as well as in embedded-resin semithin sections (Figures 7 to 10). The main advantage of autoradiography is the sensitivity of detection that it permits. Resolution depends on the isotopes: $^{3}H > {}^{35}S > {}^{32}P$. The disadvantage is the long exposure time and the difficulty of manipulating radioactive material. The absence of enzymatic and endogenous antigenic activity (e.g., biotin) turned out to be an advantage. However, the presence of autoradiographic background noise[18,19,35] inherent in the method remains a problem as regards the detection of small quantities of hybrids. Quantification is theoretically possible after the application of this method.[18,35] It is not useful to measure optical density on autoradiographic film due to the size of autoradiograms; silver grain count per μm^2 can be used instead.[36]

B. IMMUNOHISTOCHEMISTRY

Nonradioactive hybrids can be detected by various revelation methods, using an antibody (see Protocols 26, 27, 28) or an avidin-biotin system conjugated either with an enzyme (alkaline phosphatase or peroxidase) (see Protocol 25), with fluorescent reagents, (see protocol 30) or with colloidal gold (see Protocols 28, 29). The chromogenes which are most useful in revealing alkaline phosphatase are NBT/BCIP (see Protocols 25, 26, 27) and Fast red (Figure 11), while for peroxidase the best ones are AEC or DAB and 4 chloro-1-naphthol (Figure 12). Fluorescent markers can be detected directly using a fluorescent microscope (see Protocol 30). Colloidal gold particles need to be amplified (see Protocol 29) in order to be visualized (Figure 13). The same conjugates are used with an avidin-biotin system, i.e., the markers can be enzymatic, fluorescent, or colloidal gold conjugate.

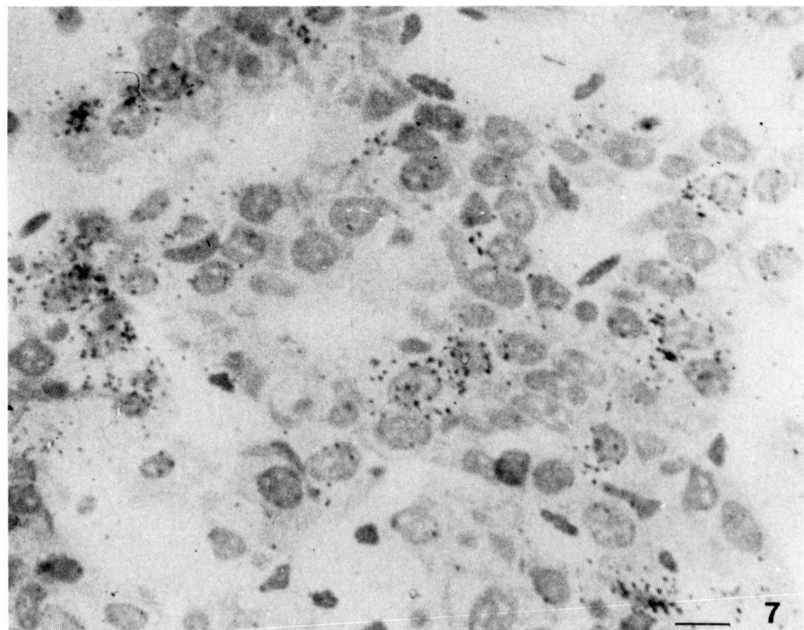

FIGURE 7. *In situ* hybridization on frozen semithin sections. Adenohypophysis was fixed with 2% paraformaldehyde and frozen as described in Protocol 4. The GH oligonucleotide probe, labeled with ^{35}S by tailing (see Protocol 12) and revealed by autoradiography, was visualized with silver grains over somatotrophs, which synthesize GH (bar: 10 μm).

An antibody detection method is useful for visualizing biotin or digoxigenin. Specific monoclonal and polyclonal antibodies are commercially available for direct and indirect (see Protocols 26, 27) immunohistological detection (Figures 14-15). Direct fluorescent methods are fast, but are of low sensitivity (see Protocol 30, Figure 16). Colloidal gold conjugate antibodies (see Protocol 28) using small particles (1 nm) amplified by silver enhancement (see Protocol 29, Figure 17) give the same sensitivity as indirect immunohistological detection. However, the resolution may be inferior.

PROTOCOL 25 — DIRECT DETECTION SYSTEM

After washing, incubate

1. Streptavidin 1:100 in Tris-sodium buffer (TBS) for 2 h at room temperature
2. Wash in TBS for 2 × 10 min
3. Biotin-alkaline phosphatase 1:100 in TBS for 2 h at room temperature
4. Wash in TBS for 10 min, then
5. Wash in distilled water for 10 min

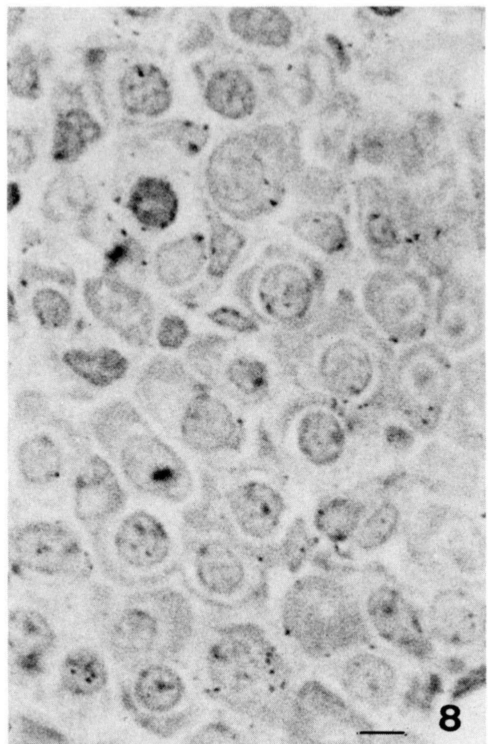

FIGURE 8. *In situ* hybridization on Lowicryl semithin sections. Adenohypophysis was fixed with 2% paraformaldehyde and frozen as described in Protocol 4. The GH oligonucleotide probe, labeled with ^{35}S by tailing (see Protocol 12) and revealed by autoradiography, was visualized with silver grains over somatotrophs, which synthesize GH (bar: 10 μm).

6. Chromogene: prepare dye solution just prior to use with: NBT 1:250 and BCIP 1:250 in TBS
7. Incubate section for 1 h to overnight at 37°C
8. Wash in distilled water for 2 × 10 min
9. Mount on a coverslip in a hydrophilic medium

Tris-sodium buffer: 50 m*M*/ml Tris-HCl (pH 7.6), 150 m*M* NaCl

PROTOCOL 26 — MONOCLONAL ANTIBODY REVELATION SYSTEM AMPLIFIED BY ABC COMPLEX (FIGURE 18)

After washing incubate with

1. Normal rabbit serum 1:5 in TBS for 20 min
2. Mouse monoclonal antibody diluted optimally in TBS for 20 to 30 min
3. Wash in TBS for 5 min

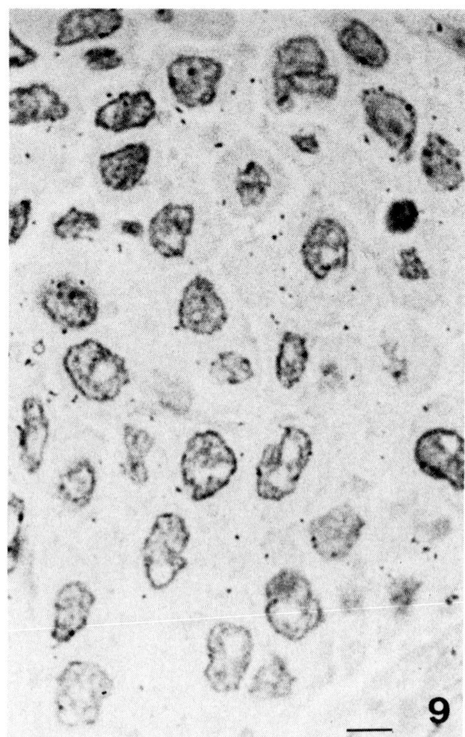

FIGURE 9. *In situ* hybridization on Lowicryl semithin sections. Breast tumor tissue was fixed with 4% paraformaldehyde and Lowicryl embedded as described in Protocol 7. The erb B2 cDNA probe, labeled with ^{35}S by random priming (see Protocol 11) and revealed by autoradiography, was visualized with silver grains in a number of epithelial cells (bar: 10 μm).

4. Biotinylated rabbit anti-mouse IgG 1:200 to 1:600 in TBS for 20 to 30 min
5. Wash in TBS for 5 min
6. Streptavidine conjugated with alkaline phosphatase for 20 to 30 min
7. Wash in TBS for 5 min
8. Alkaline phosphatase substrate solution (see Protocol 26) for 10 to 20 min
9. Rinse with distilled water
10. Counterstain and mount on a coverslip in a hydrophilic medium

PROTOCOL 27 — POLYCLONAL ANTIBODY REVELATION SYSTEM

StreptABComplex/AP using a rabbit primary antibody
After washing, incubate with

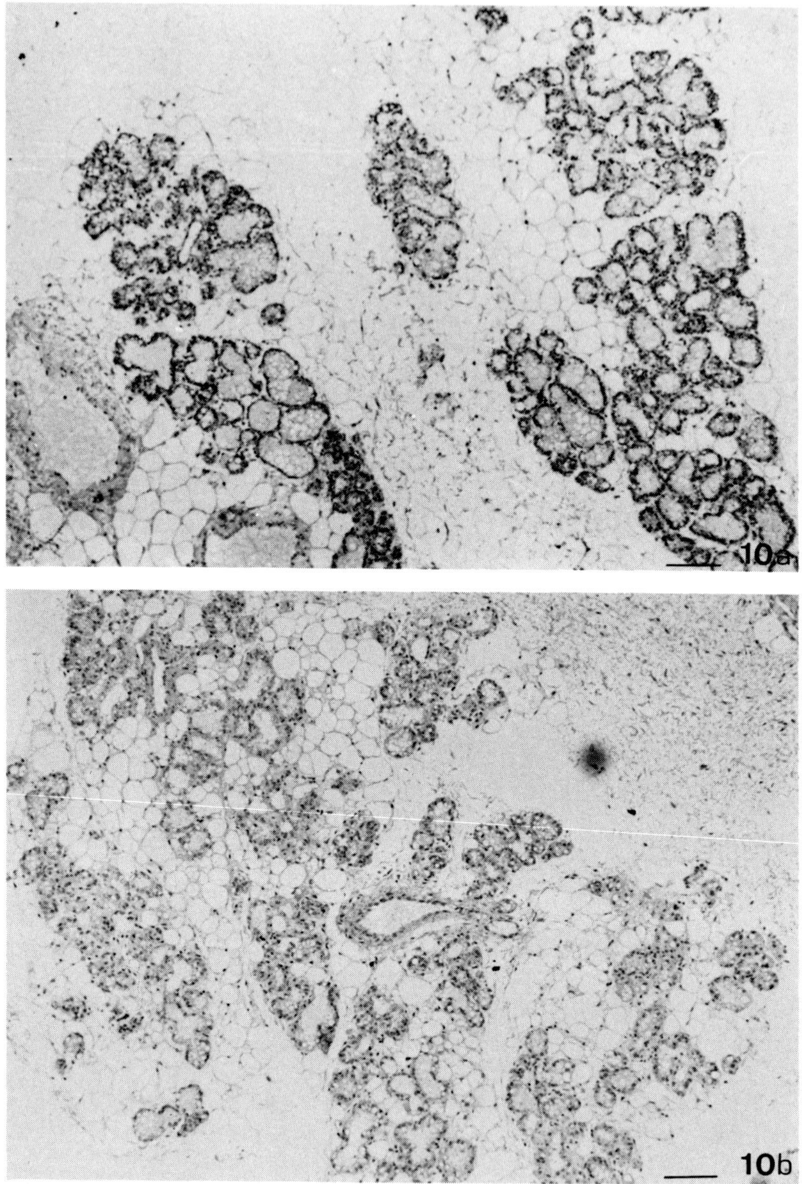

FIGURE 10. *In situ* hybridization on semithin sections of lactating mammary gland embedded in Epon-araldite, with an antisense tritiated cRNA casein probe (a), and with sense (homologous) cRNA casein probe (b) (bar: 100 μm). The photograph was a generous gift of Dr. Liscia.

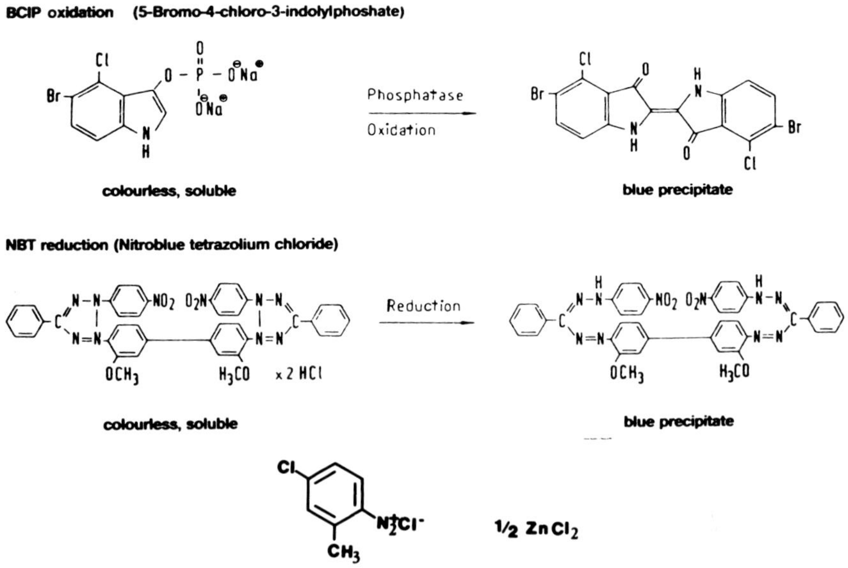

BCIP oxidation (5-Bromo-4-chloro-3-indolylphoshate)

Phosphatase

Oxidation

colourless, soluble

blue precipitate

NBT reduction (Nitroblue tetrazolium chloride)

Reduction

colourless, soluble

blue precipitate

1/2 $ZnCl_2$

FAST-RED TR-salt

FIGURE 11. Schema of NBT/BCIP reaction, Fast red salt: the chromogen used after revelation with alkaline phosphatase.

1. Normal swine serum 1:5 in TBS for 20 min
2. Rabbit antibody diluted optimally in TBS for 20 to 30 min
3. Wash in TBS for 5 min
4. Biotinylated swine anti-rabbit IgG 1:300 to 1:800 in TBS for 20 to 30 min
5. Wash in TBS for 5 min
6. StreptABComplex/AP for 20 to 30 min
7. Wash in TBS for 5 min
8. Alkaline phosphatase substrate solution for 10 to 20 min
9. Rinse with distilled water
10. Counterstain and mount on a coverslip in a hydrophilic medium

PROTOCOL 28 — REVELATION WITH ANTIBODY ANTI-DIGOXYGENIN CONJUGATED WITH COLLOIDAL GOLD

1. Wash the section in 50 m*M* Tris-HCl, 15 m*M* NaCl buffer for 10 min
2. Apply antibody antidigoxygenin conjugated with colloidal gold, 1 or 5 nm 1:30 in the same buffer for 1 to 3 h at room temperature
3. Wash in same buffer and in distilled water

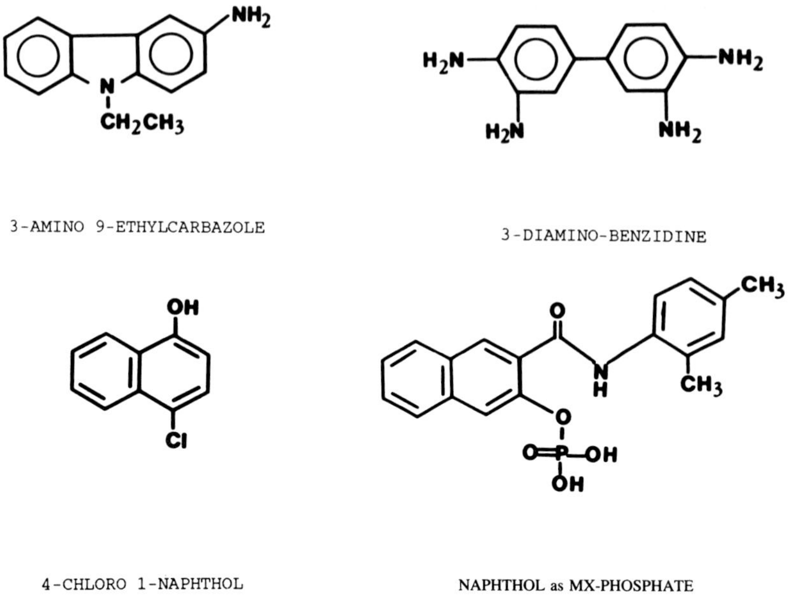

3-AMINO 9-ETHYLCARBAZOLE

3-DIAMINO-BENZIDINE

4-CHLORO 1-NAPHTHOL

NAPHTHOL as MX-PHOSPHATE

FIGURE 12. Schema of 3-diamino-benzidine, 3-amino 9-ethylcarbazole, 4-chloro 1-naphthol, and naphtho- as MX-phosphate: the chromogen used after revelation with peroxidase.

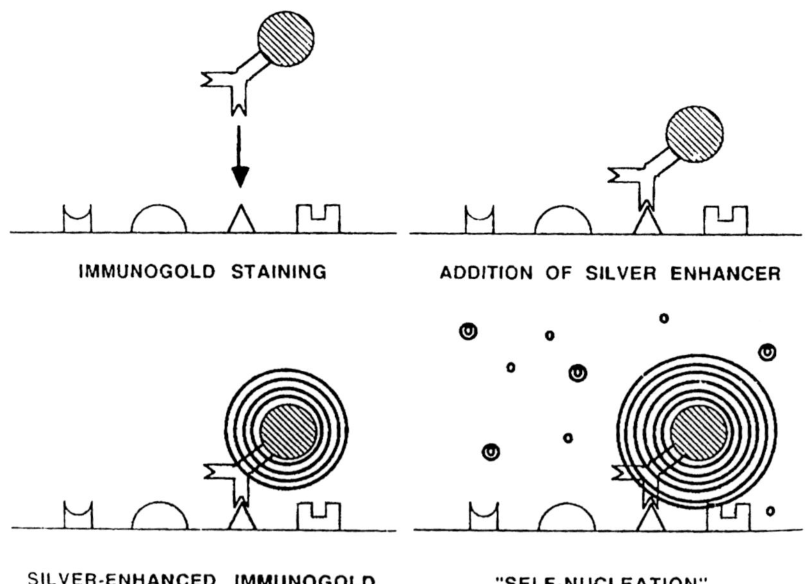

FIGURE 13. Schema of silver enhancement reaction.

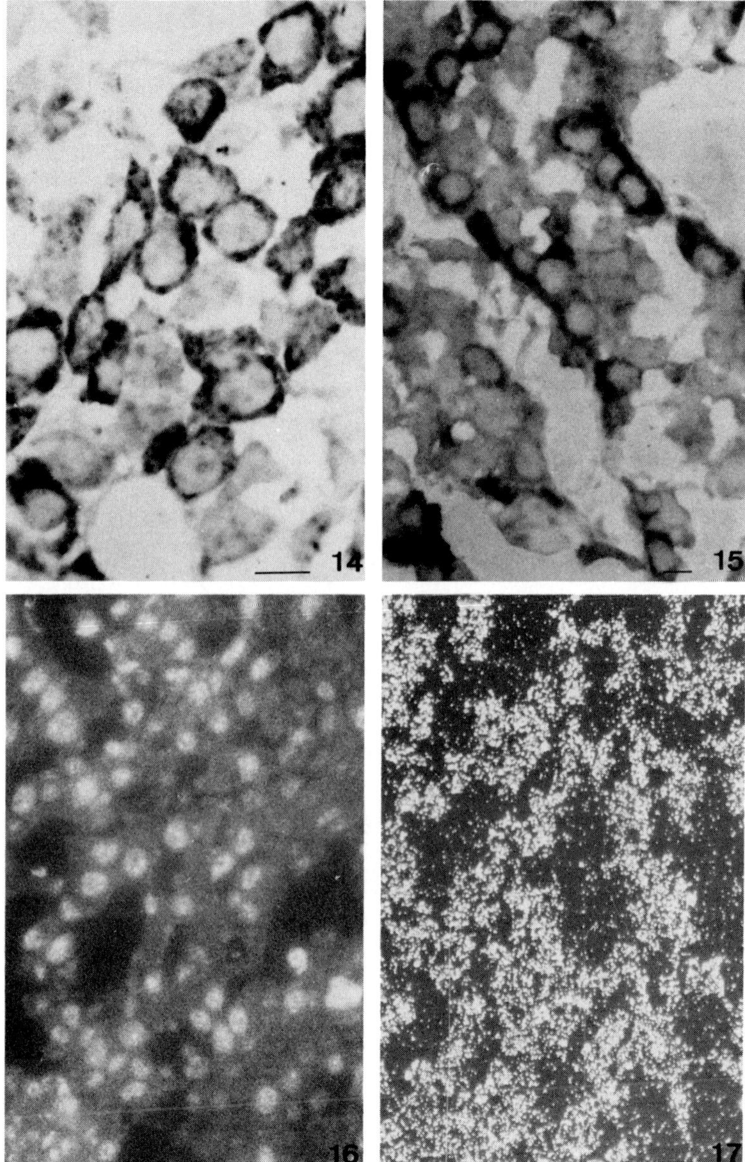

FIGURES 14, 15, 16, and 17. Visualization of mRNA coding for GH, using a 30-mer oligonucleotide probe in the anterior pituitary gland. The oligonucleotide was labeled with biotin 16-dUTP (Figure 14), digoxigenin 11-dUTP (Figure 15) by tailing (see Protocol 12). Detection of hybrids was done by immunohistochemistry (see Protocols 25 to 28, Figures 14 and 15), bar: 10 μm, and immunofluorescence (see Protocol 30, Figure 16), bar: 20 μm, or using gold particule and silver enhancement (see Protocol 29, Figure 17). Silver enhancement was visualized with dark film. bar: 20 μm.

a Mouse monoclonal antibody

b Biotinylated rabbit anti-mouse IgG

c Streptavidine alkaline-phosphatase conjugated (E)

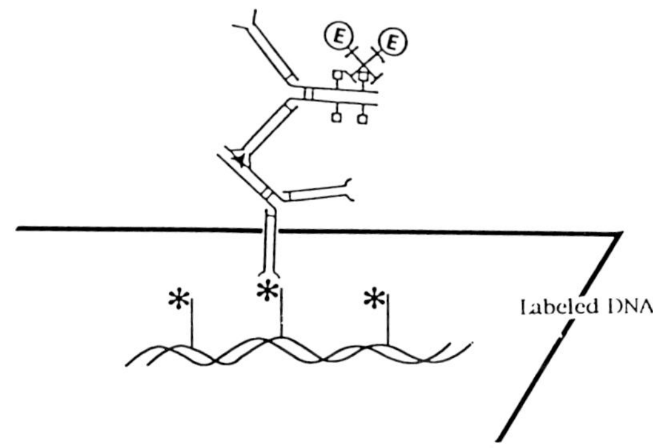

FIGURE 18. Monoclonal antibody revelation system amplified by ABC complex.

PROTOCOL 29 — SILVER AMPLIFICATION OR ENHANCEMENT[37]

1. Wash the section in 50 m*M* Tris-HCl, 15 m*M* NaCl buffer for 10 min
2. Apply antibody antidigoxygenin conjugated with colloidal gold 1 or 5 nm 1:30 in the same buffer for 1 to 3 h at room temperature
3. Wash in the same buffer and then in distilled water
4. Prepare a silver enhancement solution just before use; apply one drop of the solution to the section for 10 to 15 min
5. Wash in distilled water
6. Counterstain with Toluidine Blue
7. Mount on coverslip after dehydration

Tris buffer: 20 m*M* Tris, 650 m*M* NaCl, 0.05% Tween-20, 1% Ovalbumine, H_2O 100 ml, pH 7.6 with HCl 2 *N*, stored at 4°C

PROTOCOL 30 — DETECTION WITH FLUORESCENT REAGENT

1. Anti-biotin 1/100 in Tris-HCl buffer for 90 min
2. Wash in Tris buffer for 2 × 15 min
3. Goat anti-rabbit rhodamin conjugate 1/100 in Tris buffer for 1 h
4. Wash in Tris buffer for 2 × 15 min

5. Distilled water for 3 min
6. Mount on coverslip

C. INHIBITION OF ENDOGENOUS ACTIVITY

Some reagents used for labeling or detecting nonradioactive probes, e.g., biotin, alkaline phosphatase, and peroxidase, are present in certain mammals cells and tissues. Thus it is necessary to inhibit the endogenous activity of these reagents, which would be detected at the same time as the label and would give a nonspecific positivity to the reaction.

1. Endogenous Peroxidase Inhibition Activity

H_2O_2 — Before starting an immunological reaction, endogenous peroxidase activity was inhibited by immersing the sections in 2% peroxide in PBS for 5 to 10 min, followed by rinsing in distilled water.[38–41]

2. Endogenous Alkaline Phosphatase Inhibition Activity

Levamisol — Add levamisol 1 m*M* during all the washes and the revelation.[42,43]

Acetic acid — Treat the sections in a bath containing concentration of acetic acid from 5 to 20% in either water or ethanol at either 4°C or room temperature; then wash three times in PBS.[42,44]

3. Endogenous Biotin Inhibition Activity

Streptavidin — The endogenous biotin was competitively inhibited by preincubation with free streptavidin then saturated by addition of biotin.[44] Each streptavidin molecule possesses four binding sites for biotin, while each biotin molecule is capable of binding to only one streptavidin molecule.[45,46]

4. Inhibition of Immunoglobulin Binding Activity

Add BSA (2 to 10%) or fat-free milk (3%) during the first step in the detection procedure.

VII. CHECKING PROCEDURES

A number of essential checks are required to validate *in situ* hybridization. It is important to verify the positivity of the reaction in order to eliminate false positives and to demonstrate the specificity of the reaction.

False positives can be due to:

- The probe
- The hybridization
- The detection

Specificity is due to the formation of specific hybrids. The signal/nonspecific hybridization ratio must be maximal. Several points must be checked to confirm the specific localization of the nucleic acid studied:

- Nonspecific binding of the probe
- Background noise generated by a radioisotopic probe
- The presence of endogenous enzymatic activity in the tissue or cells studied.

A. THE PROBE

Presence of a nonspecific signal can be due to the presence of homologies between sequences of the probe or vector and sequences of other nucleic acids present in the tissues or cells studied, without relation to the target nucleic acid. This problem can be eliminated by checking the probe against the gene bank.

The specificity of the label can be verified by hybridization in the same tissue or cells using a ''sense'' oligonucleotide probe. In such cases no signal should be detected (Figure 10b).*

B. THE HYBRIDIZATION PROCEDURE

The main problem for nonspecific hybridization is that of nonspecific binding of the probe in tissues. Checks were carried out on:

- Tissue studied. Hybridization with a heterologous probe, carried out under identical conditions to those in which the test probes are used, with checks for regions of the tissue which might exhibit unusually high nonspecific binding of nucleic acids.
- Homogenous positive tissue. A large positive signal should be detected in cell lines known to contain a particularly high level of the nucleic acid under study. This technique can be cross-checked with other techniques.
- Heterologous tissue. With heterologous tissue only some cells which express the mRNA or DNA studied should be positive. The absence of a signal in the other cells or structures is a good check of specificity.
- Negative tissue. Hybridization with the probe, carried out under identical conditions to those in which a positive signal is obtained in the tissue studied. Absence of signal should be associated with absence of target nucleic acid.

Specific hybridization can be abolished by:

- Competition between a labeled probe and an excess of nonlabeled probe (100 times more concentrated than labeled probe), while a heterologous probe (in similar conditions) does not modify the signal.

* Absence of signal may be due to the absence of denaturation of double-strand cDNA, or the absence of probe labeling. Care should be taken to check the results of the labeling procedure (see Protocols 15 to 16).

- Destruction of target nucleic acid using enzymatic treatment (DNase or RNase). Pretreatment of tissue with RNase or DNase has frequently been used to demonstrate that the signal observed is dependent on the presence of RNA in the tissue and is not due to hybridization to DNA nor to adventitious binding to unknown cellular constituents. Care must be taken in this case to demonstrate that the consequent reduction in the signal is not due to destruction of the probe by residual nuclease activity.

C. THE REVELATION PROCEDURE

1. Autoradiography

Determination of background noise after autoradiography.[30]

2. Immunohistology

Hybridization is carried out without a probe in a hybridization buffer and under identical conditions to those used with a labeled probe.

A positive signal is due to the presence of a labeled probe in the tissue or revelation system used in the immunohistological procedure. Endogenous activity in the tissue or cells studied must be inhibited (see Section VI.C).

D. CORRELATION WITH OTHER TECHNIQUES

In situ hybridization detects nucleic acid in tissue. The presence of this nucleic acid can be confirmed after extraction, or by detection of the protein synthesized by the nucleic acid.

1. Blot Techniques

Dot blot, Northern blot (RNA), and Southern blot (DNA) techniques can be used to confirm the presence in cells of the nucleic acid studied, after extraction and hybridization on filters.

Addition of PCR technique may be necessary, to amplify the signal.

2. Immunohistochemistry

Immunohistochemistry can be used as a complement to *in situ* hybridization, in order to visualize protein synthesized by the cells studied.

VIII. RESULTS AND DISCUSSION

All the methods of producing semithin sections have been used with *in situ* hybridization techniques. In the pre-embedding technique, *in situ* hybridization procedure in semithin sections can be used to check *in situ* hybridization positivity, before making ultrathin sections for electron microscopy.[47] After *in situ* hybridization procedure is realized on pre-embedding tissue, it is also possible to obtain semithin sections and to treat them by immunohistochemistry technique, after removal of the resin. This technique makes it possible to compare *in situ* hybridization and immunohistochemistry

TABLE 2
Chief Characteristics of Semithin Sections: Frozen Sections Vs. Post-Embedding Sections

Method	Frozen section	Epoxy resin	LR White	Lowicryl	Spurr
Storage	After freezing	⟵	⟵	After embedding	⟶
Fixation	+	+	+	+	+
Embedding	No	Easy	Easy/rapid	Long time	Long time
Pretreatment	No	?	Possible	Possible	?
HCl			No	No	
Enzymatic			+	+	
Acetylation			+	+	
Detection					
Autoradiography	+	+		+	
Enzymatic	+		+	+	+
Gold/silver enhancement	+		+	+ +	+
Fluorescent	+				
Preservation	Medium	High	High	High	High
Sensitivity	High	Low	Medium	Medium	Medium

signals from the same section.[48] It is possible to detect simultaneously two mRNA on the same section using one radioactive label probe revealed by autoradiography and one other nonradioactive label probe revealed by immunohistochemistry.[49] Among other methods, the most useful are those which use frozen semithin sections followed by Lowicryl K4M, LR White, Spurr, and finally epoxy-resin.[50]

The first treatment of the tissues was fixation, which preserves morphology. Whatever the embedding procedure used, the tissues were fixed in 4% paraformaldehyde with or without a small percentage of glutaraldehyde. Pretreatments were possible for hydrophilic resins but not necessary,[12] since probe penetration was demonstrated using a cDNA probe without pretreatment (Figure 1). For example similar results were obtained with LR White without proteinase K treatment. With epoxy-resin it is necessary to remove resin before carrying out an *in situ* hybridization technique.

Signal detection can be, after *in situ* hybridization, carried out either autoradiographic or immunological signals. The main advantage of autoradiography is its sensitivity, but the exposure time can be long. Immunohistochemical detection can use either enzymatic or fluorescent reagents, or colloidal gold particles followed by silver-enhancement. Endogenous antigenic activity was a problem, but in all cases revelation time was short. Whatever the detection system chosen, it is now possible to quantify the reaction using an automatic analyzer (see Table 2).

Another important point concerns sensitivity. The highest sensitivity was obtained with frozen sections, but the same signal can be obtained with resin sections if the probe concentration is increased: 100 pmol/ml for resin vs. 4 pmol/ml for frozen sections. Also the quantity of the target detected was to be greater in hydrophilic resins than in frozen sections, because most of it is destroyed by the embedding procedure.

To conclude, the choice of embedding procedure depends on the quantity of target available, and on the morphological preservation required to obtain the best result, with good morphological characteristics, sensitivity, specificity, and rapidity.

IX. CONCLUSION

In situ hybridization on semithin sections and light microscopy observation allow DNA and mRNA to be detected in fixed tissues. The important points as regards morphological and nucleic acid preservation are the removal procedure, the choice of fixator, the probe (cDNA or oligonucleotide), and the label (radioactive or nonradioactive).

The best results were obtained in frozen sections and Lowicryl K4M embedded tissues with radioactive-label and nonradioactive-label oligonucleotide probes. Epoxy resin-embedded tissues give good results with radioactive-label probes.

These results show the possibility to detect by *in situ* hybridization, at light microscopic level, nucleic acid from specimens prepared for electron microscopy. In this way, all advantages of light microscopy can be added to those of electron microscopy. *In situ* hybridization on semithin sections remain an advantageous possibility to confirm the presence of target nucleic acid before *in situ* hybridization at electron microscopic level.

ACKNOWLEDGMENTS

I would like to thank the following people for their help in the preparation of this manuscript: J. Doherty (for reading), A. Beiras, T. Garcia Cabalero, G. Besteti (for preparing the embedded specimen), and D. Liscia (for the photographs).

REFERENCES

1. **Chesselet, M. F.,** *In situ Hybridization Histochemistry,* CRC Press, London, 1990, 208.
2. **Rendrop, M., Knapp, B., Winter, H., and Schweizer, J.,** Aminoalkylsilane-treated glass slides as support for *in situ* hybridization of keratin cDNAs to frozen tissue sections under varying fixation and pretreatment conditions, *Histochem. J.,* 18, 271, 1986.
3. **Weiss, L. M., and Chen, Y. Y.,** Effects of different fixatives on detection of nucleic acid from paraffin-embedded tissues by *in situ* hybridization using oligonucleotide probes, *J. Histochem. Cytochem.,* 39, 1237, 1991.
4. **Trembleau, A., Fevre-Montange, M., and Calas, A.,** Localisation ultrastructurale d'ARNm codant pour l'ocytocine par hybridation *in situ.* Etude par radioautographie à haute résolution à l'aide d'une sonde oligonucléotidique tritiée, *C. R. Acad. Sci. Paris,* 307, 869, 1988.

5. **Morel, G., Dubois, P., and Gossard, F.,** Detection ultrastructurale des ARN messagers codant pour l'hormone de croissance dans l'antéhypophyse du rat par hybridation *in situ, C. R. Acad. Sci., Paris,* 302, 479, 1986.
6. **Trembleau, A., Fevre-Montange, M., Landry, M., and Calas, A.,** Hybridation *in situ* ultrastructurale avant inclusion, in *Techniques en Microscopie Electronique: Cryométhodes, Immunocytologie, Autoradiographie, Hybridation In Situ,* Morel, G., Ed., Editions INSERM, Paris, 1991, 503.
7. **Tong, Y., Zhao, H., Simard, J., Labri, F., and Pelletier, G.,** Electron microscopic autoradiographic localization of prolactin mRNA in a rat pituitary, *J. Histochem. Cytochem.,* 37, 567, 1989.
8. **Tokyuasu, K. T.,** A technique for ultracryotomy of cell suspensions and tissues, *J. Cell Biol.,* 57, 551, 1973.
9. **Seveus, L.,** Frozen ultrathin sections, in *Microscopie Électronique,* Favard, P., Ed., 7e congrès Int., Grenoble, 423, 1970.
10. **Liscia, D. S., Doherty, P. J., and Smith, G. H.,** Localization of α-Casein gene transcription in sections of epoxy resin-embedded mouse mammary tissues by *in situ* hybridization, *J. Histochem. Cytochem.,* 36, 1503, 1988.
11. **Hogan, D. L. and Smith, G. H.,** Unconventional application of standard light and electron immunocytochemical analysis to aldehyde-fixed, araldite-embedded tissues, *J. Histochem. Cytochem.,* 30, 1301, 1982.
12. **Binder, M., Tourmente, S., Roth, J., Renaud, M., and Gehring, W. J.,** *In situ* hybridization at the electron microscope level: localization of transcripts on ultrathin sections of lowicryl K4M-embedded tissue using biotinylated probes and protein A-gold complex, *J. Cell Biol.,* 102, 1646, 1986.
13. **Roson, E. and Beiras, A.,** Immunocytologie après inclusion des tissues en résine hydrosoluble LR-White, in *Microscopie Electronique: Cryométhodes, Immunocytologie, Autoradiographie, Hybridation In Situ,* Morel, G., Ed., Editions INSERM, Paris, 1991, 211.
14. **Ichikawa, M., Sasaki, K., and Ichikawa, A.,** Immunocytochemical localization of amylase in gerbil salivary gland acinar cells processed by rapid freezing and freeze-substitution fixation, *J. Histochem. Cytochem.,* 37, 185, 1989.
15. **Carlemalm, E., Garavito, R. M., and Villiger, W.,** Resin development for electron microscopy and an analysis of embedding at low temperature, *J. Microscopy,* 126, 123, 1981.
16. **Besteti, G. E., Boujon, C. E., Reymond, M. J., and Rossi, G. L.,** Functional and morphological changes in mediobasal hypothalamus of streptozocin-induced diabetic rats, *J. Am. Diabetes Assoc.,* 38, 471, 1989.
17. **Troxler, M., Pasamontes, L., Egger, D., and Bienz, K.,** *In situ* hybridization for light and electron microscopy: a comparison of methods for the localization of viral RNA using biotinylated DNA and RNA probes, *J. Virol. Methods,* 30, 1, 1990.
18. **Cau, P.,** In *Techniques en microscopie quantitative:* Stéréologie, autoradiographie et immunocytochimie quantitatives, Editions, INSERM, Paris, 1990.
19. **Williams, M. A.,** Preparation of electron microscope autoradiographs, in *Autoradiography and Immunocytochemistry,* Glauert, A. M., Ed., North-Holland, Amsterdam, 1977, 77.
20. **Singer, R. H., Lawrence, J. B., and Rashtichian, R. N.,** Toward a rapid *in situ* hybridization methodology using isotopic and nonisotopic probes, in *In Situ Hybridization: Applications to Neurobiology,* Valentino, K. L., Eberwine, J. H., and Barchas, J. D., Eds., Oxford University Press, New York, 1987, chap. 4.
21. **Langer, P., Waldrop, A., and Ward, D.,** Enzymatic synthesis of biotin labeled polynucleotides: novel nucleic acid affinity probes, *Proc. Natl. Acad. Sci. U.S.A.,* 18, 6633, 1981.
22. **Kelly, R. G., Cozzarelli, N., Deutscher, M. P., Lehman, I. R., and Kornberg, A.,** Enzymatic synthesis of deoxiribonucleic acid, *J. Biol. Chem.,* 245, 39, 1970.
23a. **Feinberg, A. P. and Vogelstein, B.,** A technique for radiolabeling DNA restriction endonuclease fragments to high specific activity, *Anal. Biochem.,* 132, 6, 1983.

23b. **Feinberg, A. P. and Vogelstein, B.,** Addendum: a technique for radiolabeling DNA restriction endonuclease fragments to high specific activity, *Anal. Biochem.*, 137, 266, 1984.
24. **Chang, L. M. and Bollum, T. J.,** Deoxynucleotide polymerizing enzymes of calf thymus gland, *J. Biol. Chem.*, 246, 909, 1971.
25. **Maniatis, T., Fritsch, E. F., and Sambrook, J.,** in *Molecular Cloning. A Laboratory Manual,* Cold Spring Harbor Laboratory, Cold Spring Harbor, NY, 1982.
26. **Boehringer Mannheim,** Molecular biology, Gmbh, biochemica, 1989.
27. **Ausubel, F. M., Brent, R., Kingston, R. E., Moore, D. D., Seidman, J. G., Smith, J. A., and Struhl, K.,** in *Current Protocols in Molecular Biology,* Wiley Interscience, I, chap. 3, 18, 1991.
28. **Angerer, R. C., Cox, K. H., and Angerer, L. M.,** *In situ* hybridization to cellular RNAs, *Genet. Eng.*, 1984.
29. **Haase, A. T., Brahic, M., Stowring, L., and Blum, H.,** Detection of viral nucleic acids by *in situ* hybridization, *Methods Virol.*, 7, 189, 1984.
30. **Marmur, J. and Doty, P.,** Determination of the base composition of desoxyribonucleic acid from its thermal denaturation temperature, *J. Mol. Biol.*, 5, 109, 1962.
31. **McConaghy, B. L., Laird, C. D., and McCarthy, B. I.,** Nucleic acid reassociation in formamide, *Biochemistry,* 8, 3289, 1969.
32. **Wetmur, J. G., Ruyechan, W. T., and Douthart, R. J.,** Denaturation and renaturation of *Penicillium chrysogenum* mycophage double stranded ribonucleic acid in tetraalkylammonium salt solutions, *Biochemistry,* 20, 2999, 1981.
33. **Wahl, G. M., Stern, M., and Stark, G. R.,** Efficient transfer of large DNA fragments from agarose gels to diazobenzyloxymethyl paper and rapid hybridization by using dextran sulfate, *Proc. Natl. Acad. Sci. U.S.A.*, 76, 3683, 1979.
34. **Terrier, C., Chabot, J. G., Pautrat, G., Jeandel, L., Gray, D., Lutz-Bucher, B., Zing, H. H., and Morel, G.,** Arginine-vasopressin in anterior pituitary cells: *in situ* hybridization of mRNA and ultrastructural localization of immunoreactivity, *Neuroendocrinology,* 54, 303, 1991.
35. **Morel, G.,** Coupes semi-fines et ultrafines de tissus congelés: congélation, cryo-ultramicrotomie, contraste, in *Techniques en Microscopie Electronique: Cryométhodes, Immunocytologie, Autoradiographie, Hybridation In Situ,* Morel, G., Ed., Editions INSERM, Paris, 1991.
36. **Morel, G., Dihl, F., and Gossard, F.,** Ultrastructural distribution of growth hormone (GH) mRNA and GH intron I sequences in rat pituitary gland: effects of GH releasing factor and somatostatin, *Mol. Cell. Endocrinol.*, 65, 81, 1989.
37. **Holgate, C., Jackson, P., Cowen, P., and Bird, C.,** Immunogold-silver staining: new method of immunostaining with enhanced sensitivity, *J. Histochem. Cytochem.*, 31, 938, 1983.
38. **Straus, W.,** Inhibition of peroxidase by methanol or methanol nitroferricyanide for use in peroxidase procedures, *J. Histochem. Cytochem.*, 19, 682, 1971.
39. **Moriarty, G. C. and Halmi, N. S.,** Electron microscopy study of the adrenocorticotropin producing cell with the use of unlabeled antibody and the soluble peroxidase-antiperoxidase complex, *J. Histochem. Cytochem.*, 20, 590, 1972.
40. **Straus, W.,** Phenylhydrazine as inhibitor of horseradish peroxidase for use in immunoperoxidase procedures, *J. Histochem. Cytochem.*, 20, 949, 1972.
41. **Van Leeuwen, F.,** Pitfalls in immunocytochemistry with special reference to the specificity problems in the localization of neuropeptides, *Am. J. Anatomy,* 175, 363, 1986.
42. **Ponder, B. A. and Wilkinson, M. M.,** Inhibition of endogenous tissue alkaline phosphatase with the use of alkaline phosphatase conjugates in immunohistochemistry, *J. Histochem. Cytochem.*, 29, 981, 1981.
43. **Lewis, F. A., Griffiths, S., Dunnicliff, R., Wells, M., Dudding, N., and Bird, C. C.,** Sensitive *in situ* hybridization technique using biotin-streptavidine-polyalkaline phosphatase complex, *J. Clin. Pathol.*, 40, 163, 1987.

44. **Wood, G. S. and Warnke, R.,** Suppression of endogenous avidin-binding activity in tissues and its relevance to biotin-avidin detection systems, *J. Histochem. Cytochem.*, 29, 1196, 1981.
45. **Green, N. M.,** Avidin, *Adv. Protein Chem.*, 29, 85, 1975.
46. **Moss, J. and Lane, M. D.,** The biotin-dependent enzymes, *Adv. Enzymol.*, 35, 321, 1971.
47. **Le Guellec, D., Trembleau, A., Péchoux, C., Gossard, F., and Morel, G.,** Ultrastructure non-radioactive *in situ* hybridization of GH mRNA in rat pituitary gland: pre-embedding vs ultra-thin frozen sections vs post-embedding, *J. Histochem. Cytochem.*, 40, 979, 1992.
48. **Guitteny, A. F., Böhlen, P., and Bloch, B.,** Analysis of vasopressin gene expression by *in situ* hybridization and immunohistochemistry in semi-thin sections, *J. Histochem. Cytochem.*, 36, 1373, 1988.
49. **Normand, E. and Bloch, B.,** Simultaneous detection of two messenger RNAs in the central nervous system: a simple two-step *in situ* hybridization procedure using a combination of radioactive and non-radioactive probes, *J. Histochem. Cytochem.*, 39, 1575, 1991.
50. **Wenderroth, M. A. and Russel Eisenberg, B.,** Ultrastructural distribution of myosin heavy chain mRNA in cardiac tissue: a comparison of frozen and LR white embeddment, *J. Histochem. Cytochem.*, 39, 1025, 1991.

Chapter 5

PRE-EMBEDDING *IN SITU* HYBRIDIZATION

Alain Trembleau

TABLE OF CONTENTS

0-8493-4414-X/93/$0.00 + $.50

I. INTRODUCTION

In situ hybridization histochemistry using recombinant or synthetic probes became a powerful tool during the last decade for studying gene expression and its regulation within various histological preparations.[1] Some recent improvements of both the sensitivity and resolution of such techniques drawing from various approaches which had been previously developed for immunocytochemistry, such as the use of cryosections,[2,3] pre-embedding,[4-12] and post-embedding,[5,13-18] now permit the analysis of gene expression at the ultrastructural level.

During the last 4 years, we developed pre-embedding techniques using either tritiated[6,11] or biotinylated[19,21] oligonucleotide probes, allowing the detection of various mRNA coding informative molecules expressed in nervous or endocrine tissues.

These protocols were developed using the following strategies.

A. *IN SITU* HYBRIDIZATION AT THE ELECTRON MICROSCOPE (EM) LEVEL USING RADIOACTIVE PROBES

1. We first developed the technique at the light microscope level (on cryostat sections) using ^{35}S-labeled probes, which yield rapid results and facilitate the analysis of labeling specificity.

2. For a given probe, the best labeling, hybridization, and washing conditions were determined, and then the same protocol was applied to vibratome free-floating sections, except for tritium which was used as radioactive marker, yielding high resolutions of autoradiographic labeling. At this point, it was still possible to check the specificity of the technique at the light microscope level on semithin sections and to compare different fixation procedures.
3. After making sure that both the specificity and the sensitivity of the technique were not modified on vibratome and semithin sections, we performed the autoradiographic detection of the tritiated probe on ultrathin sections.

B. *IN SITU* HYBRIDIZATION AT THE EM LEVEL USING NONRADIOACTIVE PROBES

1. We checked first the specificity and sensitivity of the probe using ^{35}S as a marker on cryostat sections. The best hybridization conditions were also determined.
2. We then developed a biotinylation protocol for labeling the probe, and the hybridization was performed on cryostat sections followed by an enzymatic revelation procedure compatible with electron microscopy (peroxidase-DAB), which gave intense labeling and low background.
3. This protocol was then applied to vibratome sections. At this step, we could compare at the light level the results obtained using several permeabilization procedures and various fixatives.
4. When good results were obtained on vibratome sections, the sections were postfixed with osmium tetroxide and embedded in epoxy resin.

The aim of this chapter is to describe in detail these approaches and to discuss their advantages and disadvantages relative to other possible approaches, such as post-embedding techniques.

II. MATERIALS AND STANDARD SOLUTIONS

A. MATERIALS

Tissue fixation and sectioning — Disodium hydrogenophosphate (Prolabo); chrom alum (Merk) cryostat (Reichert-Jung); gelatin (Merk); glutaraldehyde (Fluka); paraformaldehyde (Merk); slides; sodium chloride (Sigma); sodium dihydrogenophosphate (Prolabo); sucrose (Merk); vibratome (Lancer)

Probe labeling — Acetic acid (Merk); biotin-21-dUTP (Clontech); centrifuge; ^{3}H-dCTP (69 Ci/mmol) (Amersham); ^{35}S-dATP ($>$ 1000 Ci/mmol) (Amersham); scintillation liquid (Beckman); scintillation counter (Beckman); sodium acetate (Prolabo); speed vac; sterile water (autoclaved ultrapure water, e.g., millipore); tailing buffer and $CoCl_2$ solution (provided with the terminal transferase) (Boehringer); terminal transferase (Boehringer); water bath (37°C)

Hybridization — Amberlite Biorad AG501-X 20-50 mesh; bovine serum albumin (BSA) (Sigma); coverslips; culture dishes (sterile) (12 and 24 wells); dithiothreitol (DTT) (Sigma); DNA (salmon sperm) (Calbiochem); EDTA (Sigma); ethylic alcohol (Prolabo); Ficoll (PM 400,000) (Sigma); formamide (Sigma); Owen (37°C); polyvinylpyrrolidone (PM 360,000) (Sigma); rubber cement; trisodium citrate (Prolabo); tRNA (Yeast, Sigma)

Posthybridization steps — Autoradiography boxes (Clay-Adams); avidin-biotin-peroxidase complex (ABC) (Vector); biotinylated anti-avidin D antibody (Vector); 3, 3′-diamino-benzidine (DAB) (Sigma); developer D19 (Kodak); developer Microdol X (Kodak); electron microscope (Philips and Geol); grids 200 mesh (Polyscience); hydrogen peroxide (Sigma); lead citrate (Prolabo); light microscope (Zeiss); nuclear emulsions K5 (light microscopy) and L4 (electron microscopy) (Ilford); osmium tetroxide (Pelanne Instruments); Sigmacote (Sigma); silicagel; sodium cacodylate (Prolabo); sodium thiosulfate (Prolabo); Triton X-100 (Sigma); trizma base (Sigma); trizma hydrochloride (Sigma); ultramicrotome (Reichert); uranyl acetate (Polyscience)

B. STANDARD SOLUTIONS

Deionized formamide — Add 5 g of amberlite Biorad to 50 ml formamide. Vortex 30 min and filter with Whatman paper (number 1). Aliquot and store at −20°C.

100× Denhardt — For 100-ml final volume water, add Ficoll, polyvinylpyrrolidone, and bovine serum albumin, 2 g each. Aliquot and store at −20°C.

Hybridization buffer — For a 10-ml final volume, add in a sterile tube, 5-ml deionized formamide, 2 ml 20× SSC, 0.1 ml 100× Denhardt solution, 0.125 ml tRNA solution (see below), 0.2 ml salmon sperm DNA solution (see below), and 2.575 ml sterile water. When using ^{35}S-labeled probes, add 10 m*M* dithiothreitol, in order to block the nonspecific binding of ^{35}S in the tissue.

Lead citrate — Add 1.33 g lead nitrate and 2.14 ml trisodium citrate. Mix 30 min and then add 8 ml NaOH (1 *N*). Complete with water to a final volume of 50 ml.

PBS — Prepare one 0.1 *M* Na_2HPO_4 solution and one 0.1 *M* NaH_2PO_4 solution. Adjust the pH of the first solution to 7.4 adding the second one. Then add 9 g/l NaCl. Autoclave. Store at room temperature.

PBS-Triton — Add 0.1% (vol/vol) Triton X-100 to PBS.

Prehybridization buffer — For a final volume of 10 ml, mix 2 ml 20× SSC, 100 μl 100× Denhardt, 200 μl salmon sperm DNA, and 7.7 ml sterile water.

Salmon sperm DNA — 5 mg/ml in sterile water. Sonicate 10 min, aliquot, and store at −20°C.

Sodium acetate 3*M* (pH = 4.8) — Add 40.8 g sodium acetate in 80 ml sterile water. Then adjust the pH to 4.8 with glacial acetic acid. Add water q.s.p. 100 ml. Store at room temperature.

Sodium thiosulfate solution — Add 30 g to 100 ml distilled water. Prepare just before use.

20 × SSC (NaCl 3 *M*, trisodium citrate 0.3 *M*) — For 1 l final volume, add 175.3 g NaCl and 88.2 g trisodium citrate in sterile water. Adjust the pH to 7.4. Autoclave and store at room temperature.

tRNA — 10 mg/ml in sterile water. Store at −20°C.

Tris buffer — 0.1 *M*, pH = 7.6. Store at room temperature.

Uranyl acetate — Prepare the day before using a 5% solution of uranyle acetate in distilled water. Store in the dark. Before use, filter and dilute (1/1 vol/vol) with 95% ethanol.

III. METHODS

A. OLIGONUCLEOTIDE PROBES: DESIGN, SYNTHESIS, AND LABELING

The probes used in this work were exclusively synthetic oligonucleotides, which were designed using the sequences of oxytocin, vasopressin, and growth hormones genes, respectively: a 25-mer OT probe complementary to the mRNA sequence coding the amino acids 97 to 105 of oxytocin precursor,[22] a 41-mer VP probe complementary to the mRNA sequence coding the amino acids 115 to 128 of vasopressin precursor,[22] and a 30-mer GH probe complementary to the mRNA sequence coding the amino acids 45 to 54 of growth hormone precursor.[23] These sequences were chosen in order to have specific probes which would not cross-react with other known mRNAs. For example, the VP probe was chosen in the sequence coding the C terminal glycopeptide, which does not exist for oxytocin precursor.[22] These oligonucleotides were synthesized by solid phase phosphoradimite chemistry using an automatic DNA synthesizer.

The oligonucleotide probes were labeled by tailing the 3′ end extremity with either ^{35}S-dATP, ^{3}H-dCTP, or biotin 21-dUTP. The radioactive labelings were performed using Protocol 1 and the biotinylation using Protocol 2. Following any labeling reaction, the probe was purified by ethanol precipitation (see Protocol 3).

PROTOCOL 1 — OLIGONUCLEOTIDE TAILING USING RADIOACTIVE NUCLEOTIDES

Radioactive nucleotides can be easily linked to the three prime ends of oligonucleotides using terminal transferase and alpha-labeled deoxynucleotides. Terminal transferase can incorporate several (the number depends on the tailing reaction conditions) to the 3′ extremity. The following protocol was used for labeling either with (α)^{35}S-dATP (>1000 Ci/mmol, Amersham) or (α)^{3}H-dCTP (69 Ci/mmol, Amersham).

1. Add successively in a sterile Eppendorf tube

   - 2 μl 5 × tailing buffer (Boehringer, provided with the terminal transferase)
   - 1 μl 10 × $CoCl_2$ solution (Boehringer, provided with the terminal transferase)
   - 2 picomoles of oligonucleotide
   - 15 to 20 picomoles of radioactive deoxynucleotide triphosphate*
   - 25 U of terminal transferase (Boehringer)
   - sterile water (to 10 μl)

2. Vortex and centrifuge briefly
3. Incubate 1 hour at 37°C in a water bath
4. Stop the reaction in ice by adding 100 μl 10 m*M* EDTA

PROTOCOL 2 — OLIGONUCLEOTIDE TAILING USING A BIOTINYLATED NUCLEOTIDE

Minor modifications of Protocol 1 allow for tailing oligonucleotides with biotinylated deoxynucleotides. A higher concentration of the labeled deoxynucleotide in the tailing mixture is necessary to obtain a good labeling. In addition, since the biotinylated probe is used at a higher concentration for hybridization, we always labeled 10 picomoles of the oligonucleotide (instead of 2 picomoles for radioactive labeling, see Protocol 1) in a final volume of 40 μl.

1. Add successively in a sterile Eppendorf tube

   - 8 μl 5 × tailing buffer (Boehringer, provided with the terminal transferase)
   - 4 μl 10 × $CoCl_2$ solution (Boehringer, provided with the terminal transferase)
   - 10 picomoles of oligonucleotide
   - 500 picomoles of biotin 21-dUTP (Clontech)
   - 25 U of terminal transferase (Boehringer)
   - sterile water (to 10 μl)

2. Vortex and centrifuge briefly
3. Incubate 2 h at 37°C in a water bath
4. Stop the reaction in ice by adding 100 μl 10 m*M* EDTA

* When using tritiated nucleotides provided in ethanol (e.g., ^{3}H-dCTP, Amersham), dry the right volume of the nucleotide using a speed vac first, in order to remove the ethanol. Then add in the same tube the other components of the tailing reaction.

PROTOCOL 3 — ETHANOL PRECIPITATION*

1. Add to the Eppendorf tube containing the tailing mixture (volume = V)

   - 1/10 V sodium acetate (3 *M*, pH = 4.8)
   - 3 V cold (−20°C) ethanol.

2. Vortex
3. Incubate 30 min at −80°C
4. Centrifuge (10,000 rpm, 20 min)
5. Remove the supernatant carefully
6. Dry the pellet using a speed vac (approximately 15 min)
7. Resuspend the pellet in hybridization buffer**

The specific activity of a radioactive probe was estimated by counting 5 μl of the probe in 5 ml of scintillation liquid during 1 min in a scintillation counter. The mean specific activity obtained with our tailing protocol was about 400 Ci/mmol for tritiated probes and about 5500 Ci/mmol for ^{35}S-labeled probes.

B. TISSUE PREPARATION

This study was performed on male Sprague-Dawley rats, weighing 200 to 300 g.

- When using pituitary tissue, this organ was quickly removed after sacrificing the animal and fixed by immersion in a solution of 4% paraformaldehyde (90 min).
- Brain tissue was fixed by intracardiac perfusion of fixative solution, either 4% paraformaldehyde in PBS or 4% paraformaldehyde plus 0.1% glutaraldehyde diluted in 0.1 *M* phosphate buffer (pH = 7.4) (see Protocol 4).

PROTOCOL 4 — TISSUE FIXATION BY PERFUSION

Fixation by intracardiac perfusion of an animal allows for a rapid and efficient fixation of the entire body, in particular some large organs (for example the brain) for which fixation by immersion is not satisfactory. Perfusions were done using a pump.

* Ethanol precipitation does not allow for rigorous separation of nonincorporated deoxynucleotides from labeled oligonucleotide. However, we noticed that *in situ* hybridization does not require highly purified probes: it seems that nonincorporated deoxynucleotides do not lead to a higher background. Thus we performed this easy and fast method when preparing oligonucleotides for *in situ* hybridization.

** The pellet can be resuspended in water or Tris-EDTA buffer and then the probe can be stored at −20°C. For ^{35}S-labeled probe, the addition of 10 m*M* DTT allows for longer storage (about one month).

1. Anesthetize the rat using an intraperitoneal injection of pentobarbital (50 mg/kg b.w.)
2. Open the thorax
3. Incise the left ventricle and the right auricle
4. Perfuse first 20 to 40 ml saline solution (heated to 37°C) through the left ventricle, in order to remove the blood from the circulatory system
5. Perfuse in the same way 300 to 400 ml of the ice-cooled fixative solution (75 ml/min)
6. Remove the brain, and then cut a transversal thick slice containing the hypothalamus
7. Immerse the tissue in postfixative solution (4% paraformaldehyde) for 1 to 2 h at 4°C

When using cryostat sections, the tissue was rinsed in PBS and immersed overnight in 15% sucrose diluted in PBS after the post fixation. Then the tissue was frozen at −60°C in isopentane cooled in liquid nitrogen. Twelve to twenty μm coronal hypothalamic sections were prepared and mounted on gelatine coated slides (prepared by immersion in a 1% gelatin and 0.1% chrom alum and dried at 60°C) then air dried. These sections could be processed directly or stored at −70°C.

When using vibratome slices, the tissue was rinsed in PBS and the hypothalamus was coronally cut into 50-μm slices using a vibratome. The vibratome slices were immersed in PBS in sterile culture dishes.

C. PRETREATMENTS

The sensitivity of the techniques described here was sufficient to detect vasopressin mRNA, oxytocin mRNA, and growth hormone mRNA without using any pretreatment, such as proteinase K digestion of the sections. Thus no pretreatment was used before hybridization, which helped to preserve morphology.

D. *IN SITU* HYBRIDIZATION AND WASHING

1. *In Situ* Hybridization on Slides

Slides were first immersed for 2 h at 37°C in prehybridization buffer (4 × SSC, 1 × Denhardt, 100 μg/ml salmon sperm DNA), then quickly dehydrated in absolute alcohol (twice 5 min each), and air dried.

The application of the *in situ* hybridization solution containing labeled probe onto slides was carried out according to Protocol 5.

PROTOCOL 5 — *IN SITU* HYBRIDIZATION ON SLIDES*

When using radioactive probes, the following steps should be performed with caution.

* Sometimes cryostat sections were hybridized with radioactive or nonradioactive probes as free-floating sections, in particular when *in situ* hybridization was combined with immunohistochemistry.[24] We did not notice any significant loss of hybridization signal this way and this approach yielded better immunohistochemical results.

1. Drop the hybridization solution onto sections (15 to 50 μl, depending on the number of sections per slide and/or the surface of the coverslips)
2. Cover with a coverslip making sure no air bubbles are present over the sections
3. Seal the coverslip using rubber cement around the edges of the coverslip
4. Incubate the slides in a humid chamber (overnight at 37°C)
5. Remove the coverslips by dipping the slides in 2× SSC and using forceps
6. Wash the slides in several baths of SSC buffer:

   - 2× SSC: 2 × 1 h
   - 1× SSC: 2 × 1 h
   - 0.5× SSC: 2 × 1 h

7. Dry the sections by immersing the slides in absolute ethanol (2 × 5 min)

2. *In Situ* Hybridization on Free-Floating Sections

All the following steps were performed in small volumes (0.5 to 1 ml) in sterile culture dishes with gentle shaking. The sections were transfered from one bath to the next using clean forceps. The free-floating vibratome slices were first immersed into the prehybridization buffer for 2 h at 37°C. Then they were hybridized according to Protocol 6.

PROTOCOL 6 — *IN SITU* HYBRIDIZATION ON FREE-FLOATING SECTIONS

1. Immerse the sections into the hybridization solution containing either 0.5 to 2 n*M* of the tritiated probe or 7 to 10 n*M* of the biotinylated probe.
2. Incubate overnight at 37°C*
3. Wash the slices by immersing them in several baths of SSC buffer**

   - 2× SSC: 5 × 20 min
   - 1× SSC; 2 × 1 h
   - 0.5× SSC: 2 × 1 h

4. Transfer the slices into PBS

When using radioactive probes, the slices were postfixed with osmium tetroxide (1% in 0.1 *M* sodium cacodylate, pH = 7.4) and embedded in Epon (see Protocol 7) between two siliconized slides prepared using Protocol 8.

* The slices are hybridized in sterile culture dishes (24 wells). The covers of the dishes are sealed in order to avoid any evaporation of the hybridization buffer.

** During these washes, the dishes are shaken vigorously.

PROTOCOL 7 — EMBEDDING OF VIBRATOME SLICES IN EPON

1. Dehydration: immerse the slices in several baths of alcohol as follows:

    - 70°C: 10 min
    - 80°C: 10 min
    - 90°C: 10 min
    - 95°C: 10 min twice
    - 100°C: 10 min twice

2. Immerse the slices in two baths propylene oxide (15 min each)
3. Transfer the slices into the following mixtures of propylene oxide/Epon

    - Propylene oxide/Epon 2/1 (v/v): 45 min
    - Propylene oxide/Epon 1/2 (v/v): 45 min

4. Incubate the slices in Epon (2 h to overnight at 4°C)
5. Embed the slices between two siliconized slides:

    - Drop some Epon (about 1 ml) on a siliconized slide
    - Put the flat vibratome slice in this drop of Epon
    - Cover with another siliconized slide

6. Polymerize the resin overnight at 55°C
7. Remove the siliconized slides using a razor blade

The embedded slice can be observed with light microscopy and the labeled area can be selected for preparation of semithin and ultrathin sections.

PROTOCOL 8 — PREPARATION OF SILICONIZED SLIDES

Sigmacote from Sigma was used for preparing slides. The following steps should be performed under a hood because of the high volatility and toxicity of silicone.

1. Pour the silicone into a beaker
2. Dip the slides* one by one into the silicone using forceps
3. Let the slides dry on aluminium foil in the hood (about 2 h)
4. Store the siliconized slides in aluminium foil at room temperature

* Precleaned slides should be used.

E. PROBE DETECTION

1. Radioactive Probes

a. Light Microscope Level

Two kinds of histological materials were processed for light microscopic autoradiography: cryostat sections hybridized with ^{35}S-labeled probe, and plastic semithin sections, obtained by sectioning the vibratome slices previously hybridized with tritiated probe.

Dipping — The slides were dipped into nuclear emulsion (Ilford K5) (see Protocol 9)

PROTOCOL 9 — AUTORADIOGRAPHY: DIPPING FOR CRYOSTAT OR SEMITHIN SECTIONS

All these steps should be done in a dark room under red light. This protocol using Ilford K5 emulsion was applied for autoradiography of both cryostat and semithin sections.

1. Dilute the K5 emulsion* in water (1/1, v/v) in a beaker 1 h before use
2. Melt the emulsion incubating the beaker in a water bath (40°C). During all the following steps, the beaker containing the emulsion should stay in the water bath
3. Mix 3 times the melting emulsion (every 20 min) with a glass stick
4. Dip the slides with a slow down and up vertical movement
5. Dry the slides vertically in the dark
6. Store the slides at 4°C in adequate black boxes (e.g., Clay-Adams) containing silicagel

Development — After a few days dark exposure for cryostat sections hybridized with ^{35}S-labeled probes or a few months dark exposure for semithin hybridized with tritiated probes, the slides were developed (see Protocol 10).

PROTOCOL 10 — AUTORADIOGRAPHY: DEVELOPMENT

These steps of autoradiography should be performed in a dark room, under a red light. When using sections autoradiographed with Ilford K5 emulsion, the developer used is Kodak D19. For sections autoradiographed using Ilford L4 emulsion (cf. electron microscopy), the Kodak Microdol X developer is used.

1. Remove the slides from the Clay-Adams boxes
2. Immerse them in the developer (cooled to 17°C in ice) during 4 min
3. Rinse briefly in distilled water
4. Immerse the slides in the fixer (30% sodium thiosulfate) for 10 min
5. Rinse the slides in several baths of distilled water

* For electron microscopic autoradiography, Ilford L4 emulsion is used instead of K5. The protocol of dipping is then the same but L4 emulsion is more diluted with water (1/5, v/v).

Cryostat sections were colored with cresyl violet (2 to 5 min), rinsed in water, then in 70% alcohol, dehydrated, and mounted in Eukitt. Semithin sections were colored with 1% Toluidin Blue and mounted in Eukitt.

b. Electron Microscope Level

For electron microscopic autoradiography, we used a protocol described in detail elsewhere,[25] without any modification. The time exposures were about 3 to 5 months.

2. Biotinylated Probes

Vibratome slices hybridized with biotinylated probes were revealed using a protocol[20] modified from a technique published by Arai et al.[26] (see Protocol 11).

PROTOCOL 11 — BIOTIN DETECTION IN VIBRATOME SLICES

Following the washing steps, the slices are treated in tissue culture dishes as follows:

1. Incubate the avidin-biotin-peroxidase complex (ABC) diluted in PBS-Triton (1 h, room temperature)
2. Wash in three baths of PBS (10 min each)
3. Incubate the biotinylated anti-avidin diluted in PBS-Triton (1/200, overnight [4°C])
4. Rinse in three baths of PBS (10 min each)
5. Incubate the ABC diluted in PBS-Triton (1 h, room temperature)
6. Rinse in three baths of PBS (10 min each)
7. Reveal the peroxidase in a Tris buffer containing 0.025% 3,3′-diaminobenzidine and 0.006% hydrogen peroxide
9. Check the reaction with microscopy and stop the reaction by immersing the slices in PBS
10. Rinse the slices in PBS (twice 5 min)

The slices were then postfixed with osmium tetroxide (see above), dehydrated, and embedded in Epon (see Protocol 7). Semithin and ultrathin sections were obtained. Ultrathin sections were contrasted with uranyl acetate (1 min) and/or with lead citrate (2 min).

IV. RESULTS

A. *IN SITU* HYBRIDIZATION AT THE LIGHT MICROSCOPE LEVEL

1. Detection of Oxytocin mRNA Using a Radioactive Probe

On cryostat sections (Figure 1) oxytocin mRNA was detected with the ^{35}S-labeled 25-mer oligonucleotide in magnocellular neurons in the hypotha-

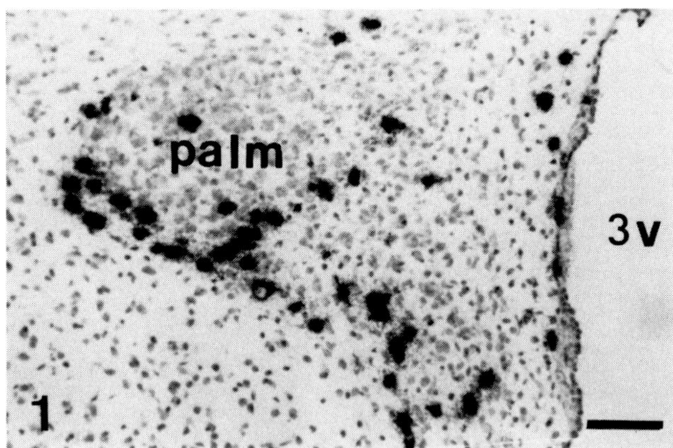

FIGURE 1. Histological detection of oxytocin mRNA using a 25-mer oligonucleotide probe labeled with ^{35}S within a 16-μm cryostat section through the hypothalamic paraventricular nucleus (PVN). The brain was fixed by perfusing 4% paraformaldehyde. A strong labeling is observed in some magnocellular neurons, most of them surrounding the lateral magnocellular nucleus of the PVN. Time exposure was about 1 week. 3V = third ventricle. Bar = 100 μm.

lamic supra-optic (SON) and paraventricular (PVN) nuclei and scattered throughout the hypothalamus. In the SON, these neurons were more abundant in the dorsal part. In the PVN, positive neurons were found mainly in the anterior commissural nucleus and in the median magnocellular nucleus. Figure 1 shows a section of the PVN where positive neurons labeled by the OT probe surround the lateral magnocellular part of the PVN. This particular localization at this level of the PVN demonstrates the specificity of the labeling and the absence of cross-hybridization with vasopressin mRNA, which is mainly localized in the lateral magnocellular part of the PVN[27,28] (see Figure 2).

OT mRNA detected on semithin sections after hybridization of the tritiated OT probe in vibratome sections was localized in the same areas of the hypothalamus. Figure 3 shows a high magnification of an autoradiographed semithin section through the SON. This section was cut transversally to the main plane of the vibratome slice. The presence of the positive neurons in the median part of the section (arrows) demonstrates that oligonucleotide probes can penetrate the entire depth of at least 50-μm vibratome sections, without the addition of any detergent. The resolution obtained with the tritiated probe was much better than with the ^{35}S-labeled probe and permitted us to notice the absence of labeling over the nucleus of positive cells.

2. Detection of Vasopressin mRNA Using a Biotinylated Probe

In both cryostat (not shown) and vibratome sections (Figure 2), vasopressin mRNA could be visualized in some magnocellular neurons of the hypothalamus using the biotinylated 41-mer oligonucleotide detected with peroxidase-DAB. These labeled neurons were localized in the SON and PVN and scattered throughout the hypothalamus. In the SON they occupied mainly

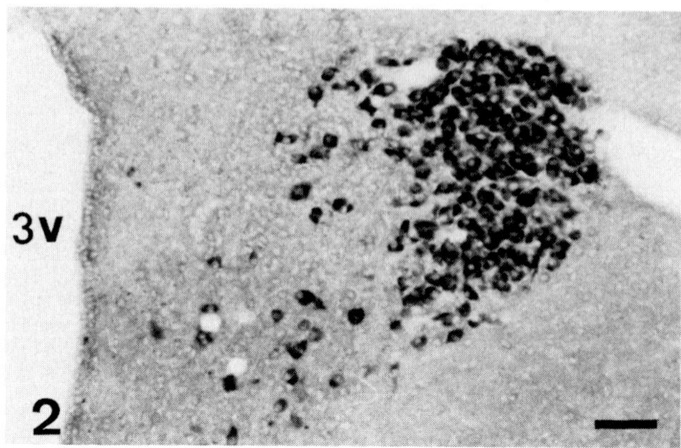

FIGURE 2. Histological localization of vasopressin mRNA using a biotinylated 41-mer oligonucleotide probe within a 50-μm vibratome section through the hypothalamic PVN. The brain was fixed by perfusion of 4% paraformaldehyde. A cellular labeling is found in many magnocellular neurons within the lateral magnocellular nucleus of the PVN. 3V = third ventricle. Bar = 100 μm.

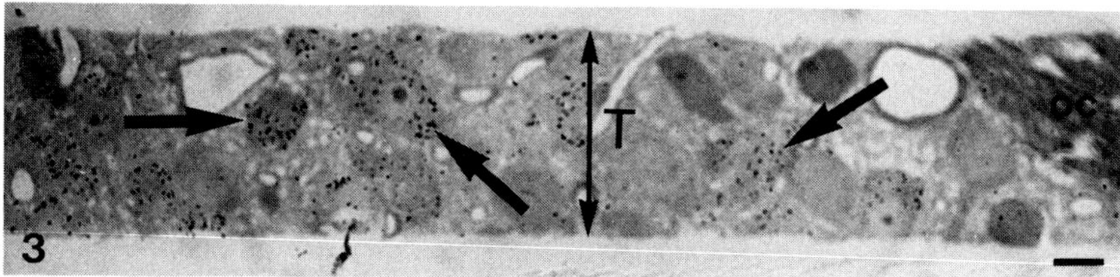

FIGURE 3. Autoradiographic detection of oxytocin mRNA using a tritiated 25-mer oligonucleotide probe in a 1-μm semithin section through the supraoptic nucleus. The probe was first hybridized to a 50-μm vibratome slice. Then the slice was embedded in Epon and transversal semithin sections were cut and autoradiographed. Some labeled magnocellular neurons are observed through the entire depth (T = 50 μm) of the vibratome slice. The labeling of some neurons localized in the median part of the vibratome slice (arrows) demonstrates that radioactive oligonucleotide probes can diffuse through the entire depth of the vibratome slice. oc = optic chiasma. Time exposure = 5 months. Bar = 10 μm.

the ventral area. In the PVN they were very abundant in the lateral magnocellular nucleus (Figure 2). Within the positive neurons, labeling was restricted to the cytoplasm and proximal dendrites.

B. *IN SITU* HYBRIDIZATION AT THE ELECTRON MICROSCOPE LEVEL

1. Detection of Oxytocin mRNA Using a Tritiated Probe

The morphology obtained using tissue fixed with both 4% paraformaldehyde and 0.1% glutaraldehyde was satisfactory despite the hybridization procedure steps and allowed the identification of the main subcellular struc-

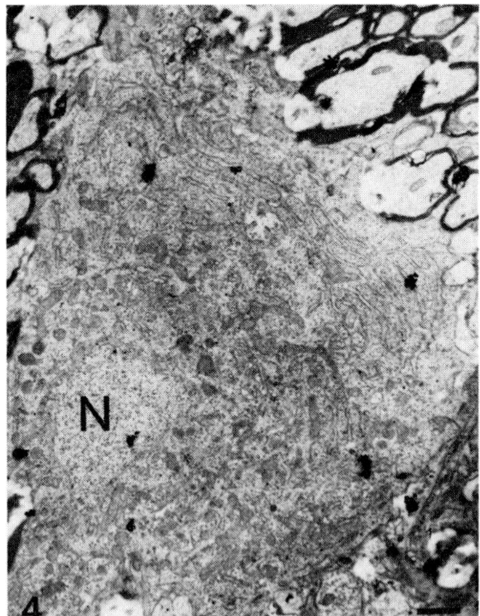

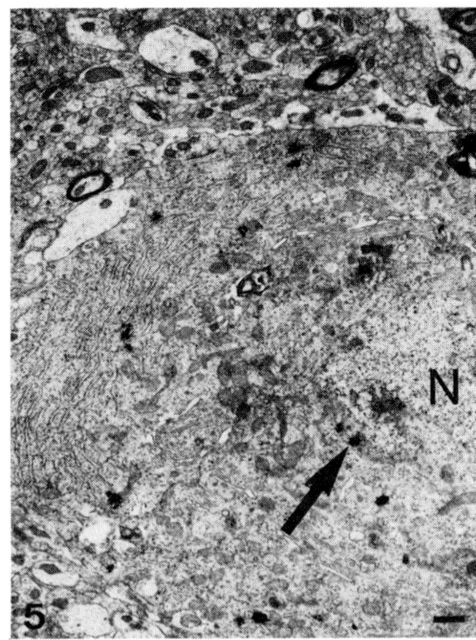

FIGURES 4 and 5. Ultrastructural localization in the supraoptic nucleus of oxytocin mRNA using the tritiated 25-mer oligonucleotide. The tissue was fixed by perfusion of 4% paraformaldehyde and 0.1% glutaraldehyde and hybridized prior to being embedded. The labeling (silver grains) is mainly localized in some areas of the cytoplasm containing the Nissl bodies. A weak but significant labeling is found at the level of the nuclear envelope (arrow) either on the cytoplasmic compartment on the nuclear matrix. N = nucleus. Time exposure = 3 months. Bar = 1 μm.

tures (Figures 4 and 5). The cellular labeling (silver grains) was quite weak, even after three months exposure, but it was easy to find because background was always extremely low (Figures 4 and 5). Within the positive cell bodies, silver grains were mainly found in areas containing the rough endoplasmic reticulum (RER) of the Nissl bodies. In addition, significant labeling was observed close to the nuclear envelope, sometimes on the cytoplasmic compartment, sometimes over the nuclear matrix (Figure 5). Very few silver grains were observed in areas of the cytoplasm containing either a lot of mitochondria or the Golgi apparatus.

2. Detection of mRNAs Using Biotinylated Probes

a. Vasopressin mRNA

At the ultrastructural level, although the morphology was not very well preserved, the high signal-to-noise ratio allowed the visualization of positive neurons (Figure 6). The labeling was very high in neurons close to the surface of the vibratome slice and lower in its median part (Figure 6). Within the positive neurons the labeling was very strong in some peripheral areas of the cytoplasm containing Nissl bodies and very weak in some large perinuclear areas containing a lot of mitochondria and the Golgi apparatus but no RER (Figure 6). At the level of the Nissl bodies, the DAB precipitate was associated

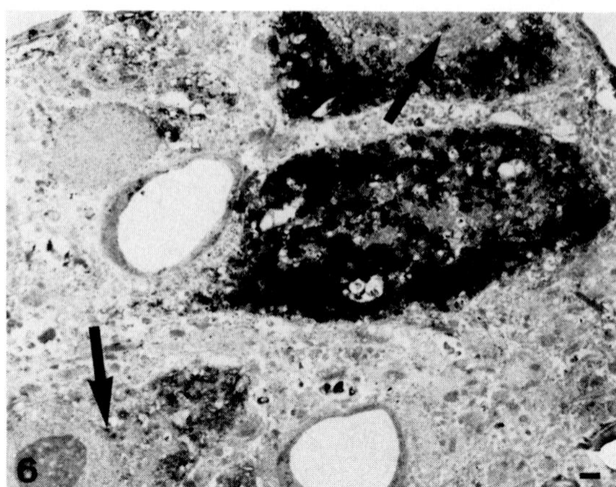

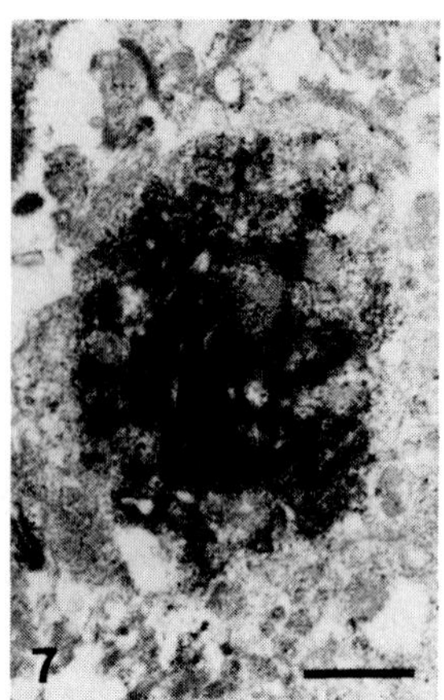

FIGURES 6 and 7. Ultrastructural localization of vasopressin mRNA in the supraoptic nucleus using the biotinylated 41-mer oligonucleotide probe. The tissue was fixed by perfusion with 4% paraformaldehyde and permeabilized with triton when revealing the biotin using an avidin-biotin peroxidase complex and a biotinylated anti-avidin antibody. Within the positive cell bodies, the peroxidase-DAB labeling is mainly associated with the endoplasmic reticulum of the Nissl bodies (Figures 6 and 7), but not in the lumen. A significant labeling is also observed in the vicinity of the nuclear envelope, in the cytoplasmic compartment (Figure 6, arrows). The labeling is strong close to the vibratome slice surface (Figure 6, top) and weaker in the middle part of the vibratome slice (Figure 6, bottom). Nuclei, mitochondria, as well as Golgi apparatus, are devoid of labeling. Bar = 1 μm.

with the RER membrane, but it was not present inside the lumen (Figure 7). This labeling was often more intense in discrete areas along the RER. A discrete labeling was also observed in many cases close to the nuclear envelope, always in the cytoplasmic compartment (Figure 6, arrows). The nucleus, mitochondria (Figure 6), and the Golgi apparatus were never labeled.

b. Growth Hormone mRNA

On vibratome sections, we observed that a subpopulation of anterior pituitary cells only were labeled (not shown). At the ultrastructural level, these positive cells could be identified as somatotrophs, using morphological criteria such as the lamellar endoplasmic reticulum and the type of secretory granules[29] (Figure 8). The morphology was satisfactory (Figures 8 and 9). Within positive cells, the labeling was mainly localized in the cytoplasmic areas containing the lamellar RER (Figure 8). At this level, discrete areas between the cisternae were strongly labeled, but no DAB precipitate was found inside the reticular lumen (Figure 9). Additional weak and discrete labeling was sometimes found in others areas of the cytoplasm. Nuclei, Golgi, mitochondria, and secretory granules were never labeled.

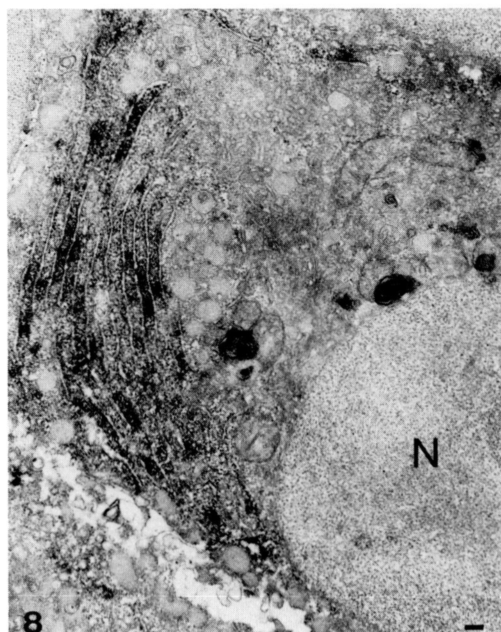

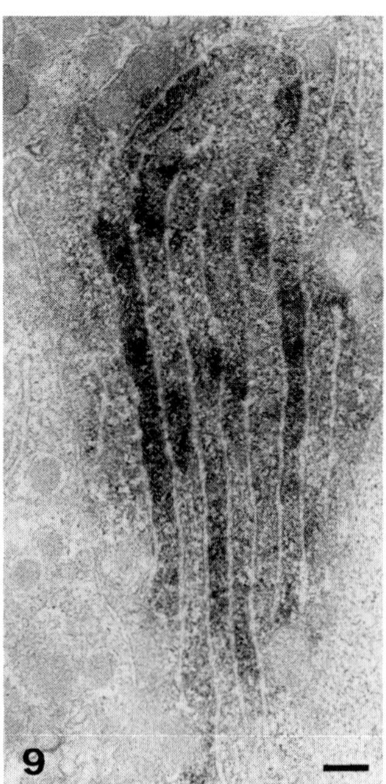

FIGURES 8 and 9. Ultrastructural detection of growth hormone mRNA in the anterior lobe of the pituitary using a 30-mer oligonucleotide probe. The pituitary was fixed by immersion into 4% paraformaldehyde and permeabilized with triton. The labeling is found in somatotrophs only and it is restricted to the lamellar endoplasmic reticulum, but not in its lumen (Figure 8). Nuclei, mitochondria, Golgi apparatus as well as secretory granules, are devoid of labeling (Figure 9). Within the reticulum, the labeling is intense in some discrete areas only, suggesting a possible compartmentalization of growth hormone mRNA at this level (Figures 8 and 9). N = nucleus. Bar = 0.1 μm.

V. COMMENTARIES

A. FLEXIBILITY, mRNA PRESERVATION, SPECIFICITY, SENSITIVITY, TISSUE PRESERVATION, AND RESOLUTION OF OUR TWO PRE-EMBEDDING TECHNIQUES

Since the hybridization step precedes the embedding of the vibratome slices, the mRNA preservation and accessibility are not impaired in such approaches. In fact, with pre-embedding techniques the tissue does not suffer any strong modification of classical protocols used for light microscopic studies and can be hybridized in almost the same conditions. Thus, when a protocol allowing the detection of a given mRNA on cryostat or paraffin sections has been developed, the same protocol can be used with minor modifications for pre-embedding electron microscopic *in situ* hybridization. However, it should be mentioned that pretreatments such as HCl treatment or proteinase K digestion, which are very often used on cryostat or paraffin sections, should be limited for vibratome sections because of the destruction of cellular morphology.

Regardless of the type of marker used for *in situ* hybridization, pre-embedding techniques permit inspection of experimental results at the light microscope level, without any requirement for electron microscopy. This can be done using film autoradiography when a ^{35}S-labeled probe has been used, or by dipping in nuclear emulsion either the vibratome section[11] or some semithin sections[30] (Figure 3) when tritiated or ^{35}S-labeled probes have been used. When using biotinylated probes, the sections can be mounted on glass slides and inspected at the light microscope. Such possibilities avoid loss of time at the electron microscope looking for labeling when one does not know whether the experiment has worked.

Another advantage of pre-embedding techniques is that the specificity of the labeling can be worked out at the light microscope before moving to the electron microscope. For example, when studying localization of neuropeptide mRNA in the brain, one can check on vibratome slices or on semithin sections that labeling is localized in the nuclei where the transcripts are known to be expressed. For example, in the present study we checked that the lateral magnocellular nucleus of the paraventricular nucleus contained mainly vasopressin mRNA expressing cells (Figure 2) but only a few oxytocin mRNA expressing neurons (Figure 1). In the same way, the specificity of the hybridization can be checked at the light level using all the classical controls including RNAse treatment, competition tests,[11] heterologous probes[11] as well as combination with immunohistochemistry.[30]

We demonstrated using transversal semithin sections and tritiated oligonucleotides that these probes can penetrate the entire depth of at least 50-μm vibratome slices without addition of any detergent to the hybridization buffer. This property of oligonucleotides makes them ideal for pre-embedding techniques.

The sensitivity of protocols using either radioactive or nonradioactive oligonucleotides appeared to be similar, although very long time exposures were required to visualize hybridized tritiated probes. In addition it was necessary to enhance the avidin-peroxidase detection of biotin using a biotinylated anti-avidin antibody as previously described[26] to obtain strong labeling of vasopressin mRNA: when we revealed DAB after the first incubation of the ABC, only a very weak labeling was obtained. The addition of glutaraldehyde to the fixative seemed to decrease labeling intensity, as previously reported,[31] but still allowed the visualization of neuropeptide mRNAs with tritiated probes at the ultrastructural level. When hybridizing with biotinylated probes, we chose to fix the tissue with paraformaldehyde only to avoid decreasing labeling. In our experiments pre-embedding techniques either using radioactive or biotinylated oligonucleotides were sufficiently sensitive to detect vasopressin and oxytocin mRNAs in the hypothalamus and growth hormone encoding mRNA in the pituitary at the electron microscope level. Many other mRNAs were also detected at the electron microscope level using similar techniques in other groups.[4-9,11,12] Although these mRNAs are known to be expressed at moderate or high abundance with corresponding cells, we recently detected at the ultrastructural level less abundant expressed transcripts (for example vasopressin mRNA in parvocellular neurons of the hypothalamus and tyrosine hydroxylase mRNA in the brainstem) using digoxigenin-labeled probes and pre-embedding *in situ* hybridization on vibratome slices fixed with both paraformaldehyde and glutaraldehyde (unpublished data). Thus the sensitivity of such approaches can be quite high.

Tissue preservation was excellent when hybridizing a tritiated oligonucleotide on vibratome slices fixed with 4% paraformaldehyde and 0.1% glutaraldehyde. In this case, it was not necessary to use some detergent because of the good diffusibility of such a probe. In contrast, the morphology obtained with tissues processed with biotinylated probes was not as good, because they were fixed with paraformaldehyde only, and also because Triton X-100 was used during the immunoenzymatic steps of biotin detection. Despite these treatments, the main cellular ultrastructures including nucleoli, mitochondria, Golgi apparatus, endoplasmic reticulum, and secretory granules were, however, sufficiently preserved to allow their identification in both nervous and endocrine tissues.

The resolution of tritium is low because of uncertainty concerning the precise localization of the isotope and because of the large size of silver grains. When silver grains are found over a membrane, such as a membrane belonging to the RER, it is not possible to determine whether the corresponding mRNA is localized inside or outside the cisternae. In the same way, it is impossible to determine whether the significant autoradiographic labeling observed close to the nuclear envelop with the tritiated oxytocin probe corresponds to some intra- or extranuclear mRNA.

However, although this isotope does not allow for precise localization of mRNAs, it can indicate cellular areas or large structures, such as Nissl bodies, where mRNAs are particularly abundant. Although the peroxidase labeling is sometimes diffuse, its resolution appears to be better, since it can be determined with this technique that there is no labeling inside the lumen of the endoplasmic reticulum and that the nuclear envelop associated labeling is localized in the cytoplasmic compartment only. Furthermore, we found, using peroxidase labeling, that the localization of growth hormone mRNA seems to be discrete along the endoplasmic reticulum. However, enzymatic labeling cannot be used for very fine localization of cellular molecules, and it will be hard in the future to improve the resolution of pre-embedding techniques because markers of choice such as colloidal gold cannot be used easily without very strongly permeabilizing the tissue (and thus significantly disrupting the morphology). Hence, we consider the poor resolution of labelings to be the main drawback of pre-embedding techniques.

B. ADVANTAGES AND DISADVANTAGES OF PRE-EMBEDDING VS. POST-EMBEDDING APPROACHES

1. Advantages

The pre-embedding approach presents the distinct advantage that the same protocols designed for light microscopy are likely to work for electron microscopy with only minor modifications. This is also true for hybridization on ultrathin frozen sections but not for post-embedding. Indeed, this latter approach necessitates hybridization of the probe on tissue embedded in resin which is very different from nonembedded tissue in terms of mRNA preservation and accessibility.

Another appreciable advantage of pre-embedding is that the result of such experiments can be checked directly on vibratome sections, before any embedding and electron microscopic observation. Thus, when experiments do not work, a lot of time can be saved by stopping tissue processing. In addition, as noted above, many control experiments can be done at the light microscope level when using pre-embedding techniques. Post-embedding techniques, however, do not share this advantage. Moreover, for nervous system studies, it is very often useful to spot some labeled cells, in particular when they are rare or scattered (e.g., somatostatin expressing cells in the cortex) before embedding, in order to locate them more easily at the electron microscope. For this purpose, we embedded flat vibratome slices between glass slides allowing, after the resin polymerization to deliniate and cut the right area spotted using light microscopy. Such light microscopic selection of labeled areas cannot be done when post-embedding techniques are used.

The morphology obtained using pre-embedding technique is quite good, mostly because the tissue can be postfixed with osmium tetroxide following hybridization, and much more well-preserved compared to frozen sections or post-embedding techniques. Indeed, post-embedding *in situ* hybridization has been applied in almost all cases to tissues embedded in hydrophilic resins such as lowicryl, which are not compatible with osmium tetroxide fixation. One work only used epoxy resin for post-embedding hybridization.[16] In this latter work, resin was removed before the hybridization step; thus, the morphology was very poor.

Finally, although hybridization on frozen sections appears most sensitive,[21] the sensitivity of pre-embedding techniques seems to be higher than that of post-embedding methods.[21] So far, post-embedding studies permit the ultrastructural localization of very abundant nucleic acids only, such as ribosomal DNA and RNA,[13,14] viral DNAs and RNAs,[15] and a few very abundant cellular mRNAs.[5,16-18] Our recent localization of vasopressin mRNA in some parvocellular neurons (Trembleau et al., unpublished data), as well as in the axonal compartment of magnocellular neurons[32] using pre-embedding *in situ* hybridization, demonstrates that the sensitivity of this approach can be extremely high.

2. Disadvantages

First of all, as it was already mentioned above, the main limit of our pre-embedding approaches is their poor resolution of either the radioactive or enzymatic markers.

Another disadvantage of pre-embedding techniques vs. ultrathin frozen sections and post-embedding is the necessity of processing the tissue (preparation of the sections, hybridization, embedding) immediately following fixation. This could be a major limitation when working on valuable specimens such as pathological tissue. With post-embedding and frozen sections, it is possible to store the tissue after embedding it in Lowicryl or after freezing, respectively. However, we recently noticed that adequately cryoprotected brain or pituitary vibratome slices can be stored at $-70°C$ without significant loss of either reactivity or morphology (unpublished data). Thus, this limitation can be alleviated at least for brain and pituitary tissues.

C. ULTRASTRUCTURAL LOCALIZATION OF HYPOTHALAMIC NEUROPEPTIDES AND PITUITARY HORMONE ENCODING mRNAs AS DETERMINED USING PRE-EMBEDDING *IN SITU* HYBRIDIZATION

Our results showing that mRNA encoding growth hormone is localized in discrete areas along the endoplasmic reticulum suggest a possible compartmentalization of this mRNA within the endoplasmic reticulum. Such a localization with respect to the reticulum was also observed for vasopressin and oxytocin mRNAs in magnocellular neurons, in particular in some experiments in which the labeling was weaker than the one shown on the pictures in the present chapter.[33] This kind of compartmentalization of mRNAs within the endoplasmic reticulum fits well with an hypothesis recently proposed by van Leeuwen et al.[34] postulating a possible functional (translational) compartmentalization of the endoplasmic reticulum. According to this view, different mRNAs could be translated at different sites of the endoplasmic reticulum. Our observations could constitute morphological evidence supporting such an hypothesis.

We also identified vasopressin and oxytocin mRNAs at the level of the nuclear envelope, inside the cytoplasm. Although the functional signification of this localization remains unclear, it could represent a transitory localization on the pathway between the nucleus and the endoplasmic reticulum. The absence of any nuclear labeling regardless of the probe used suggests a rapid turnover of oxytocin, vasopressin, and growth hormone mRNAs within the nucleus and/or fast nucleo-cytoplasmic transfer.

VI. CONCLUSION

Pre-embedding techniques appear to be sensitive tools for studying the subcellular compartmentalization of mRNAs in various models. Further development of such approaches will most probably yield new information concerning the cellular and molecular biology of various nucleic acids including viral nucleic acids (DNAs and/or RNAs) and cellular mRNAs with respect to the subcellular structures. In the field of Neuroscience, recent studies documented a dendritic[35,36] but also, more surprisingly, an axonal transport of some mRNAs in the brain.[16,32,37,38] The axonal localization of mRNAs raises many fundamental questions including the mechanisms of such regulated transport and the functional significance of the presence of these molecules in the axonal compartment. Electron microscopic *in situ* hybridization and, more particularly, pre-embedding approaches (due to their high sensitivity) appear to be among the available tools to answer some of these questions.

ACKNOWLEDGMENTS

A.T. Is a Fellow of International Brain Research Organization and of the Ministère des Affaires Etrangères (French Government). Kathleen R. Melia is acknowledged for kindly reviewing the English style.

REFERENCES

1. **Bloch, B.,** L'hybridation *in situ* appliquee a la detection des ARN messagers. Situation actuelle en 1991 et perspectives, in *Hybridation In Situ: Methodes Pratiques,* Calas, A., Bloch, B., Fournier, J.-G., and Trembleau, A., Eds., SFME, Paris, 1991, 1.
2. **Morel, G., Dubois, P., and Gossard, F.,** Detection ultrastructurale des ARN messagers codant pour l'hormone de croissance dans l'antehypophyse du rat par hybridation *in situ, C. R. Acad. Sci., Paris,* 302(Série III), 479, 1986.
3. **Morel, G., Dihl, F., and Gossard, F.,** Ultrastructural distribution of GH mRNA and GH intron I sequences in rat pituitary gland: effect of GH releasing factor and somatostatin, *Mol. Cell Endocrinol.,* 65, 81, 1989.
4. **Harris, N. and Croy, R. R. D.,** Localization of mRNA for pea legumin: *in situ* hybridization using a biotinylated cDNA probe, *Protoplasma,* 130, 57, 1986.
5. **Webster, H. de F., Lamperth, L., Favilla, J. T., Lemke, G., Tesin, D., and Manuelidis, L.,** Use of a biotinylated probe and *in situ* hybridization for light and electron microscopic localization of Po mRNA in myelin-forming Schwann cells, *Histochemistry,* 86, 441, 1987.
6. **Trembleau, A., Fevre-Montange, M., and Calas, A.,** Localisation ultrastructurale de l'ARNm codant l'ocytocine par hybridation *in situ.* Etude par radioautographie à haute résolution à l'aide d'une sonde oligonucléotidique tritiée, *C. R. Acad. Sci., Paris,* 307(Série III), 869, 1988.
7. **Guitteny, A. F. and Bloch, B.,** Ultrastructural detection of vasopressin messenger RNA in the normal and Brattleboro rat, *Histochemistry,* 92, 227, 1989.
8. **Brangeon, J., Nato, A., and Fochioni, A.,** Ultrastructural detection of ribulose-1,5-biphosphate carboxylase protein and its subunit mRNAs in wild type and holoenzyme deficient Nicotiana using immunogold and *in situ* hybridization techniques, *Planta,* 177, 151, 1989.

9. **Tong, Y., Zhao, H. F., Simard, J., Labrie, F., and Pelletier, G.,** Electron microscopic autoradiographic localization of prolactin mRNA in the rat pituitary, *J. Histochem. Cytochem.*, 37, 567, 1989.
10. **Wolber, R. A., Beals, T. F., and Massaab, H. F.,** Ultrastructural localization of Herpes simplex virus RNA by *in situ* hybridization, *J. Histochem. Cytochem.*, 37, 97, 1989.
11. **Trembleau, A., Calas, A., and Fevre-Montange, M.,** Ultrastructural localization of oxytocin mRNA in the rat hypothalamus by *in situ* hybridization using a synthetic oligonucleotide, *Mol. Brain Res.*, 8, 37, 1990.
12. **Pomeroy, M. E., Lawrence, J. B., Singer, R. H., and Billings-Gagliardi, S.,** Distribution of myosin heavy chain mRNA in embryonic muscle tissue visualized by ultrastructural *in situ* hybridization, *Dev. Biol.*, 49, 99, 1991.
13. **Binder, M., Tourmente, S., Roth, J., Renaud, M., and Gerhing, W. J.,** In situ hybridization at the electron microscope level: localization of transcripts on ultrathin sections of Lowicryl K4M-embedded tissue using biotinylated probes and protein A-gold complexes, *J. Cell. Biol.*, 102, 1646, 1986.
14. **Thiry, M. and Thiry-Blaise, L.,** In situ hybridization at the electron microscope level: an improved method for precise localization of ribosomal DNA and RNA, *Eur. J. Cell Biol.*, 50, 235, 1989.
15. **Puvion-Dutilleul, F. and Puvion, E.,** Ultrastructural localization of viral DNA in thin sections of herpes simplex virus type 1 infects cells by *in situ* hybridization, *Eur. J. Cell Biol.*, 49, 99, 1989.
16. **Jirikowski, G. F., Sanna, P. P., and Bloom, F. E.,** mRNA coding oxytocin is present in axons of the hypothalamo-neurohypophyseal tract, *Proc. Natl. Acad. Sci. U.S.A.*, 87, 7400, 1990.
17. **Le Guellec, D., Frappart, L., and Wilhems, R.,** Ultrastructural localization of fibronectin mRNA in chick embryo by *in situ* hybridization using ^{35}S or biotin labeled cDNA probes, *Biol. Cell.*, 70, 159, 1990.
18. **Escaig-Haye, F., Grigoriev, V., Peranzi, G., Lestienne, P., and Fournier, J.-G.,** Analysis of human mitochondrial transcripts using electron microscopic *in situ* hybridization, *J. Cell Sci.*, 100, 851, 1991.
19. **Trembleau, A., Fevre-Montange, M., Landry, M., and Calas, A.,** Techniques d'hybridation *in situ* ultrastructurale avant inclusion, in *Techniques en Microscopie Electronique: Cryomethodes, Immunocytologie, Autoradiographie, Hybridation In Situ,* Morel, G., Ed., Editions INSERM, Paris, 1991, 503.
20. **Trembleau, A., Landry, M., Fevre-Montange, M., and Calas, A.,** Comparative ultrastructural localization of neuropeptide mRNAs using tritiated and biotinylated oligonucleotide probes, in *Neurocytochemical Methods,* NATO ASI Series, Vol. H58, Calas, A. and Eugene, D., Eds., Springer-Verlag, Paris, 1991, 313.
21. **Le Guellec, D., Trembleau, A., Pechoux, C., Gossard, F., and Morel, G.,** Ultrastructural non-radioactive *in situ* hybridization of GH mRNA in rat pituitary gland: pre-embedding vs ultra-thin frozen sections vs post-embedding, *J. Histochem. Cytochem.*, 40, 379, 1992.
22. **Ivell, R. and Richter, D.,** Structure and comparison of the oxytocin and vasopressin genes from rat, *Proc. Natl. Acad. Sci. U.S.A.*, 81, 2006, 1984.
23. **Cooke, N. E., Coit, D., Welner, R. I., Baxter, J. D., and Martial, J. A.,** Structure of cloned DNA complementary to rat prolactin messenger RNA, *J. Biol. Chem.*, 255, 6502, 1980.
24. **Trembleau, A. and Roche, D.,** Couplages de l'hybridation *in situ* a d'autres techniques cytologiques, in *Hybridation In Situ, Methodes Pratiques,* Calas, A., Bloch, B., Fournier, J.-G., and Trembleau, A., Eds., SFME, Paris, 1991, 57.
25. **Boyenval, J. and Fischer, J.,** Dipping technique, *J. Microscop. Biol. Cell,* 27, 115, 1976.
26. **Arai, H., Emson, P. C., Agarwal, S., Cristodoulou, C., and Gait, M. J.,** *In situ* hybridization histochemistry: localization of vasopressin mRNA in rat brain using a biotinylated oligonucleotide probe, *Mol. Brain Res.*, 4, 63, 1988.

27. **Hou Yu, A., Lamme, A. T., Zimmerman, E. A., and Silverman, A. J.,** Comparative distribution of vasopressin and oxytocin neurons in the rat brain using a double-labeled procedure, *Neuroendocrinology,* 44, 235, 1986.
28. **Trembleau, A., Calas, A., and Fevre-Montange, M.,** Combination of immunohistochemistry and *in situ* hybridization with a synthetic oligonucleotide probe to localize simultaneously vasopressin, oxytocin and their mRNAs in hypothalamic magnocellular neurons, *Bull. Assoc. Anat.,* 72, 101, 1988.
29. **Nakane, P. K.,** Classification of anterior pituitary cell types with immunoenzyme histochemistry, *J. Histochem. Cytochem.,* 18, 9, 1970.
30. **Guitteny, A. F., Bohlen, P., and Bloch, B.,** Analysis of vasopressin gene expression by *in situ* hybridization and immunohistochemistry in semi-thin sections, *J. Histochem. Cytochem.,* 36, 1373, 1988.
31. **Bloch, B., Popovici, T., Le Guellec, D., Normand, E., Chouham, S., Guitteny, A. F., and Bohlen, P.,** *In situ* hybridization histochemistry for the analysis of gene expression in the endocrine and central nervous system tissues: a three-year experience, *J. Neurosci. Res.,* 16, 183, 1986.
32. **Trembleau, A., Morales, M., and Bloom, F.,** Compartmentalization of vasopressin mRNA within hypothalamic magnocellular perikarya and their projections to the median eminence: *in situ* hybridization at the ultrastructural level, 22nd annual meeting Society For Neuroscience, Abst. 414.6.
33. **Trembleau, A.,** Detection d'ARNm en microscopie electronique: hybridation *in situ* d'un oligonucleotide biotinyle en pre-inclusion, *Pathol. Biol.,* in press.
34. **van Leeuwen, F. W., van der Beek, E. M., van Heerikhuize, J. J., Sluiter, A. A., Felix, D., and Imboden, H.,** Vasopressin and angiotensin II are absent but spontaneously reappear in solitary hypothalamic neurons of the homozygous Brattleboro rat, *Neurosci. lett.,* 127, 207, 1991.
35. **Bloch, B., Guitteny, A. F., Normand, E., and Chouham, S.,** Presence of neuropeptide messenger RNAs in neuronal processes, *Neurosci. Lett.,* 109, 259, 1990.
36. **Kleinman, R., Banker, G., and Steward, O.,** Differential subcellular localization of particular mRNAs in hippocampal neurons in culture, *Neuron,* 5, 821, 1990.
37. **Mohr, E., Fehr, S., and Richter, D.,** Axonal transport of neuropeptide encoding mRNAs with in the hypothalamo-hypophyseal tract of rats, *EMBO J.,* 10, 2419, 1991.
38. **Wensley, C. H., Kauer, J. S., Margolis, D. M., and Chikaraishi, D. M.,** OMP mRNA is transported to the axon terminals of olfactory receptor cells, *Soc. Neurosci. (Abstr.)* 17, #252.9, 637, 1991.

Chapter 6

IN SITU HYBRIDIZATION ON ULTRATHIN FROZEN SECTIONS

Gérard Morel

TABLE OF CONTENTS

0-8493-4414-X/93/$0.00 + $.50

I. INTRODUCTION

The easiest way to transpose *in situ* hybridization from light microscopy to electron microscopy is by the use of ultrathin frozen sections, which have the advantage of excluding all the steps of embedding in hydrophylic or epoxy resin. And denaturation of nucleic acid as a result of dehydration does not seem to be a significant risk in this type of study. At this moment, however, the fact of embedding in epoxy resin makes it impossible to detect hybrids at the ultrastructural level.

The main disadvantage of *in situ* hybridization on ultrathin frozen sections is the loss of ultrastructural morphology, and perhaps also the lability of the target nucleic acid, which may be lost during processing. The advantages are the ease of penetration of the probe into the sections, the sensitivity of the *in situ* hybridization method, and the rapidity of specimen preparation.

II. PREPARATION OF ULTRATHIN FROZEN SECTIONS

All tissues, cells (in cultured monolayers harvested after enzymatic treatment, or in suspension, either cultivated or not, and pelleted after washing), and clinical biopsies can be used as sources of specimens. Optimal sample size is about 1 mm^3.

Preparation of specimens for *in situ* hybridization on ultrathin frozen sections is the crucial step for ultrastructural preservation. Before freezing, specimens are fixed, and then cryoprotected. Ultrathin frozen sections are made using a cryoultramicrotome technique. To limit the undesirable consequences of drying, sections were stained and embedded in methylcellulose before observation.

A. FIXATION

The aim of fixation was to preserve ultrastructural aspects and nucleic acids. We have previously shown that a strong *in situ* hybridization signal is of little interest if cellular integrity is lost.[1] Thus, it is necessary to reach a compromise between the preservation, on the one hand, of the ultrastructures and the target nucleic acid, and, on the other hand, the fixation, the probe, and the revelation system. At the light microscopic level a wide variety of fixatives are used, but at the present time aldehyde generally seems to be the preferred type of fixative.[2-9] We have never obtained an *in situ* hybridization signal with osmium tetroxide, which we have used up to now.[9] Paraformaldehyde, either with or without the addition of glutaraldehyde (concentration $<0.5\%$), is the most widely used fixative (see Protocol 1). Aldehyde fixative seems to limit RNA loss. High glutaraldehyde concentration ($>0.5\%$) increases ultrastructural preservation, but also eliminates the *in situ* hybridization signal by limiting the access of the probe to the target nucleic acid and increasing the background signal, whether with or without treatment with sodium borohydride (see Protocol 2), by decreasing the efficiency of the washing steps.[10-11]

PROTOCOL 1 — FIXATION

1. Fix fresh tissue by vascular perfusion (15 min) and/or by immersion (2 h) in 2 to 4% freshly prepared paraformaldehyde, in a 0.1 *M* Sorensen phosphate buffer[a] (pH 7.4), with or without 0.2% glutaraldehyde, at 4°C.
2. Rinse in the same buffer for 30 min at 4°C
3. Incubate in cryoprotectant at 4°C

[a] The composition of phosphate buffer is given in the appendix.

PROTOCOL 2 — SODIUM BOROHYDRIDE PRETREATMENT

1. Prepare a 1% sodium borohydride solution containing 0.3% H_2O_2 in sterile distilled water
2. Incubate tissue after the postfixation washing step for 30 min, or tissue section for 10 min, before the *in situ* hybridization

B. FREEZING

To obtain ultrathin sections of a biological specimen, the tissue must be hardened. For hydrated biological specimens to retain water, the most convenient way to harden them is by freezing, i.e., the conversion of water from a liquid to a solid state. This apparently simple phase transition is, in fact, a complex series of events which is governed by such factors as tissue structure and composition, pretreatment, and cooling rate. Depending on the characteristics of the tissue, the questions are what is the optimal cooling rate, and over what temperature range; what cryoprotectant can be used to improve the quality of the frozen specimen without producing unacceptable consequences for the study?

When pure water is cooled below its melting temperature (Tm), it does not freeze immediately but remains in a metastable supercooled state. Supercooling cannot continue indefinitely, and eventually spontaneous crystallization occurs, accompanied by the release of latent heat of fusion which warms the water back up to its Tm.[12-13] By ultrarapid freezing, it is possible to obtain "vitrified ice", "vitreous ice", or "amorphous ice", though, in fact, the word "ice" as such implies a crystalline state.[14] However, critical cooling rates which give true vitrification have rarely been reported.[15-17] These cooling rates are not sufficient to vitrify untreated hydrated biological specimens at normal atmospheric pressure. The rate of cooling is absolutely critical in determining the properties of the final frozen tissue. Biologists have frequently used the term "vitrification" to describe cases of freezing where the ice crystals formed are smaller than the resolution of the measuring technique employed.

Crystallization may occur after freezing if the tissue has been stored, or during thawing. A recrystallization process will allow ice crystals to form and grow at all temperatures between the recrystallization temperature and the Tm. And when normal crystalline specimens are warmed to above the recrystallization temperature, ice crystals continue to grow at increasingly high rates. Finally, the ice crystals undergo a phase transition, with concomitant absorption of latent heat, and revert to the original liquid state before drying out.

The best way to overcome the problem of harmful ice crystal formation in specimens which have to be frozen in 1 mm^3 pieces is by the use of a cryoprotectant.[18-20] Substances which allow cells to be frozen with the formation of smaller ice crystals, and with less damage, than would otherwise be the case are now referred to as cryoprotectants. For ultrathin frozen sections, cryoprotectants are used only to improve in the preservation of ultrastructural and physiological features in the solid hydrated state, as opposed to cryoprotectants intended to facilitate freeze-thaw survival. However this is not a useful distinction, since the same compounds may serve either function. The purpose of such compounds is to protect tissue during the freezing process. In general, the ways in which this may be done are either to lower the Tm, to lower the temperature of homogeneous ice nucleation (increase the supercooling capacity), or to raise the recrystallization temperature and the critical cooling rate. Different cryoprotectants may have different combinations of these properties. In addition to its specific cryoprotection properties, it is necessary that the cryoprotectant should not be unacceptably injurious to the tissues, nor cause excessive artifacts. There should also be no interference with the material or molecule studied.

The range of available cryoprotective substances is quite large, although those which have been used for the ultrastructural detection of molecules by immunocytology or by *in situ* hybridization are relatively few in number: glycerol, dimethylsulfoxide, polyvinylpyrrolidone, and sucrose. There may be no difference between penetrating and nonpenetrating cryoprotectants as regards the cryoprotection of fixed tissue, where the plasma membranes are partially destroyed within tissue. However, a high concentration of a nonpenetrating cryoprotectant may produce hyperosmolarity artifacts.[21] Among these cryoprotectants, the most widely used is the nonpenetrating cryoprotectant, sucrose, whose advantage is its nonreactivity with nucleic acid and with many antigens.

The cryoprotective step is carried out between the fixation and freezing steps. Tissue can be cryoprotected with sucrose, in either (1) a hyperosmolar 2.3 *M* sucrose solution (see Protocol 3) or a 1.6 *M* sucrose solution added to 0.025 *M* polyvinylpyrrolidone neutralized with 0.055 *M* Na_2CO_3,[22] or (2) an iso-osmolar 0.4 *M* sucrose solution (see Protocol 4).

PROTOCOL 3 — CRYOPROTECTION WITH HYPER-OSMOLAR SUCROSE

1. Wash after fixation in Sorensen phosphate buffer[a] 0.1 *M*, pH 7.4, for 30 min at 4°C
2. Incubate in gradient of saccharose from 0.5, 1, 1.5, 2 *M*, for >1 h in each bath, at 4°C
3. Incubate overnight in a 2.3 *M* sucrose solution at 4°C
4. Place the sample on a cryoultramicrotome support
5. Lower into liquid nitrogen or solid nitrogen[b] (see Protocol 5)
6. Store the sample at −196°C in liquid nitrogen

[a] The composition of phosphate buffer is given in the appendix.
[b] The advantage of this freezing procedure is the absence of chelefactor effects.

PROTOCOL 4 — CRYOPROTECTION WITH ISO-OSMOLAR SUCROSE

1. Wash after fixation in Sorensen phosphate buffer[a] 0.1 *M*, pH 7.4, for 30 min at 4°C
2. Incubate in 0.4 *M* sucrose phosphate buffer (0.1 *M*; pH 7.4) solution for 30 to 90 min at 4°C
3. Place the sample on a cryoultramicrotome support
4. Freeze (see Protocol 6)
5. Store the sample in liquid nitrogen

[a] The composition of phosphate buffer is given in the appendix.

PROTOCOL 5 — PREPARATION OF SOLID NITROGEN

1. Place the liquid nitrogen in a vacuum evaporator
2. Pump with a rotary pump until the nitrogen becomes solid
3. Open the evaporator rapidly
4. Shoot the sample into the melting nitrogen

PROTOCOL 6 — FREEZING WITH ISO-OSMOLAR CRYOPROTECTANT

The following material is required:

- Digital thermometer with thermocouple (sensitivity 0.1°C)
- Cooling system, nitrogen gradient (see Protocol 6), LM10 (Air Liquide), programmable cooling system (Minicool, Air Liquide)

1. Immerse the specimen in saturated or 2.3 *M* sucrose solution for 1 min[a]
2. Place the sample on a cryoultramicrotome support

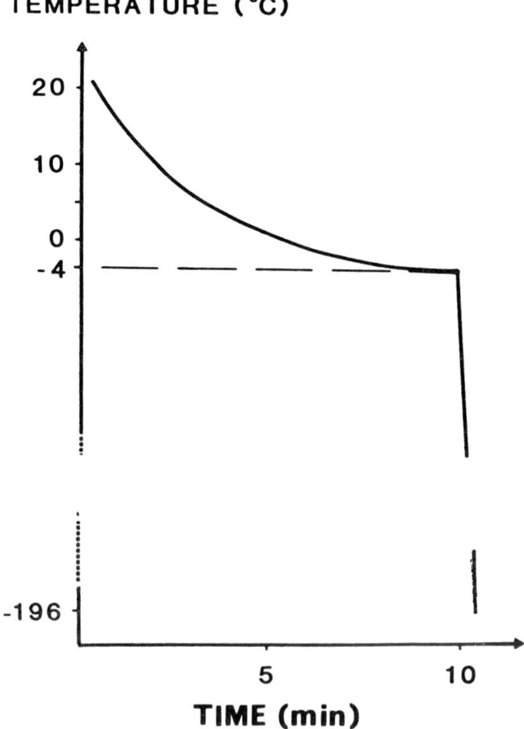

FIGURE 1. Freezing curve temperature. The specimens are cryoprotected in an iso-osmolar cryoprotective agent (sucrose 0.4 *M* in phosphate buffer), placed on a cryoultramicrotome support, and introduced into a cold gradient of fuming nitrogen, to −4°C, then immersed in liquid nitrogen for storage.

3. Place it in the cooling system
4. Lower the temperature slowly to −4°C[b] (Figure 1)
5. Lower the temperature quickly from −4°C to −130°C[c] or −196°C[d] (Figure 1)
6. Store tissue at −196°C

[a] The aim of this step is to protect the tissue from desiccation during cooling; the sucrose solution must be reduced to a thin film.

[b] At this temperature (temperature between Tm and the supercooling temperature) the liquid in the tissue is always in liquid phase and remains so to freezing point if the cooling rate is <1°C/min.

[c] This temperature was obtained with the programmable freezing system or LM 10 and a cooling rate >30°C/min.

[d] This temperature was obtained with the cold gradient by immersion in liquid nitrogen.

C. CRYOULTRAMICROTOMY

If the freezing of the tissue was done without the formation of large ice crystals or the destruction of tissue or cell structure, this tissue sectioning or

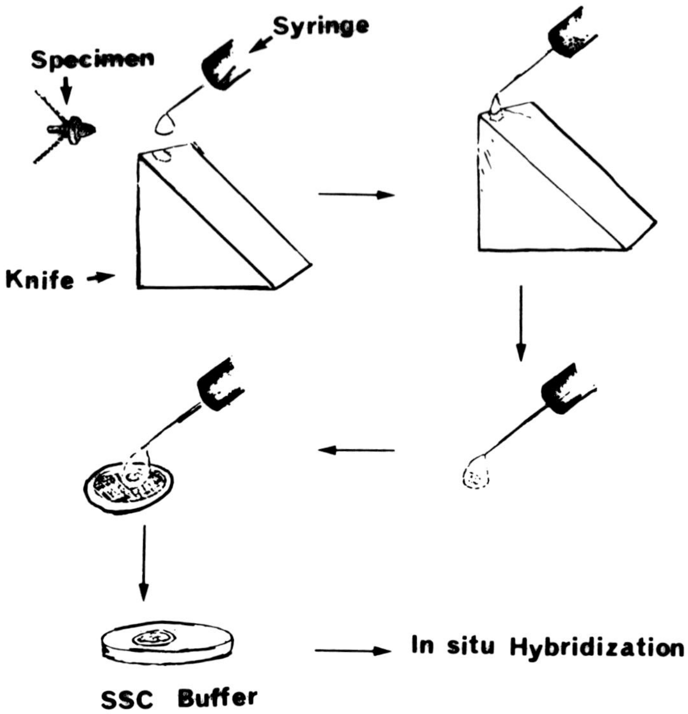

FIGURE 2. Dry sectioning method.

cryoultramicrotomy step is easy to perform. The method used by most people has been described by Tokuyasu and is called the ''dry sectioning method'' (see Protocol 7).[23]

PROTOCOL 7 — DRY SECTIONING METHOD (FIGURE 2)

1. Gold sections are on the knife
2. Put a drop of satured sucrose over the sections on the knife
3. Transfer the drop and the sections onto a carbon-coated film on a grid (see Protocol 8)
4. Wash in SSC[a] × 2
5. Hybridize

[a] The SSC composition is given in the appendix.

PROTOCOL 8 — PREPARATION OF COATED GRIDS

The following materials are required:

- A tall beaker
- A dropper

- An acetate film
- 200- or 400-mesh nickel grids
- 1% collodion solution in isoamyl acetate

1. Put one drop of collodion solution on the water in the beaker.[a] The collodion film floats to the surface of water
2. Air-dry
3. Put the grids on the collodion film
4. Put an acetate film over the collodion film and grids and take out
5. Dry
6. Deposit a 10-nm layer of carbon in a vacuum by evaporation

[a] The thickness of the collodion film can be reduced by floating the grids on the solvent.

The sectioning parameters are the knife, angle of the knife, specimen temperature, knife temperature, chamber temperature, and sectioning speed. The quality of the knife is important, if regular sections are to be obtained. A diamond knife can be used for making ultrathin frozen sections, but great care must be taken in making glass knives. A method described by Griffiths gives 45° angle knives of excellent quality.[24] The life of a knife can be prolonged by evaporation of a thin film of tungsten onto it.[24] A knife angle of 6° seems to be the best. The temperature of the specimen must be below the recrystallization temperature (if the cryoprotectant used was sucrose 0.4 *M*, this temperature must be < -100°C) to avoid all artifacts of recrystallization during sectioning and after, if tissue was restored after sectioning. The chamber temperature and that of the knife must be about -100°C, for the same reasons. The sectioning speed is generally around 5 mm s^{-1}, and can be reduced if the tissue is soft. For particular types of tissue, such as bone, sectioning must be done manually and at a low temperature (< -130°C).[25,26]

Grids with sections can be stored at 4°C for a long period.

D. STAINING

The aim of staining is to increase the contrast of sections, as is the case for embedding sections, but it must avoid the destruction of structure and ultrastructure during drying.[27,28] Stierhof et al. and Keller et al. have shown that the ultrastructure of frozen sections of fixed tissue was perfect after embedding in epoxy resin and classical staining (see Protocol 9).[27,28]

PROTOCOL 9 — EMBEDDING IN EPON[a]

1. Fix by contact with a drop of osmium tetroxide for 10 min
2. Wash in phosphate buffer[b] for 3 × 5 min
3. Wash in distilled water for 5 min
4. Dehydrate in ethanol 50%, then 70%, for 2 min each
5. Stain in 2% uranyl acetate in ethanol 70%, for 20 min

6. Dehydrate in ethanol 85%, 90%, 95%, 100% for 2 × 2 min each
7. Embed in 2% Epon solution in ethanol 100%
8. Dry the grids until they turn a gold-purple color
9. Dry in an oven at 60°C overnight
10. Stain the grids, if desired, with lead citrate (5 to 15 min)

[a] The grid must be coated with a film of Formvar, since collodion is soluble in alcohol.
[b] The composition of phosphate buffer is given in the Appendix.

It is necessary to define optimal conditions of embedding and staining in order to reveal ultrastructures in similar conditions to classical embedding methods. However, the main advantage of using freezing is that it preserves all the soluble substances, which are lost during classical dehydration, and embedding must be preserved during the staining step. There are several protocols which can be used to preserve some (see Protocol 10) or all (see Protocol 11) of the soluble molecules during staining.[22,24,29-32]

PROTOCOL 10 — EMBEDDING IN LOWICRYL[a]

1. Fix by contact with a drop of glutaraldehyde 1 to 2% for 10 min
2. Wash in phosphate buffer[b] for 3 × 5 min
3. Wash in distilled water for 5 min
4. Dehydrate in ethanol 50%, then 70%, for 2 min each
5. Stain in 2% uranyl acetate in ethanol 70% for 20 min
6. Dehydrate in ethanol 85%, 90%, 95%, 100% for 2 min each
7. Embed in 30% Lowicryl solution in ethanol 100% for 5 min
8. Embed in 60% Lowicryl solution in ethanol 100% for 5 min
9. Embed in 100% Lowicryl solution in ethanol 100% for 5 min
10. Dry the grids until they turn a gold-purple color
11. Dry the sections overnight at room temperature or at 60°C in an oven

[a] The grid must be coated with a film of Formvar, since collodion is soluble in alcohol.
[b] The composition of phosphate buffer is given in the appendix.

PROTOCOL 11 — EMBEDDING IN METHYLCELLULOSE

Prepare 2% methylcellulose (25 centipoises) in distilled water, by heating at 95°C

- Agitate overnight at 4°C
- Centrifuge at 100,000 *g* for 1 h
- The preparation can be stored at 4°C or −20°C for a long period

1. Fix by contact with a drop of glutaraldehyde 1 to 2% for 10 min
2. Wash in phosphate buffer[a] for 3 × 5 min

3. Wash in distilled water for 5 min
4. Stain on a drop of 4% neutralized uranyl acetate[b] for 10 min
5. Incubate sections on a drop of methylcellulose, with or without 10 to 20% of 4% neutralized uranyl acetate, placed on parafilm of ice for 10 min
6. Dry the grids until they turn a gold-purple color[c]
7. Observe after 1 h

[a] The composition of phosphate buffer is given in the appendix.
[b] A mixture of 4% uranyl acetate and 0.3 *M* oxalic acid (v/v) was neutralized with ammonia to pH 7.5.
[c] If the methylcellulose is too dense it can be diluted to 1% in distilled water, and the uranyl acetate omitted.

The choice of staining method depends on the ultrastructural aspect desired. Embedding in resin shows ultrastructures similar to classical electron microscopy observations, while methylcellulose gives results with negative or positive-negative staining. In the field of *in situ* hybridization the use of methylcellulose is the best way to preserve the mRNA present in a cytoplasmic matrix. The detection of viral nucleic acid in its capsid is better with resin embedding.

III. PREPARATION OF THE PROBE

A. CHOICE OF PROBE

Only DNA probes (complementary DNA [cDNA] and oligonucleotide) were used for *in situ* hybridization on ultrathin frozen sections. Some tests were performed with antisense RNA, without success.

The cDNA used was always limited to the insert. The plasmid pBR322 which served to amplify the probe had never shown an *in situ* hybridization signal on ultrathin sections, whatever the choice of labeling (radioactive vs. nonradioactive label). The main advantage of cDNA, compared to an oligonucleotide probe, is the density of labeling. After labeling, a cDNA probe was cut in 200 to 300 bp labeled fragments. A target nucleic acid can hybridize several labeled fragment from cDNA.

An oligonucleotide must be synthesized, and this can be done if the sequence of the target nucleic acid has been published. The length of the most commonly used oligonucleotide is between 20 and 30 mer. Shorter oligonucleotides can give specificity problems, and longer ones are more costly, but no more effective. The choice of the sequence is the main factor determining specificity. The sequence must be checked against a gene bank, since it is known that some oligonucleotide probes cross-hybridize with other gene sequences. The stability of a hybrid depends on the percentage of G-C bases; about 50 to 60% seems to be best. The inconvenience of oligonucleotides compared to cDNA probes, as far as the density of labeling is concerned, can be compensated for by the use of several oligonucleotides for the same target nucleic acid; but a single *in situ* hybridization signal per target nucleic acid is the only possibility for quantification. The standardization of the *in situ* hybridization protocol for "a labeled" oligonucleotide of a given length can be used, with minor variations, for others of the same length. Labeling by tailing at the 3′ end, using nonisotopic nucleotides, does not modify the sequence which hybridizes with target nucleic acid.

B. CHOICE OF LABEL

This choice determines the revelation system: the use of radioactive nucleotides involves autoradiographic detection, while that of antigenic-conjugated nucleotides involves immunocytological detection.

The main radioisotopes used were tritium (^{3}H) and sulfur-35 (^{35}S) (see Chapter 2), whose resolutions, expressed in terms of the half distance (HD) of the isotope origin, are 250 to 310 nm and 390 nm, respectively. Exposure time is proportional to the half-life of the isotope.[33,34]

Nonisotopic labeled nucleotides or antigen-conjugated nucleotides are more frequently used at the electron microscopic level. The most successful of these nonradioactive nucleotides use biotin- and digoxigenin-conjugated nucleotides, respectively (see Chapter 2). The advantages of these antigen-conjugated nucleotides result from their revelation: immunocytology. They give high cellular and subcellular resolution, if colloidal gold particles are used as the mark. The rapidity of the revelation in the case of immunocytology, compared to autoradiography, is an important factor of choice. Disadvantages result from the revelation method, namely in the quantification process (gold particle count), the presence of endogenous antigens such as endogenous biotin in tissue, and aspecific binding of digoxigenin with a "receptor". Low sensitivity can result from a low level of nucleotide labeling.

C. LABELING OF PROBE

The same methods of labeling can be used for *in situ* hybridization on ultrathin frozen sections (see Chapter 2). Nick translation (see Protocol 12) and random priming (see Protocol 13) were used to label cDNA with radioactive and nonradioactive labeled nucleotides. The tailing 3′ end labeling method was performed with oligonucleotide probes (see Protocol 14) and also with isotopic and nonisotopic labeled nucleotides.

PROTOCOL 12 — NICK TRANSLATION

The following materials and reagents are required:

- Water bath at 15°C
- 1 μg of cDNA
- dNTP mixture (0.2 m*M* of each dNTP other than the labeled nucleotide, in 500 m*M* Tris-HCl (pH 7.8), 50 m*M* $MgCl_2$, 100 μg/ml nuclease-free BSA, and 100 m*M* 2-mercaptoethanol if the labeled nucleotide is radioactive)

- Enzymes (DNA pol I, 4 U/ml; DNase I, 40 pg/ml)
- Labeled nucleotide: 0.3 m*M*/l biotin 16-dUTP, or 0.3 m*M*/l digoxigenin 16-dUTP, or 50 μCi (α^{35}S) dATP
- Sterilized distilled water, RNase-free

1. Mix

DNA	1 μg
Buffer	2 μl
dNTP	3 μl
Labeled nucleotide	x μl
Enzyme	1 μl
Water to	20 μl

2. Incubate at 15°C for 2 h
3. Precipitate (see Protocol 15)
4. Dilute in water
5. Store at −20°C
6. Denature before use by heating to 95°C for 5 min

PROTOCOL 13 — RANDOM PRIMING

The following materials and reagents are required:

- Bath at 37°C
- 25 ng of cDNA
- dNTP (0.5 m*M* of dNTP other than the labeled nucleotide)
- Reaction mixture with hexanucleotides (×10): 130 mM K_2PO_4, 6.5 m*M* $MgCl_2$, 33 μ*M* dTTP, 1 m*M* DTT (if the labeled nucleotide is radioactive), 32 μg/ml BSA (pH 7.4)
- Enzyme (Klenow fragment 2 U/μl)
- Labeled nucleotide: 0.3 m*M*/l biotin 16-dUTP, or 0.3 m*M*/l digoxigenin 16-dUTP, or 50 μCi (α^{35}S) dATP
- Sterilized distilled water, RNase-free

1. Denature cDNA
2. Mix

DNA	25 ng
Reaction mixture	5 μl
dNTP mixture	5 μl
Labeled nucleotide	x μl
Enzyme	5 μl
Water to	50 μl

3. Incubate at 37°C for 30 min
4. Precipitate (see Protocol 15)
5. Dissolve in water
6. Store at −20°C
7. Denature before use by heating to 95°C for 5 min

PROTOCOL 14 — TAILING 3′ END

The following materials and reagents are required:

- Bath at 37°C
- 10 p*M* of oligonucleotide (2 p*M*/μl)
- Buffer (×5): 1 *M* potassium cacodylate, 125 m*M* Tris-HCl, 1.25 mg/ml BSA, 1 m*M* DTT (if the labeled nucleotide is radioactive, pH 6.6)
- $CoCl_2$ 1.5 m*M*
- Enzyme (Terminal deoxynucleotidyl Transferase [TdT], 25 U/μl)
- Labeled nucleotide: 0.3 mM/l biotin 11-dUTP, or 0.3 m*M*/l digoxigenin 11-dUTP, or 50 μCi ($\alpha^{35}S$) dATP
- Sterilized distilled water, RNase-free

1. Mix

Oligonucleotide	5 μl
Buffer	10 μl
$CoCl_2$	5 μl
Labeled nucleotide	x μl
Enzyme	1 μl
water to	50 μl

2. Incubate at 37°C for 1 h
3. Precipitate (see Protocol 15)
4. Dissolve in water
5. Store at −20°C

PROTOCOL 15 — PRECIPITATION OF PROBE

The following reagents are required:

- tRNA 10 mg/ml
- Ethanol 100% at −20°C
- Ammonium acetate 7.5 *M* or sodium acetate 3 *M*

1. Add to 1 V of reaction reagent
   - 2 μl of tRNA
   - 3 V of ethanol
   - 1/10 V ammonium acetate or sodium acetate
2. Mix
3. Incubate at −20°C overnight or at −80°C for 1 h
4. Centrifuge for at least 30 min at 14,000 *g*
5. Dry
6. Dissolve in water
7. Store at −20°C

D. LABELING CHECK

Before use, the labeled probe must be checked to determine the density of the label introduced into the nucleic acid. For the radioactive labeled probe it is necessary to determine the specific activity. For other nonradioactive labeled probes, the incorporation of biotin or digoxigenin is more difficult to check, requiring as it does a large amount of the labeled probe (see Chapter 4 for details). This incorporation is compared to a standard by spotting on a nitrocellulose filter and visualization by immunohistochemistry (see Chapter 4).

E. STORAGE OF THE PROBE

Radioactive probes labeled with ^{35}S (half-life: 84 days) must be used within 1 month, and probes labeled with ^{3}H (half-life 12.3 years) within 2 months, while nonradioactive probes can be stored for up to 1 year.

IV. PRETREATMENT OF ULTRATHIN FROZEN SECTIONS

The aim of pretreatment is to bring about the contact between the probe and the nucleic acid.

In the absence of an embedding procedure, such as the ultrathin frozen section method, pretreatments are reduced to a minimum. Treatment aimed at partially removing cellular proteins using protease (proteinase K) (see Protocol 16) or HCl, which causes substantial morphological deterioration, is used when the level of the signal is extremely low.

To reduce background noise during *in situ* hybridization by saturating nonspecific adsorption sites, two possibilities exist: acetylation and prehybridization. Acetylation cannot be performed without causing serious damage to sections. Prehybridization consists of incubating sections on a hybridization buffer for 1 h, then hybridizing without a washing step.

If the target nucleic acid is a double-strand nucleic acid, it is necessary to denature it by heating without formamide to 95°C for 5 min, or with 50% formamide at 65°C for 5 min. The latter gives better preservation of ultrastructures.

PROTOCOL 16 — PROTEINASE K PRETREATMENT

1. Wash sections in phosphate buffer for 10 min
2. Fix ultrathin frozen sections in 4% paraformaldehyde in phosphate buffer for 10 min
3. Wash section in phosphate buffer for 2 × 5 min
4. Wash sections in TE buffer (×1) for 5 min at room temperature
5. Incubate in proteinase K solution[a] (1 μg/ml) for 5 to 10 min at room temperature
6. Wash in TE buffer (×1) for 2 × 5 min at room temperature

7. Postfix in 4% paraformaldehyde in phosphate buffer for 10 min
8. Wash sections in phosphate buffer for 2 × 5 min
9. Wash sections in SSC (×2) for 2 × 5 min
10. Prehybridize or hybridize

[a] Proteinase K is stored at −20°C in aliquot at a concentration of 1 μg/μl; before use it must be heated to 37°C for 30 min

V. HYBRIDIZATION

The hybridization step consists of putting the probe (complementary nucleic acid) in contact with the ultrathin frozen sections (which contain the target nucleic acid) to form hybrids by complementarity. The presence of an isotope or antigen in hybrids permits their detection by autoradiography and immunocytology, respectively.

A. GENERAL CONSIDERATIONS

The detailed nature of the *in situ* hybridization procedure is influenced by: (1) the buffer, (2) the probe, (3) the incubation, and (4) the washes, since the tissue specimens were all similar to one another (ultrathin frozen sections of fixed tissue). In each of these elements the main parameters were

1. Buffer
   - Saline concentration
   - Formamide concentration (see Protocol 17)
   - Macromolecules (tRNA, DNA, Denhardt's solution)
2. Probe
   - Nature (oligonucleotide vs. cDNA)
   - Label (radioactive vs. nonradioactive)
   - Density of labeling
   - Concentration
3. Incubation
   - Temperature
   - Time
4. Wash
   - Saline concentrations
   - Temperatures
   - Time

The content of the buffer must be given optimal conditions to form hybrids at the incubation temperature. Saline concentration (from SSC $\times 20$) and deionized formamide concentration have similar effects in conventional hybridization (see Chapter 1). Macromolecules are added to reduce nonspecific adsorption between the probe and the tissue section.

The choice of the probe determines the hybridization protocol (see below), while the choice of label defines the revelation and the choice of label, like the density of labeling, influences probe concentration (radioactive probes are used at lower concentrations than nonradioactive probes).

Incubation is carried out for a shorter period than in the case of light microscopy because ultrathin frozen sections are destroyed by long incubation periods (overnight incubation destroys about 90% of these sections). As previously shown, an incubation period of 3 h is adequate for *in situ* hybridization.[9] The incubation temperature, which is about 20 to 30°C below the hybridization temperature, is that which gives optimal hybridization, though with the possibility of mismatches, which are eliminated during the washing step and/or the revelation procedure.

These washes are intended to eliminate nonspecific adsorption between the probe and the tissue section, as well as all mismatches between the probe and other nucleic acids. This is done by incubation of sections on drops of saline solution at decreasing concentrations (SSC $\times 5$ to $\times 0.5$), for 10 min to 1 h each, at room temperature to 40°C. All these parameters — decreasing concentration of ions, increasing incubation period, and temperature — increase stringency of the washing step. If this step is too stringent, the *in situ* hybridization signal disappears. This can be used as a check of hybridization.

PROTOCOL 17 — PREPARATION OF DEIONIZED FORMAMIDE

The reagents required are formamide and amberlite resin (Biorad AG 501-X8(D), 20 to 50 mesh).

1. Add 3 g of resin to 100 ml formamide
2. Mix for 30 min
3. Filter
4. Aliquot and store at -20°C[a]

[a] Formamide can be stored at -20°C, if it remains frozen, for 1 to 2 months.

B. HYBRIDIZATION PROTOCOL WITH cDNA

1. Probe

The growth hormone (GH) cDNA probe consisted of the complete 800 bp *Hind* III fragment from the plasmid prGHI.[35,36] An *Eco* RI to *Xba* I fragment of the rat GH gene containing approximately 2 kb of 5′ untranscribed DNA, the first exon and the first intron, was subcloned at the appropriate restriction sites in pGEMI (Promega). After digestion with *Pst* I and *Acc* I, a 186-bp fragment was obtained that corresponded to intron I of the GH gene.

Probes were labeled by nick translation (see Protocol 12) or random priming (see Protocol 13) with a mixture of ^{3}H-dCTP and ^{3}H-dTTP, or with ^{35}S-dCTP (1200 Ci/m*M*, Amersham). The specific activities obtained with ^{35}S-labeled nucleotide were of the order of 10^7 cpm/μg and 10^8 cpm/μg for nick translation and random priming, respectively. Unlabeled pBR322 DNA used for prehybridization was nicked with DNase I at the same enzyme-to-DNA ratio. Probes were boiled for 5 min and then chilled before being added to the hybridization buffer.

2. Pretreatment

Ultrathin frozen sections were preincubated for 1 h at 37°C in a hybridization buffer without probe but containing 2 μg/ml of denatured pBR322 DNA. No proteolytic pretreatment was done.

3. Hybridization

The buffer used was

Components	Storage conc	Final conc
Deionized formamide	100%	50%
SSPC[a]	×20	×5
Denhardt's solution[a]	×50	×1
DNA	1 mg/ml	2 μg/ml
RNasin	10,000 U/ml	300 U/ml

[a] Their compositions are given in the appendix.

The probe was added to the hybridization buffer at a concentration of 10^6 cpm/ml of hybridization buffer and 3×10^5 cpm/ml of hybridization buffer for ^{35}S-labeled probe and ^{3}H-labeled probe, respectively (about 0.1 to 0.2 μg/ml of hybridization buffer). Dithiotreithol was added to the hybridization buffer at a final concentration of 10 m*M* in the case of ^{35}S-labeled probes. The grids were floated on a drop of hybridization buffer (3 grids per 100 μl).

The hybridization incubation can be performed at 37°C or 40°C, without significant differences in results, in a humid chamber containing SSC buffer (concentration ×5) for 3 h.

4. Washing

Grids were washed at room temperature in SSC buffer (the composition of the SSC buffer is given in appendix) as follows:

SSC ×4	for 5 min
SSC ×2	2 to 6× for 10 min
SSC ×0.2	2× for 3 min

The grids were then postfixed with 1% osmium tetroxide in a 0.1 *M* sodium phosphate buffer (pH 7.4) washed in the same buffer, and air-dried overnight before autoradiographic revelation.

```
rGH mRNA probe:  5' TGA GAA GCA GAA CGC AGC CTG GGC ATT CTG  3'

        50
        Ala Lys Ala Ile  Asn  Asp                           Cys Pro Thr
rPRL    GCC AAG GCC AUC AAU GAC                             UGC CCC ACU
rGH         UAU UCC AUU CAG AAU GCC CAG GCU GCG UUC UGC UUC UCA
            Tyr Ser  Ile Gln Asn Ala Gln Ala Ala Phe Cys Phe Ser
                                             50

 60                                          70
Ser Ser Leu Ala Thr Pro Glu Asp Lys Glu Gln Ala Gln Lys Val
UCU UCC CUA GCU ACU CCU GAA GAC AAG GAA CAA GCC CAG AAA GUC
GAG ACC AUC CCA GCC CCC ACC GGC AAG GAG GAG GCC CAG CAG AGA
Glu Thr  Ile Pro Ala Pro Thr Gly Lys Glu Glu Ala Gln Gln Arg
                     60

                                                   100
Pro  Glu  Val  ------------------------------- Leu  Gly  Gly
CCU  CCG  GAA--------------------------------CUA  CGU  GGA
ACU  GAC  AUG--------------------------------UUU  ACC  AAC
Thr  Asp  Met  ------------------------------- Phe  Thr  Asn
70
                               110
Ile His Glu Ala Pro Asp Ala Ile Ile Ser Arg Ala Lys Glu
AUC CAU GAA GCU CCU GAU GCU AUC AUA UCA AGA GCC AAA GAG
AGC CUG                     AUG UUU GGU ACC UCG GAC CGC
Ser Leu                     Met Phe Gly Thr Ser Asp Arg
   100

              120                                   130
Ile Glu Glu Gln Asn  Lys Arg Leu Leu Glu Gly  Ile Glu Lys
AUU GAG GAA CAA AAC AAG CGG CUU CUC GAA GGC AUC GAA AAG
GUC UAU GAG AAA CUG AAG GAC CUG GAA GAG GGC AUC CAG GCU
Val Tyr Glu Lys Leu Lys Asp Leu Glu Glu Gly Ile Gln Ala
        110                                 120

rPRL mRNA probe:  5' TGA TAT GAT AGC ATC AGG AGC TTC ATG GAT  3'
```

FIGURE 3. Choice and preparation of oligonucleotide probe. The sequences of rat growth hormone (rGH) and rat prolactin (rPRL) cDNA show homologies. Probes are selected in cDNA sequence which are not expressed in the other hormone. Then complementary sequences are synthesized.

C. HYBRIDIZATION WITH OLIGONUCLEOTIDES

1. Probes

The probes used are 30-mer synthetic oligonucleotides synthesized by solid phase phosphoramidite chemistry. To detect GH mRNA or Prolactin (PRL) mRNA in the pituitary gland, where these hormones are synthesized, we used complementary probes from a sequence absent in the mRNA of the other hormone (Figure 3). The problem of distinguishing between these two mRNAs is due to the high degree of homology of their sequences, resulting from a common ancestral gene.[37]

The oligonucleotide was labeled by tailing the 3′ end' (see Protocol 14) using a radioactive or nonradioactive nucleotide. For ultrastructural studies we used for preference nonradioactive labeling methods as well as biotin- or digoxigenin-labeled dUTP. The length of the carbon chain (11 or 16) does not affect labeling methods or detection.

2. Pretreatment

No pretreatment was carried out before hybridization on ultrathin frozen sections using oligonucleotide probes.

3. Hybridization

The buffer used was

Components	Storage conc	Final conc
Deionized formamide	100%	30% – 50%
SSC[a]	×20	×4
Denhardt's solution[a]	×50	×1
tRNA	10 mg/ml	250 μg/ml

[a] Their compositions are given in the appendix.

The oligonucleotide probe was added to the hybridization buffer at a concentration of 3 to 5 p*M*/ml of hybridization buffer for the radioactive labeled probe, and 10 to 30 p*M*/ml of hybridization buffer for the nonradioactive labeled probe. Dithiotreithol was added to the hybridization buffer at a final concentration of 10 m*M*, for the radioactive labeled probe. The grids were floated on a drop of hybridization buffer (3 grids per 100 μl of hybridization buffer).

The hybridization incubation was performed at between 20 and 40°C, depending on the oligonucleotides used, in a humid chamber containing SSC buffer (concentration ×5) for 3 h.

4. Washing

Grids were washed at room temperature in SSC buffer (the composition of SSC is given in appendix) as follows:

SSC ×4	for 5 min
SSC ×2	3 to 6× for 10 min
SSC ×1	2× for 5 min

Then the grids were postfixed with 2% glutaraldehyde or 4% paraformaldehyde in a 0.1 *M* sodium phosphate buffer (pH 7.4) and washed in the same buffer.

For radioactive labeled probes, the grids were air-dried overnight before autoradiographic revelation.

For nonradioactive labeled probes, the grids remained in 0.1 *M* sodium phosphate buffer (pH 7.4) for immunocytochemical detection of label.

D. CHECK OF HYBRIDIZATION PROCEDURE

Nonspecific hybridization is due to a nonspecific adsorbtion of the probe in tissues. This can be checked on different tissues and by abolition of hybridization.

1. Tissue

The tissue studied must be hybridized with a heterologous probe (cDNA or oligonucleotide probe similar to those used). The hybridization procedure was carried out under identical conditions to those in which the test probes are used, checking for regions of the tissue which might exhibit unusually high nonspecific adsorbtion of nucleic acid.

Positive signals must be detected with identical hybridization procedures in cell lines or tissues known to contain particularly high levels of the nucleic acid under study which could be correlated with other techniques: this is the homogenous positive tissue control.

In the same way, with heterologous tissue only some cells which express the mRNA or DNA studied should be positive. The absence of a signal in the other cells or structures is a good check of specificity: this is the heterologous tissue control.

In the opposite way, hybridization with the probe, carried out under identical conditions to those in which the positive signal is obtained in tissues studied, must show an absence of the signal in cell lines or tissues known to not contain the nucleic acid with other techniques: this is the negative tissue control.

2. Abolition of Hybridization Signal

Specific hybridization can be abolished by:

- Competition between a labeled probe and an excess of a nonlabeled probe (100 times more concentrated than labeled probe), while an heterologous probe (in similar conditions) does not modify the signal.
- Destruction of target nucleic acid using enzymatic treatment (DNase or RNase). Pretreatment of tissue with RNase[9] or DNase has frequently been used to demonstrate that the signal observed is dependent on the presence of RNA in the tissue and is not due to hybridization to DNA nor to adventitious adsorption to unknown cellular constituents.

VI. REVELATION

Visualization of hybrids remains the main aim of the microscopist. The choice of method (autoradiography vs. immunocytology) depends on the type of probe labeling (isotope vs. antigen label), but the resolution and the sensitivity can be modified at this stage.

A. AUTORADIOGRAPHIC DETECTION

1. General Considerations

Hybrids which contain isotopic molecules emit particles during radioactive disintegration. The isotopes used (^{3}H or ^{35}S) emit electrons (β-particles). Autoradiographs are produced by these particles using a photographic emulsion in close contact with the section. This photographic emulsion consists of a suspension of silver bromide crystals in gelatin. The emission of particles causes the formation of a ''latent image'' in some crystals. The ''latent image'' is subsequently developed to give grains (filamentous silver bodies or silver particles: the size and morphology of silver grains depends on the developer used and the development time) which are visualized at the electron microscopic level.[33,34,38,39]

2. Photographic Emulsion Application

This step will entirely determine the quality of hybrid visualization. The photographic emulsion (Ilford L4 or Amersham EM1) consists of a suspension of silver bromide crystals in gelatin, which must be prepared (see Protocol 18) for application as a monolayer film over the sections. This monolayer film application, which is carried out by the use of a loop technique[34,40] (see Protocol 19) or dipping technique[41] (see Protocol 20), is the guarantee of resolution of the detection. The monolayer film can be checked with an electron microscope without treatment, since AgBr crystals are electron-dense.[34,42]

PROTOCOL 18 — PREPARATION OF NUCLEAR EMULSION

The emulsion is prepared the day before the coating session. The concentration of the emulsion must be checked for each new batch.

The following materials and reagents are required:

- Water bath at 45°C
- Jar to dilute emulsion
- Clean histological slide
- Nuclear emulsion: Ilford L4 or Amersham EM1
- Sterilized distilled water

1. Place the jar of emulsion in the water bath and allow it to stand until it melts
2. Place sterilized distilled water in the water bath
3. Dilute the emulsion with sterilized distilled water (Ilford L4: to 2 volumes of emulsion add 1 volume of water)
4. Mix without bubbles for at least 30 min
5. Check the dilution of the emulsion by making a monolayer film over a coated grid[a] (see Protocol 19). If the film is too dense add water. If

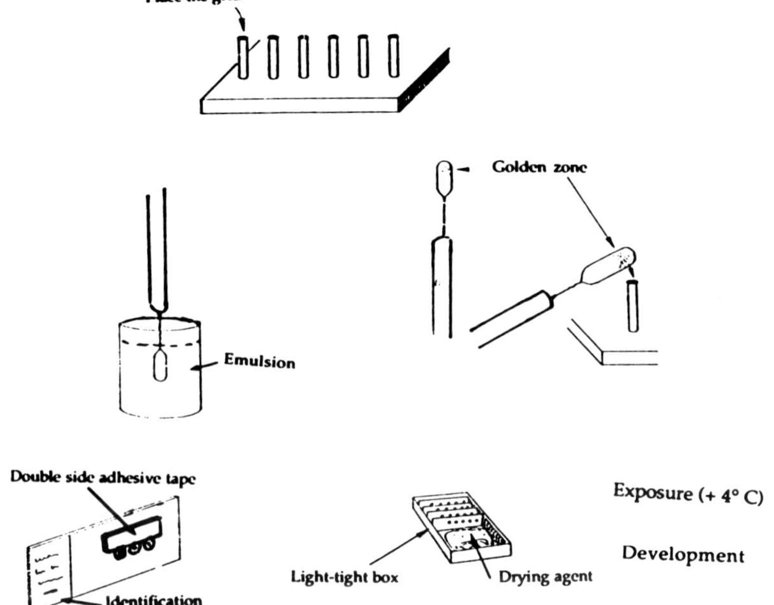

FIGURE 4. Emulsion application by the loop technique for autoradiography (as described in Protocol 19).

it is not dense enough, add emulsion mix, and check until a monolayer film of the required density is obtained.

6. Store the emulsion in aliquot at 4°C in the dark

[a] Without treatment AgBr crystals are electron-dense.

PROTOCOL 19 — EMULSION APPLICATION BY THE LOOP TECHNIQUE (FIGURE 4)

Check the darkroom equipment:

- Water bath at 45°C
- Jar of diluted emulsion (see Protocol 18)
- Clean histological slide
- Histological slide with a short section cut from one of its ends and stuck to its surface, and a piece of double-sided adhesive tape stuck to the top of the section
- Wire loop of tungsten, platinum, or nichrome[a]
- Support, with a line of posts 3 mm in diameter[b]
- Small movable safelight
- Small light-tight boxes, containing a drying agent wrapped in paper tissue
- Adhesive tape

The darkroom must be completely light-tight (with a suitable safelight of the nuclear emulsion used); temperature 35 to 40°C; humidity >70%.

1. Place the jar of diluted emulsion in the water bath and allow it to stand there until it melts
2. Place a grid over the pole
3. Immerse the whole loop in the diluted emulsion
4. Withdraw it carefully, edge-first, covered with a thin membrane of emulsion
5. Rotate the loop vertically in front of the safelight to see if the emulsion still flows. When this is no longer the case, i.e., when the membrane becomes a gelled film, the emulsion is ready to use[c]
6. Apply the film over the grids. The best area is generally the upper third of the loop
7. Blow on the film to close the contact between film and grid
8. Replace the grid and attach it by an edge to the adhesive tape over the histological slide (3 to 4 grids per slide)
9. Put the specimen away in a light-tight box
10. Store at 4°C to expose

[a] The loop is obtained by winding the wire around two hemolytic tubes to form an ellipse and fusing the ends into a support.

[b] The poles are 2 cm apart; the end of each pole is polished and cleaned before placing it on the grid.

[c] If the film emulsion is applied ungelled to a collodion-coated grid, AgBr crystals accumulate near the grid bar.

PROTOCOL 20 — EMULSION APPLICATION BY DIPPING IN LIQUID EMULSION (FIGURE 5)

- Water bath at 45°C
- Jar of diluted emulsion (see Protocol 18)
- Clean histological slide
- Small movable safelight
- Tissue paper
- Small light-tight boxes, containing a drying agent wrapped in paper tissue

The darkroom must be completely light-tight (with a suitable safelight for the nuclear emulsion used); temperature 20°C; humidity 20 to 40%.

1. Prepare the slide
   a. Mark 2 crosses on the back of the slide, 2.5 cm from the bottom and 0.5 cm from the edge
   b. Put the slide into collodion or formvar solution
   c. Drain briefly

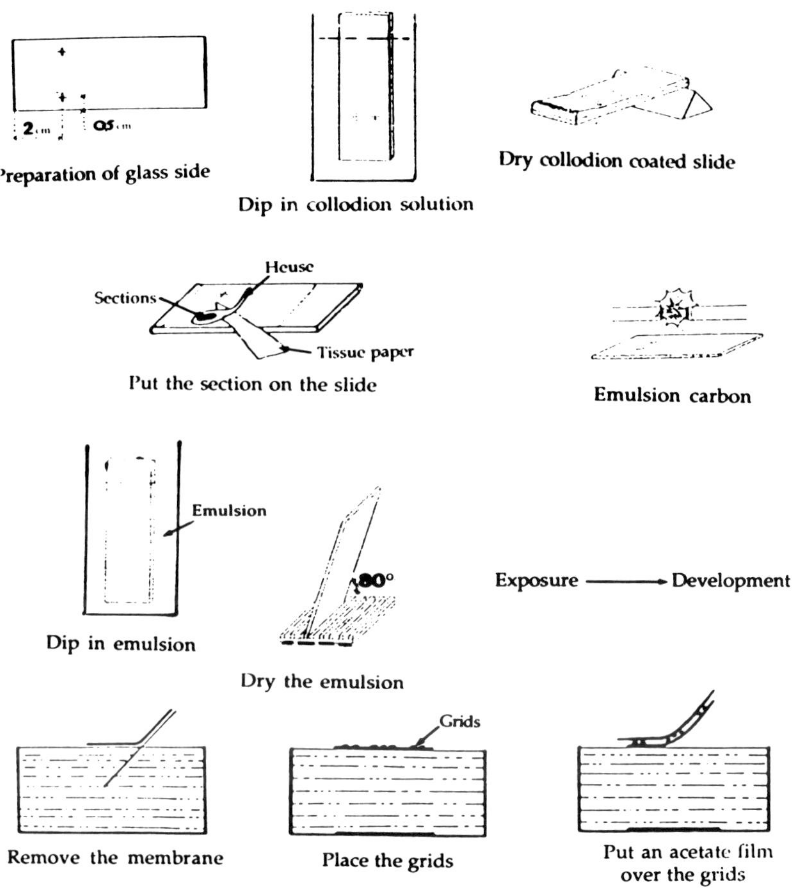

FIGURE 5. Emulsion application by dipping in liquid emulsion for autoradiography (as described in Protocol 20).

   d. Dry at a shallow angle to the vertical
   e. Cover the drying slides with an inverted jar, allowing a crack for air, but avoiding dust
   f. After the slides are thoroughly dry, redip the lower end of each to a depth of about 0.5 cm, then dry them as before

2. Place ultrathin sections, with a Heuse, on the front of the slide over the cross, and dry with a filter paper. Do not move the Heuse during this step, so as not damage the film
3. Counterstain with lead citrate or uranyl acetate and lead citrate
4. Evaporate a 10 to 20 nm layer of carbon
5. Dip the slide in diluted liquid emulsion[a]
7. Air-dry the slide vertically for several hours

8. Put the slides away in a light-tight box, containing a drying agent wrapped in paper tissue
9. Expose at 4°C

[a] As in Protocol 19, it is necessary to check the layer of emulsion at the electron microscopic level, then take the membrane off the slide and put the grid over it at the place where the sections are.

3. Exposure

Grids are stored in a light-tight box at 4°C in the presence of a drying agent (see Figures 4 and 5).

Exposure time is usually long, never less than 1 month. According to the radioisotope used, it can reach 12 months. It is possible to gauge exposure time from semithin preparations (see Chapter 4). It is necessary, in order to avoid this problem, to prepare some grids which will develop at different times, so as to estimate the presence of silver grains over sections.

4. Development

The aim of the development process is to cause the precipitation of enough silver metal on to the latent image for it to be visible by electron microscopy.[34,38] The latent image is composed of a few atoms. This nucleus acts as a site where silver precipitates out from the solution. This results in the structure known as silver grain. The size of the silver grain which is formed depends on the size of the latent image, and on the nature of developer and time of development. Development conditions also need to be closely controlled so as to achieve silver grain formation from as many exposed crystals as possible, while at the same time generating no silver grains from unexposed crystals.

Two standard developers are widely used: D19 and Microdol X, which are highly effective when used to develop Ilford L4 and Amersham EM1 (see Protocol 21). The differences between them are that: (1) D19 develops more latent images than Microdol X,[43] and (2) D19 has a higher pH, and may cause more contamination on the sections than Microdol X.[44] The fine grain developers, with or without gold latensification (see Protocol 22), and also paraphenylenediamine and phenidon developers, have to be made up from their individual chemicals components and do not keep (see Protocol 23). The lower degree of reproducibility of treatment which is possible with these fine grain developers than with D19 or Microdol X is compensated for by a slight increase in the resolving power of the autoradiographic technique due to reduced photographic error.[34] However, it is sometimes difficult to recognize what constitutes a single silver grain, since several small silver granules may be developed from one bromide crystal.[45]

PROTOCOL 21 — DEVELOPMENT PROCESS

1. Dip the slide in diluted developer for 2 min
2. Wash in water
3. Fix in 30% sodium thiosulfate for 1 to 2 min
4. Wash in running water and then in distilled water
5. Stain with uranyl acetate 4% for 5 min
6. Air-dry

PROTOCOL 22 — GOLD LATENSIFICATION[42,46,47]

When using fine grain development, the loss of sensitivity incurred can be offset by prior treatment of the emulsion with gold thiocyanate. This treatment increases sensitivity by a factor of three or four.

1. Prepare 2% $AuCl_4$
2. Mix 1 ml of $AuCl_4$ and 99 ml of distilled water, and adjust the pH to 7 using NaOH
3. Add 250 mg of KCNS and 300 mg KBr, bringing up to 500 ml with sterile distilled water
4. Soak the autoradiographs in this latensifying solution at 20°C for 5 min
5. Rinse briefly in distilled water
6. Develop (see Protocol 23)

PROTOCOL 23 — SPECIAL DEVELOPMENT PROCESS

1. Paraphenylenediamine development[a,48]
    a. Prepare a solution containing sodium sulfite 0.1 *M* and paraphenylenediamine 0.01 *M*. Dissolve the sodium sulfite at 50°C, then add the paraphenylenediamine
    b. Filter and use while still fresh
    c. Develop at 20°C for 1 or 2 min
    d. Rinse briefly in 3% acetic acid and then in distilled water
    e. Fix in 30% sodium thiosulfate for 1 to 2 min
    f. Wash in distilled water
2. Phenidon development[49]

    Dissolve in 200 ml of distilled water, in the following order

Ascorbic acid	6 g
Phenidon(1-ϕ-3 pytazolidone)	1 g
KBr	2.4 g
K_2CO_3	5.2 g
Na_2SO_3	40 g
NaCNS	24 g

a. Make the solution up to 300 ml. At this stage the solution must be clear[b]
b. Filter and use while fresh
c. Develop at 17°C for 1 or 2 min
d. Rinse briefly in 2% acetic acid (stop bath)
e. Fix in 30% sodium thiosulfate for 1 to 2 min
f. Wash in distilled water

[a] The developer is unstable, and the results are not reproducible, unlike those of the standard developers. The relative efficiency is also rather low.[34]

[b] Turbidity in the solution indicates poor dissolution of the reagents, and the procedure must then be carried out again.

5. Observations

The specimens must be recovered from their supports, if the loop technique has been used; grids may be easily removed from the slide, after the final washing; if ultrathin sections have been placed on a membrane-coated slide it is more difficult to put them on grids (see Protocol 24). These grids can then be observed in the same conditions as before.

PROTOCOL 24 — RECOVERY OF SECTIONS FROM MEMBRANE-COATED SLIDE

This protocol to recover the sections differs according to whether the slide was recovered with collodion or formvar.

1. Recovery of section from collodion-coated slide
   a. Scratch with a razor blade round the thick bottom edge of the collodion on the slide
   b. Remove the membrane
   c. Place a 200-mesh grid over each group of sections on the collodion film
   d. Put an acetate film over the grids and take them out
   e. Dry in an oven at 40°C for 1 to 2 h
   f. Detach the grids from the acetate film[a]
2. Recovery of sections from a formvar-coated slide
   a. Scratch around the ribbons of sections, ensuring that each one is completely circumscribed
   b. Introduce a little hydrofluoric acid solution[b] into the scratch
   c. Allow the preparation to stand until the carbon layer begins to lift
   d. Repeat until the layer bearing the sections is floating free
   e. Transfer the floating layer to a watch-glass containing distilled water

f. Rinse by transfer until free of acid
g. Pick the layer up on a 200-mesh grid

[a] If necessary the collodion can be thinned by immersing the grids in isoamyl acetate for 3 min.
[b] 3% of hydrofluoric acid solution (dilute 40 to 48% hydrofluoric acid solution).

B. IMMUNOCYTOLOGICAL DETECTION

1. General Considerations

The revelation of hybrids is obtained by immunocytology after hybridization using a probe labeled with a dUTP-biotin or dUTP-digoxigenin, which can be considered as antigen molecules. Antigen localization is based on specific antibody binding. If an animal is injected repeatedly with a foreign molecule, after a time its serum is found to contain specific antibodies which will form an antigen-antibody complex when brought into contact with antigen. Hybrids containing antigen molecules (biotin or digoxigenin) are revealed by classical immunocytological reactions.

An immunological reaction may be applied in two ways — either the antibody may be conjugated to the label and used to stain the antigen directly (the ''direct'' method), or a primary unconjugated antibody may be applied to the antigen, and the antibody thus bound can then be stained with a labeled secondary antibody against the primary antibody (the ''indirect'' method). The primary antibody can be a molecule with high affinity to the antigen (i.e., streptavidin and biotin). The antibody can bind two antigens with a Fab fragment, while streptavidin can bind four molecules of biotin.[50]

The marker used must be electron-dense. Colloidal gold is the most widely used marker since it permits multiple detections (the diameter of these particles is 5, 10, or 15 nm). It is conjugated to an antibody (primary or secondary antibody), to Protein A (used as the secondary antibody), or to streptavidin. It is possible to use an enzymatic marker (peroxidase or others), but these are less electron-dense[51,52] and the resolution is poor (except for peroxidase revealed by 4-chloro-1-naphthol, as described by Stenberger).[53]

2. Immunocytological Detection Using the Direct Method

After the formation of hybrids, detection is carried out directly using an antibody conjugated with a marker, which can be colloidal gold particles or an enzyme. If the hybrids contained biotin, two methods are available: an antibody against biotin (see Protocol 25) or streptavidin (see Protocol 26). If the antigen was digoxigenin, only antibodies can be used (see Protocol 25). The main advantage of this method is its rapidity.

PROTOCOL 25 — IMMUNOLOGICAL DETECTION WITH LABELED ANTIBODY (FIGURE 6)

1. Prepare the buffer (100 m*M* mono-di-phosphate buffer;[a] 650 m*M* NaCl; 0.5% Ovalbumine; pH 7.4)

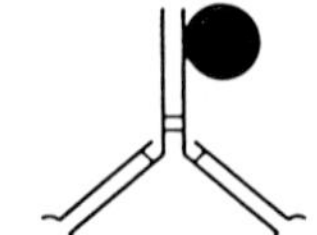

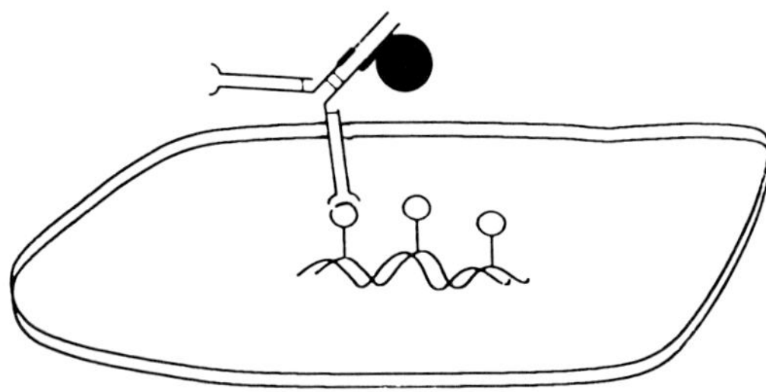

FIGURE 6. Principle of detection of hybrids by a direct immunocytological method, using an anti-probe-label serum conjugated with colloidal gold particles (as described in Protocol 25).

2. Wash the grids on buffer, with gentle agitation, for 10 min
3. Wash the grids on buffer containing 1% Ovalbumine, with gentle agitation, for 10 min
4. Incubate the grids on labeled antibody diluted 1/20 to 1/50 in buffer containing 1% Ovalbumine, with gentle agitation, for 60 min
5. Wash the grids on buffer, with gentle agitation, for 20 min
6. Stain the grids (see Protocols 9 to 11)

PROTOCOL 26 — IMMUNOLOGICAL DETECTION WITH LABELED STREPTAVIDIN (FIGURE 7)

1. Prepare the buffer (as described in Protocol 25)
2. Wash the grids on buffer A, with gentle agitation, for 10 min
3. Wash the grids on buffer containing 1% Ovalbumine, with gentle agitation, for 10 min
4. Incubate the grids on labeled steptavidin diluted 1/20 to 1/30 in buffer containing 1% Ovalbumine, with gentle agitation, for 60 min
5. Wash the grids on buffer, with gentle agitation, for 20 min
6. Stain the grids (see Protocols 9 to 11)

3. Immunocytological Detection Using the Indirect Method (see Protocol 27)

The advantage of this method is its sensitivity: the binding of unlabeled molecules (antibodies or steptavidin) is better than that of labeled molecules. Other detection methods, like the protein A, can be used with all labels

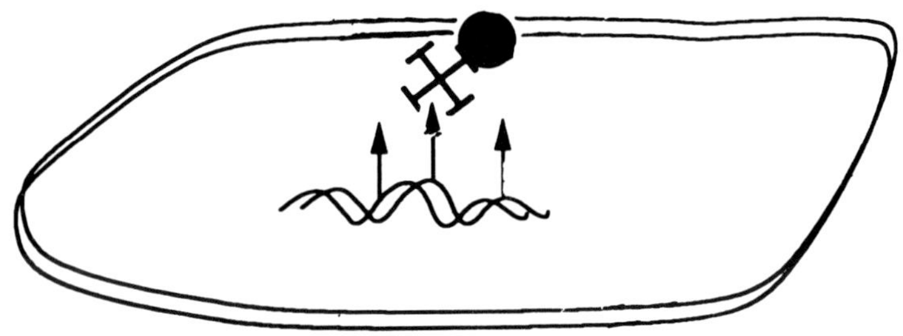

FIGURE 7. Principle of detection of hybrids labeled with biotin by a direct immunocytological method, using streptavidin conjugated with colloidal gold particles (as described in Protocol 26).

(enzymes, colloidal gold particules, other electron-dense particles). Any of the amplification systems (PAP, biotinylated antibodies, labeled avidin-streptavidin complexes) can be used, but the quantification is difficult.

PROTOCOL 27 — IMMUNOLOGICAL DETECTION BY THE INDIRECT METHOD (FIGURE 8)

1. Prepare buffer A (20 m*M* tris-HCl buffer; 650 m*M* NaCl; 0.05% Tween-20; 0.5% Ovalbumine; pH 7.6) and the buffer B (as described in Protocol 25)
2. Wash the grids on buffer A, with gentle agitation, for 10 min
3. Wash the grids on buffer A containing 1% Ovalbumine, with gentle agitation, for 10 min
4. Incubate the grids on unlabeled antibody diluted 1/50 to 1/200 in buffer A containing 1% Ovalbumine, with gentle agitation, for 60 min
5. Wash the grids on buffer A, with gentle agitation, for 10 min
6. Wash the grids on buffer B, with gentle agitation, for 10 min
7. Incubate the grids on labeled antibody diluted 1/30 to 1/50 in buffer B containing 1% ovalbumin, with gentle agitation, for 30 min
8. Wash the grids on buffer B, with gentle agitation, for 10 min
9. Stain the grids (see Protocols 9 to 11)

4. Simultaneous Detection of Two Nucleic Acids

The possibility of labeling probes with two different antigens permits their revelation with specific antibodies (direct method) (see Protocol 28), or

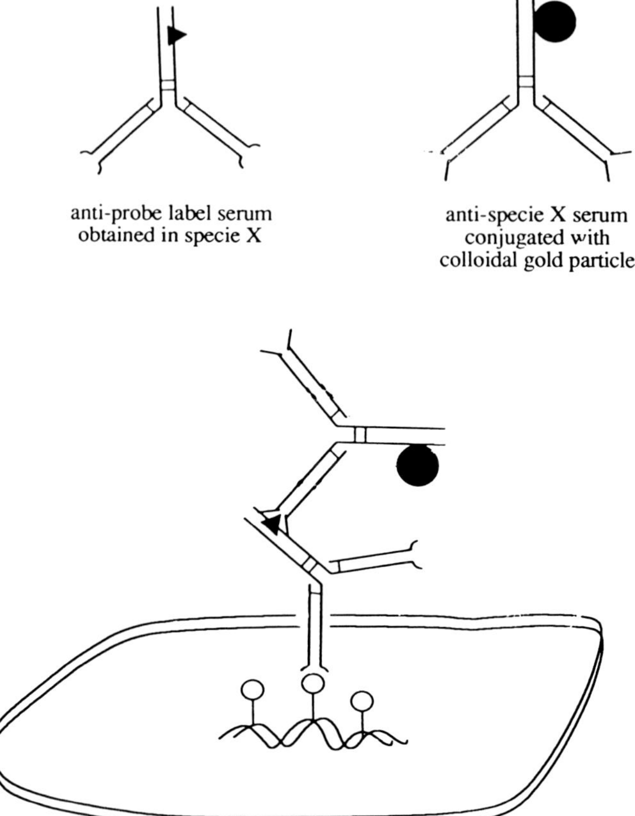

FIGURE 8. Principle of indirect immunocytological detection of hybrids labeled with biotin or digoxigenin using an anti-probe-label serum revealed by an anti-species serum conjugated with colloidal gold particles (as described in Protocol 27).

by a combination of direct and indirect methods (see Protocol 29), or by indirect methods alone (see Protocol 30). This is made possible by combining different types of marker, such as enzyme and colloidal gold particles, or a model using two different sizes of colloidal gold particles, which is more common. The size of these colloidal gold particles must be checked for their Gauss distribution, and the distinction between large and small particles must be clear. It has been reported that small particles increase the level of staining.[54] It was also necessary to check for double labeling. The use of 1 nm colloidal gold particles, which improves the signal detection, must be accompanied by silver enhancement (see Protocol 31).

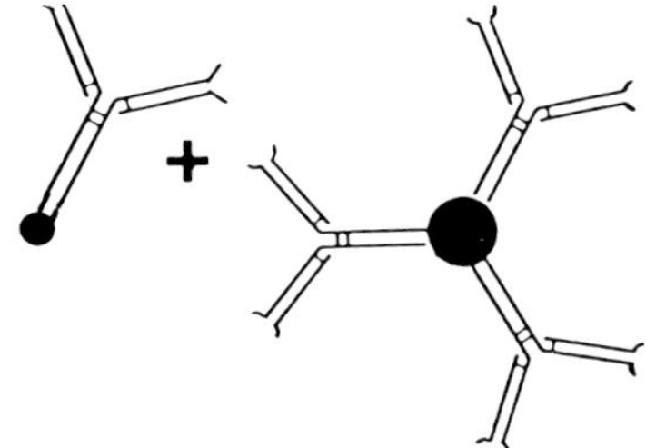

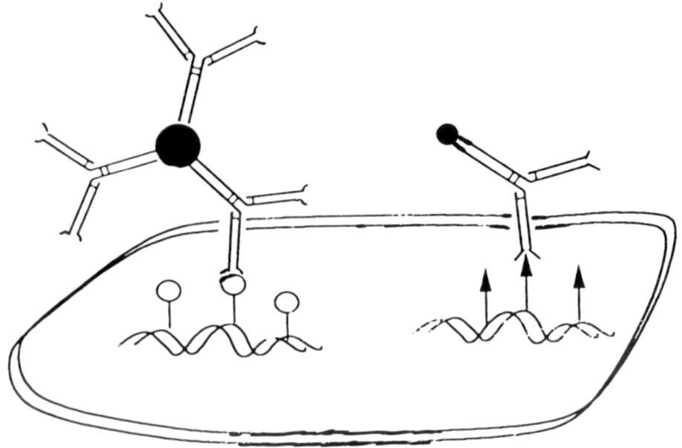

FIGURE 9. Principle of simultaneous direct immunocytological detection of two distinct hybrids labeled with biotin or digoxigenin using an anti-biotin serum and an anti-digoxigenin serum conjugated with two sizes of colloidal gold particles (as described in Protocol 28).

PROTOCOL 28 — REVELATION OF TWO HYBRIDS LABELED WITH BIOTIN AND DIGOXIGENIN, RESPECTIVELY, BY DIRECT IMMUNOLOGICAL REACTIONS (FIGURE 9)

The following reagents are used: streptavidin labeled with 5 and 15-nm colloidal gold particles for biotin detection and anti-digoxigenin labeled with 15 and 5-nm colloidal gold particles for digoxigenin detection.

The 5 and 15-nm colloidal gold particles were used with streptavidin and antidigoxigenin, respectively, and then inversely so as to avoid the problem of colloidal gold particle size.

1. Prepare the buffer (as described in Protocol 25)
2. Wash the grids on buffer, with gentle agitation, for 10 min
3. Wash the grids on buffer containing 1% Ovalbumine, with gentle agitation, for 10 min

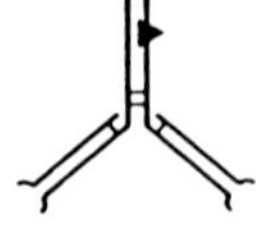

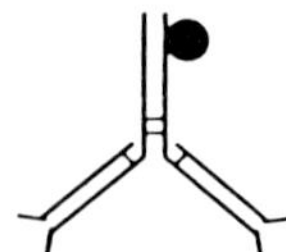

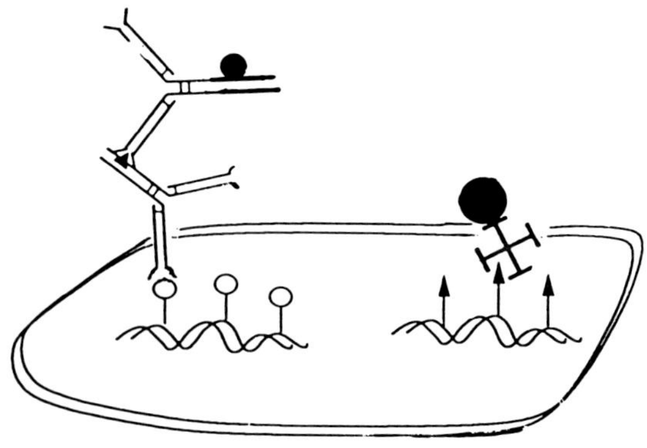

FIGURE 10. Principle of simultaneous detection of two distinct hybrids: (1) a hybrid labeled with biotin, by direct immunocytological reaction, using streptavidin conjugated with the first size of colloidal gold particles, and (2) a hybrid labeled with digoxigenin, by indirect immunocytological reaction, using an anti-digoxigenin serum and then a second antibody against the species of anti-digoxigenin, conjugated with the second size of colloidal gold particles (as described in Protocol 29).

4. Incubate the grids on a mixture of labeled steptavidin diluted 1/30 and labeled anti-digoxigenin diluted 1/20 in buffer containing 1% Ovalbumine, with gentle agitation, for 60 min
5. Wash the grids on buffer, with gentle agitation, for 20 min
6. Stain the grids (see Protocols 9 to 11)

PROTOCOL 29 — SIMULTANEOUS REVELATION OF TWO HYBRIDS LABELED WITH BIOTIN AND DIGOXIGENIN, RESPECTIVELY, BY DIRECT AND INDIRECT IMMUNOLOGICAL REACTIONS (FIGURE 10)

The following labeled reagents can be used:

1. Streptavidin labeled with 5- and 15-nm colloidal gold particles for biotin detection, and unlabeled mouse anti-digoxigenin and sheep anti-mouse labeled with 15 and 5-nm colloidal gold particles for digoxigenin detection

2. Unlabeled rabbit anti-biotin and goat anti-rabbit labeled with 15- and 5-nm colloidal gold particles for biotin detection, and mouse anti-digoxigenin labeled with 5- and 15-nm colloidal gold particles for digoxigenin detection.

The 5- and 15-nm colloidal gold particles were used with streptavidin and anti-digoxigenin, respectively, and then inversely so as to avoid the problem of colloidal gold particle size.

1. Prepare the buffers A (as described in Protocol 27) and B (as described in Protocol 25)
2. Wash the grids on buffer A, with gentle agitation, for 10 min
3. Wash the grids on buffer A containing 1% Ovalbumine, with gentle agitation, for 10 min
4. Incubate the grids on unlabeled serum (anti-digoxigenin [see above 1] or anti-biotin [see above 2]) diluted 1/50 to 1/200 in buffer A containing 1% of Ovalbumine
5. Wash the grids on buffer A, with gentle agitation, for 20 min
6. Wash the grids on buffer B, with gentle agitation, for 10 min
7. Incubate the grids on a mixture of labeled reagents (steptavidin diluted 1/30 and anti-mouse diluted 1/50 [see above 1], or anti-rabbit diluted 1/50 and anti-digoxigenin diluted 1/20 [see above 2] in buffer B containing 1% Ovalbumine, with gentle agitation, for 30 min
8. Wash the grids on buffer B, with gentle agitation, for 10 to 20 min
9. Stain the grids (see Protocols 9 to 11)

PROTOCOL 30 — REVELATION OF TWO HYBRIDS LABELED WITH BIOTIN AND DIGOXIGENIN, RESPECTIVELY, BY INDIRECT IMMUNOLOGICAL REACTIONS (FIGURE 11)

The following labeled reagents can be used:

1. Unlabeled streptavidin, and biotin labeled with 5- and 15-nm colloidal gold particles, for biotin detection, and unlabeled mouse anti-digoxigenin, and sheep anti-mouse labeled with 15- and 5-nm colloidal gold particles, for digoxigenin detection
2. Unlabeled rabbit anti-biotin, and anti-rabbit labeled with 15- and 5-nm colloidal gold particles, for biotin detection, and unlabeled mouse anti-digoxigenin, and anti-mouse labeled with 5- and 15-nm colloidal gold particles, for digoxigenin detection

The 5- and 15-nm colloidal gold particles were used with streptavidin and anti-digoxigenin, respectively, and then inversely so as to avoid the problem of colloidal gold particle size.

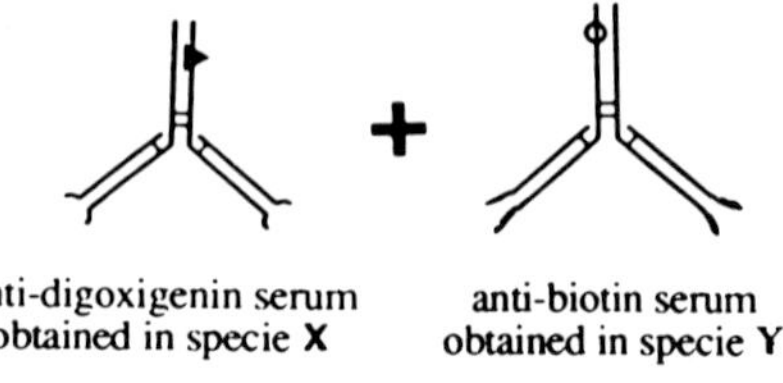

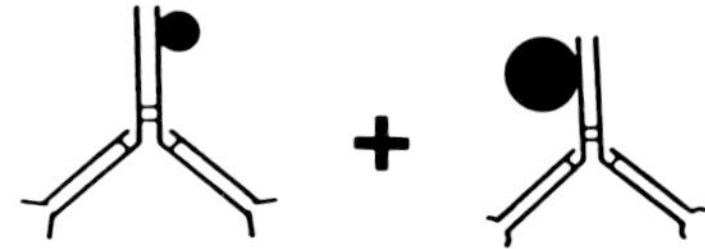

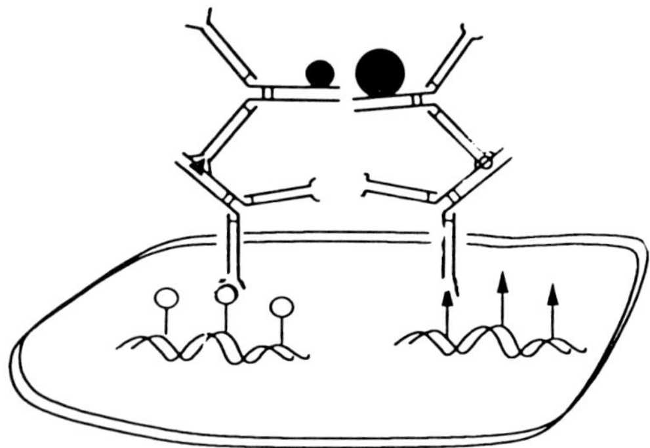

FIGURE 11. Principle of simultaneous detection of two distinct hybrids labeled with biotin and digoxigenin, respectively, using two indirect immunocytological reactions. In the first step an anti-biotin and an anti-digoxigenin serum are incubated, and then in the second step, incubation of sera against the species of the anti-probe-label sera conjugated with two different sizes of colloidal gold particles (as described in Protocol 30).

1. Prepare the buffers A (as described in Protocol 27) and B (as described in Protocol 25)
2. Wash the grids on buffer A, with gentle agitation, for 10 min
3. Wash the grids on buffer A containing 1% Ovalbumine, with gentle agitation, for 10 min
4. Incubate the grids on unlabeled serum (streptavidin and anti-digoxigenin [see above 1] or antibiotin and anti-digoxigenin [see above 2]) diluted 1/50 to 1/200 in buffer A containing 1% of Ovalbumine
5. Wash the grids on buffer A, with gentle agitation, for 20 min
6. Wash the grids on buffer B, with gentle agitation, for 10 min
7. Incubate the grids on a mixture of labeled reagents (biotin diluted 1/40 and anti-mouse diluted 1/50 [see above 1], or anti-rabbit diluted and

anti-mouse diluted 1/50 [see above 2]) in buffer B containing 1% Ovalbumine with gentle agitation, for 30 min

8. Wash the grids on buffer B, with gentle agitation, for 10 to 20 min
9. Stain the grids (see Protocols 9 to 11)

PROTOCOL 31 — AMPLIFICATION OF 1-NM COLLOIDAL GOLD PARTICLES BY SILVER ENHANCEMENT

1. Prepare a silver enhancement solution just before use
2. Wash in distilled water
3. Apply one drop of the solution to the section for 3 min[a]
4. Wash in distilled water
5. Stain the grids (see Protocols 9 to 11)

[a] The time, which is important, depends on temperature and agitation, and must be checked.

5. Simultaneous Detection of Nucleic Acid and Antigen

A combination of the detection of a nucleic acid by *in situ* hybridization and of an antigen by immunocytology is used to characterize the product synthesized by the nucleic acid, or the cell which contains this nucleic acid. This technique is similar to that of the detection of two nucleic acids by immunocytology, since the antigen can be detected by antibodies (rabbit or mouse antibodies mainly). If the hybrids are labeled with biotin, there are two possibilities: (1) streptavidin to detect biotin, or (2) an anti-biotin serum obtained from rabbit or mouse; if the hybrids are labeled with digoxigenin they can be revealed by antibodies (obtained from mouse or goat). The possibilities for simultaneous detection are the following:

- If the hybrids are labeled with biotin, the antigen can be detected by any antibody (see Protocol 32)
- If the hybrids are labeled with digoxigenin, the antigen can be detected by an antibody obtained in other species than the anti-digoxigenin serum (see Protocol 33)
- If the hybrids are labeled with biotin, and can be detected only by a rabbit anti-biotin serum, the antigen can be detected only by an antibody obtained in other species than rabbit (see Protocol 34)
- If the hybrids are labeled with digoxigenin, and can be detected only by a mouse anti-digoxigenin serum, the antigen can be detected only by an antibody obtained in other species than mouse (see Protocol 35)

PROTOCOL 32 — SIMULTANEOUS DETECTION OF NUCLEIC ACID LABELED WITH BIOTIN AND AN ANTIGEN (FIGURE 12)

The following labeled reagents are used: streptavidin labeled with 5- and 15-nm colloidal gold particles for biotin detection and anti-species labeled with 15- and 5-nm colloidal gold particles for antigen detection.

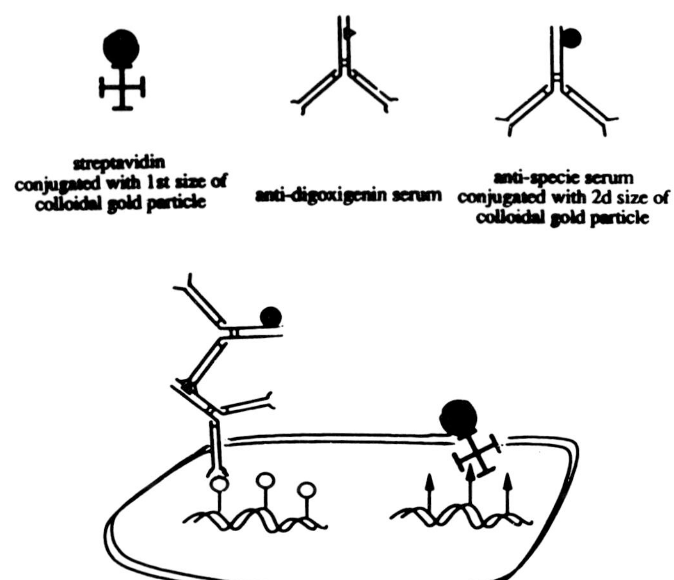

FIGURE 12. Principle of simultaneous detection of hybrid labeled with biotin and antigen using a direct immunocytological method (streptavidin conjugated with colloidal gold particles) for biotin revelation, and an indirect immunocytological reaction for antigen revelation (as described in Protocol 32).

The 5- and 15-nm colloidal gold particles were used with streptavidin and anti-antigen, respectively, and then inversely so as to avoid the problem of colloidal gold particle size.

1. Prepare the buffers A (as described in Protocol 27) and B (as described in Protocol 25)
2. Wash the grids on buffer A, with gentle agitation, for 10 min
3. Wash the grids on buffer A containing 1% Ovalbumine, with gentle agitation, for 10 min
4. Incubate the grids on unlabeled serum anti-antigen diluted 1/50 to 1/200 in buffer A containing 1% of Ovalbumine
5. Wash the grids on buffer A, with gentle agitation, for 20 min
6. Wash the grids on buffer B, with gentle agitation, for 10 min
7. Incubate the grids on a mixture of labeled reagents streptavidin diluted 1/50, and anti-species of serum anti-antigen diluted 1/50 in buffer B containing 1% Ovalbumine, with gentle agitation, for 30 min
8. Wash the grids on buffer B, with gentle agitation, for 10 to 20 min
9. Stain the grids (see Protocols 9 to 11)

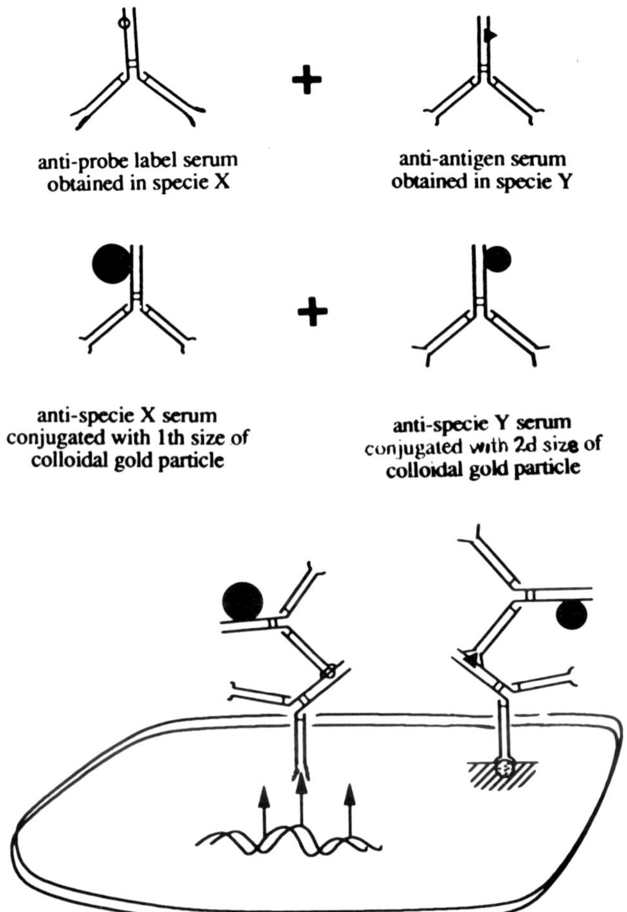

FIGURE 13. Principle of simultaneous detection of hybrids and antigen using two indirect immunocytological reactions. In the first step an anti-probe-label and anti-antigen sera are incubated, and then in the second step, incubation of sera against the species of the previous sera conjugated with two different sizes of colloidal gold particles (as described in Protocol 33).

PROTOCOL 33 — SIMULTANEOUS DETECTION OF NUCLEIC ACID AND ANTIGEN BY INDIRECT IMMUNOLOGICAL REACTIONS (FIGURE 13)

The following reagents can be used:

1. For hybrid detection: unlabeled mouse or goat anti-probe-label, and sheep anti-mouse labeled with 5- and 15-nm colloidal gold particles, for digoxigenin detection

2. For antigen detection: unlabeled rabbit (or species other than mouse or goat) anti-antigen, and anti-rabbit labeled with 15- and 5-nm colloidal gold particles

The 5- and 15-nm colloidal gold particles were used for hybrid and antigen binding, respectively, and then inversely so as to avoid the problem of colloidal gold particle size.

1. Prepare buffers A (as described in Protocol 27) and B (as described in Protocol 25)
2. Wash the grids on buffer A, with gentle agitation, for 10 min
3. Wash the grids on buffer A containing 1% Ovalbumine, with gentle agitation, for 10 min
4. Incubate the grids on first unlabeled serum diluted 1/50 to 1/200 in buffer A containing 1% of Ovalbumine
5. Wash the grids on buffer A, with gentle agitation, for 20 min
6. Wash the grids on buffer B, with gentle agitation, for 10 min
7. Incubate the grids on a mixture of labeled reagents anti-mouse, and anti-rabbit diluted 1/50 in buffer B containing 1% Ovalbumine, with gentle agitation, for 30 min
8. Wash the grids on buffer B for 10 to 20 min with gentle agitation
9. Stain the grids (see Protocols 9 to 11)

PROTOCOL 34 — SIMULTANEOUS DETECTION OF NUCLEIC ACID AND ANTIGEN USING PROTEIN A[a,55,56]

The following reagents can be used:

- Unlabeled rabbit anti-biotin diluted 1/200
- Unlabeled rabbit anti-antigen diluted 1/200
- Unlabeled protein (100 μg/ml)
- Protein A labeled with 5- and 15-nm colloidal gold particles, diluted 1/40
- PBS (mono-di-phosphate buffer 10 m*M*, NaCl 150 m*M*, pH 7.0)
- Bovin serum albumin (BSA) 5% in PBS
- PBS-Gly (PBS added of 10 m*M* glycine)[b]
- Glutaraldehyde 1% in PBS

1. Wash the grids on PBS, with gentle agitation, for 4 × 2 min
2. Incubate the grids on 5% BSA, with gentle agitation, for 10 min
3. Wash the grids on PBS-Gly, with gentle agitation, for 2 min
4. Incubate the grids on unlabeled rabbit anti-biotin for 60 min
5. Wash the grids on PBS-Gly, with gentle agitation, for 6 × 1 min
6. Incubate the grids on protein A labeled with 5-nm gold particles, with gentle agitation, for 30 min
7. Wash the grids on PBS-Gly, with gentle agitation, for 6 × 1 min

8. Incubate the grids on unlabeled protein A, with gentle agitation, for 5 min
9. Wash the grids on PBS-Gly, with gentle agitation, for 6 × 1 min
10. Incubate the grids on unlabeled rabbit anti-antigen for 60 min
11. Wash the grids on PBS-Gly, with gentle agitation, for 6 × 1 min
12. Incubate the grids on protein A labeled with 15-nm gold particles, with gentle agitation, for 30 min
13. Wash the grids on PBS-Gly, with gentle agitation, for 2 × 1 min
14. Wash the grids on PBS, with gentle agitation, for 4 × 1 min
15. Fix in 1% glutaraldehyde in PBS for 5 min
16. Wash the grids on PBS, with gentle agitation, for 2 × 1 min
17. Wash the grids on water, with gentle agitation, for 4 × 1 min
18. Stain the grids (see Protocols 9 to 11)

[a] It can be modified to the method described by Thorpe.[57]
[b] Glycine is used to neutralize the aldehyde function

PROTOCOL 35 — SIMULTANEOUS DETECTION OF NUCLEIC ACID AND ANTIGEN USING PROTEIN G

The following reagents can be used:

- Unlabeled mouse anti-biotin diluted 1/200
- Unlabeled mouse anti-antigen diluted 1/200
- Unlabeled protein G (100 μg/ml)
- Protein G labeled with 5 and 15-nm colloidal gold particles, diluted 1/40
- PBS (mono-di-phosphate buffer 100 m*M*, NaCl 500 m*M*, pH 7.4)
- BSA 5% in PBS

1. Wash the grids on PBS, with gentle agitation, for 4 × 2 min
2. Incubate the grids on 5% BSA, with gentle agitation, for 10 min
3. Wash the grids on PBS with gentle agitation for 2 min
4. Incubate the grids on unlabeled mouse anti-biotin for 60 min
5. Wash the grids on PBS, with gentle agitation, for 6 × 1 min
6. Incubate the grids on protein G labeled with 5-nm gold particles, with gentle agitation, for 30 min
7. Wash the grids on PBS, with gentle agitation, for 6 × 1 min
8. Incubate the grids on unlabeled protein G, with gentle agitation, for 5 min
9. Wash the grids on PBS, with gentle agitation, for 6 × 1 min
10. Incubate the grids on unlabeled mouse anti-antigen for 60 min
11. Wash the grids on PBS, with gentle agitation, for 6 × 1 min
12. Incubate the grids on protein G labeled with 15-nm gold particles, with gentle agitation, for 30 min
13. Wash the grids on PBS, with gentle agitation, for 5 × 1 min

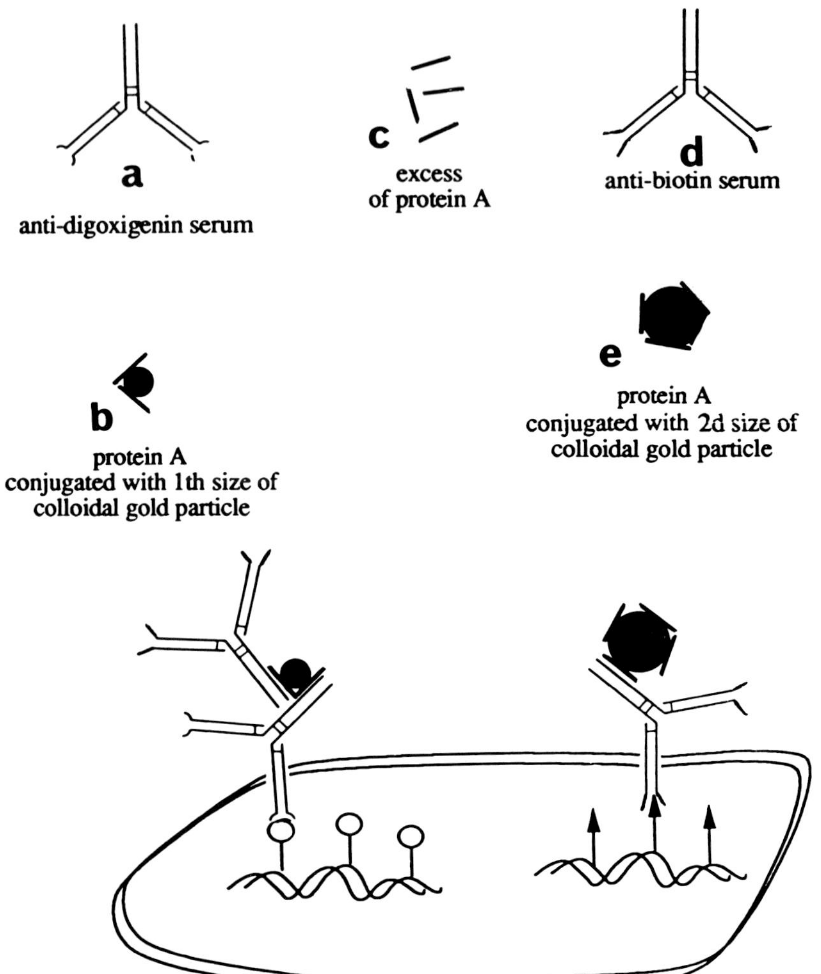

FIGURE 14. Principle of simultaneous detection of two distinct hybrids labeled with biotin and digoxigenin, respectively, using antibodies from the same species, by immunocytological reaction with protein A. To begin with, the incubation is carried out using an anti-digoxigenin serum (a), then protein A conjugated with the small size of colloidal gold particles (b). The binding sites of protein A are saturated by incubation with an excess of unlabeled protein A (c). Then the second hybrid is detected by incubation with an anti-biotin serum (d), then with protein A conjugated with the large size of colloidal gold particles (e) (as described in Protocol 34).

14. Fix in 1% glutaraldehyde in PBS for 5 min
15. Wash the grids on PBS, with gentle agitation, for 2 × 1 min
16. Wash the grids on water, with gentle agitation, for 4 × 1 min
17. Stain the grids (see Protocols 9 to 11)

The Protocols 34 (see Figure 14) and 35 can be used for simultaneous detection of two hybrids labeled, respectively, using biotin and digoxigenin.

To check the specificity of the immunocytological reaction (described in Protocols 34 and 35) the antigen detection must be carried out before the hybrid detection both with protein A and with protein G.

6. Observations

These observations are obtained more quickly and with better resolution, using this form of detection than after autoradiographic revelation. Sensitivity is lower, but can be increased by the use of small colloidal gold particles.[58]

C. CHECKING OF REVELATION PROCEDURE

The method of revelation must not introduce a signal which may mask the hybridization signal.

1. Autoradiography

After autoradiography there is always background noise, for a variety of different reasons.[34,39] It is determined by taking a series of randomized micrographs of the support film around the tissue section. If the background grain frequency/unit area exceeds 5% of grain frequency over labeled cells, it is best to abandon the specimen.

2. Immunocytology

Hybridization is carried out without a probe in a hybridization buffer and under identical conditions to those used with a labeled probe, or else the revelation procedure alone is carried out. In the latter case a nonspecific signal is more abundant than in the former. Constituents of hybridization buffer (such as formamide or Denhardt solution) limit nonspecific adsorption.

A positive signal can be due to:

- The presence of nonspecific adsorption of a labeled probe in tissue
- The presence in tissue of a probe label (endogenous biotin or digoxigenin)
- Endogenous enzymatic activity in the tissue or cells studied, if the marker used is an enzyme (for details of inhibition of endogenous enzymatic activity see Chapter 4)

The presence of nonspecific adsorption of a labeled probe in the tissue can be checked by examining heterologous tissue, where only the cells which are known to express the target nucleic acid will give a positive result. The disappearance of this nonspecific adsorption is obtained by increasing the period and the stringency of the washing steps. Increasing the washing temperature is difficult with ultrathin frozen sections, if ultrastructures are to be preserved.

Endogenous biotin activity is inhibited by competition with streptavidin. Sections are preincubated with free streptavidin, then saturated by addition of biotin.[59] Each streptavidin molecule possesses four binding sites for biotin, while each biotin molecule is capable of binding to only one streptavidin molecule.[60,61]

However, if biotin or digoxigenin is present in the tissue or cells studied, it is better to label the probe with the antigen which is absent than to inhibit the endogenous activity of the one which is present. A hybridization procedure can reveal an inhibited endogenous biotin.

VII. CHECKING PROCEDURE

To validate an *in situ* hybridization reaction, it is essential to check all the elements of the procedure which might introduce false positivities into the reaction signal. In order to eliminate background and false positives, to increase the ratio signal/nonspecific signal, and to demonstrate the specificity of the reaction, a number of checks are required.

Background is due to nonspecific adsorption of the probe and the revelation procedure.

False positives can be caused by the probe, the hybridization reaction, and the revelation procedure.

Specificity is due to the formation of specific hybrids. The signal/nonspecific hybridization ratio must be maximized by decreasing the background and increasing the formation of specific hybrids. Several points must be checked to confirm the specific localization of the nucleic acid studied.

- The probe (see part: preparation of the probe)
- The hybridization reaction (see part: checking of hybridization procedure)
- The revelation procedure (see part: checking of revelation procedure)

The simultaneous detection of mRNA and the protein synthesized by the nucleic acid, using a combination of hybridization and immunocytology, and the correlation of this method with other techniques (e.g., blot techniques, amplified or not by PCR), constitute strong evidence for the presence of nucleic acid in the tissue, but not for subcellular localization of nucleic acid. Variations of this localization after treatment may argue in favor of the specificity of this localization.

VIII. RESULTS

A. AUTORADIOGRAPHIC LOCALIZATION (FIGURE 15)

Hybridization of ultrathin frozen sections of rat anterior pituitary gland with a rat growth hormone (GH) cDNA, labeled by nick translation (see Protocol 12) with ^{35}S dCTP, was carried out for 3 h at 40°C. After washing, the hybrids were detected by autoradiography. Silver grains were present over cell populations with the cytological characteristics of somatotrophs (secretory granules of 350 nm in diameter and endoplasmic reticulum in parallel cisternae)[62] and absent over gonadotrophs, corticotrophs, and thyrotrophs (Figure 15).[9] This was the first check of the localization of this mRNA.

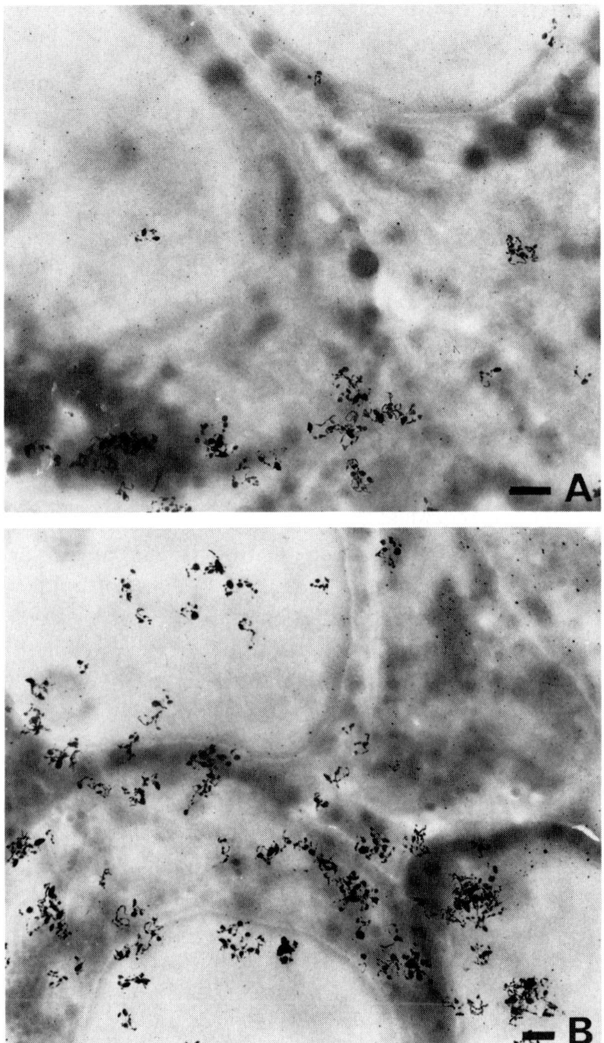

FIGURE 15. Ultrastructural localization of growth hormone mRNA in rat anterior pituitary gland using a cDNA labeled by nick translation with ^{35}S-dCTP on ultrathin frozen sections. Silver grains were observed in cells showing characteristic features of somatotroph: secretory granules around 350 nm in diameter and abundant parallel endoplasmic reticulum cysternae. Two populations of somatotrophs were distinguished: (A) shows cells from the population with silver grains localized in the cytoplasmic matrix containing few secretory granules, while (B) shows somatotrophs from the other population with silver grains both in the nuclei and in the cytoplasmic matrix containing a large number of secretory granules. Silver grains were absent from gonadotrophs (LH-FSH). Bar = 1 μm.[9]

The second was a physiological check. The synthesis of growth hormone is dependent on a neuropeptide, growth hormone-releasing factor, which increases the release of growth hormone,[1,9,63,64] whereas somatostatin did not change the transcription rate of the growth hormone gene,[64] but decreased the growth hormone release.[65] After injection of either peptide, localization of growth hormone mRNA was restricted to somatotrophs, and was thus similar to mock-treated animals. As in light *in situ* hybridization experiments, an increase in labeling with the GH cDNA probe was observed at the ultrastructural level after growth hormone-releasing factor injection, as compared to mock-treated animals. However, this increase was lower than in light *in situ* hybridization experiments, but the distribution of silver grains was different (see below). After a somatostatin injection, no modification of signal was observed, as compared to mock-treated animals, but the distribution was modified (see below).

The third check was another probe containing the first intron of the GH gene, which is specific to the precursor of growth hormone mRNA and the spliced intron. This was a cDNA probe, labeled by nick translation with ^{35}S-dCTP, like the growth hormone probe, and hybridized in similar conditions. The signal was mainly localized in the nuclei of somatotrophs.[9] Some silver grains were present in the cytoplasmic matrix.

As for the autoradiographic studies, it was necessary to quantify the signal (see below) to determine if silver grains were randomly distributed.

B. IMMUNOCYTOLOGICAL LOCALIZATION (FIGURES 16 AND 17)

Detection of the same growth hormone mRNA using an oligonucleotide probe labeled with a nonradioactive nucleotide showed the same cellular and subcellular localizations (Figure 16).[9,66] The probe was a 30-mer oligonucleotide complementary to the codons of the mRNA from Gln 45 to Ser 54.[67] The sequence of this oligonucleotide is given in Figure 3. The oligonucleotide was labeled by tailing the 3′ end using biotin 16-dUTP as described in Protocol 14. The ultrathin frozen sections were incubated for 3 h at 37°C in a hybridization buffer containing 4 p*M* of biotinylated probe per ml. Revelation of hybrids was carried out as described in Protocol 26, using streptavidin conjugated with colloidal gold particles, and as described in Protocol 27 using anti-biotin serum and then anti-rabbit immunoglobulin conjugated with colloidal gold particles. The size of the colloidal gold particles was important, since the signal decreases as their size increases.[68] With 5-nm gold particles, labeling density was 3 times greater than with 15-nm gold particles.[66]

Gold particles indicating growth hormone mRNA hybrids were mainly localized in anterior pituitary gland cells containing dense, spherical secretory granules of about 350 nm in diameter and in lamellar endoplasmic reticulum, i.e., the somatotrophs. As previously noted, in gonadotrophs, corticotrophs, and thyrotrophs, only gold particles corresponding to background were observed (1 to 3 gold particles per cell) (first check).

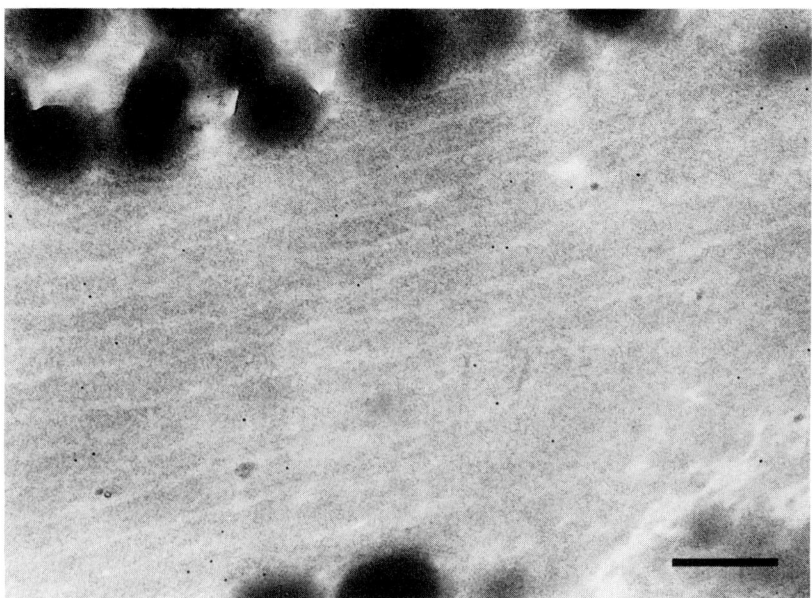

FIGURE 16. Ultrastructural localization of growth hormone mRNA in rat anterior pituitary gland using a 30-mer oligonucleotide labeled by tailing a 3′ end with biotin 16-dUTP on ultrathin frozen sections. Biotin was revealed using an indirect immunocytological reaction (first: rabbit anti-biotin serum, second: anti-rabbit serum conjugated with 5-nm colloidal gold particles). Gold particles were present in the cytoplasm and nucleus of somatotrophs characterized by their ultrastructural cytology. In the cytoplasm they were present in the matrix and in the lamellar endoplasmic reticulum area. The sections were stained with uranyl acetate and embedded in methylcellulose. Bar = 0.5 μm.[66]

No signal was detected where there was an excess of nonlabeled growth hormone oligonucleotide probe concomitantly with the labeled probe (second check), or with the hybridization buffer, without a probe, followed by any of the revelation processes (third check). When an excess of nonlabeled prolactin oligoprobe was added to the hybridization buffer, labeling was observed in the somatotrophs (fourth check).

Using the probe characterized in Figure 3, which encodes for prolactin, the *in situ* hybridization signal was localized in exactly the same conditions in a cell population characterized by a lamellar endoplasmic reticulum, identical to somatotrophs, and polymorphous secretory granules of 600 to 700 nm in diameter, i.e., the lactotrophs (Figure 17).

In order to characterize the presence of a nucleic acid in one particular cell type of heterogenous tissue, such as the pituitary gland, it is necessary to identify the hormone produced and stored by the cell simultaneously with the nucleic acid. The atrial natriuretic factor (ANF) was uptaken in the anterior pituitary gland by three cell populations: the gonadotrophs, the corticotrophs, and the lactotrophs.[69] The presence of an ANF-receptor is suggested by this

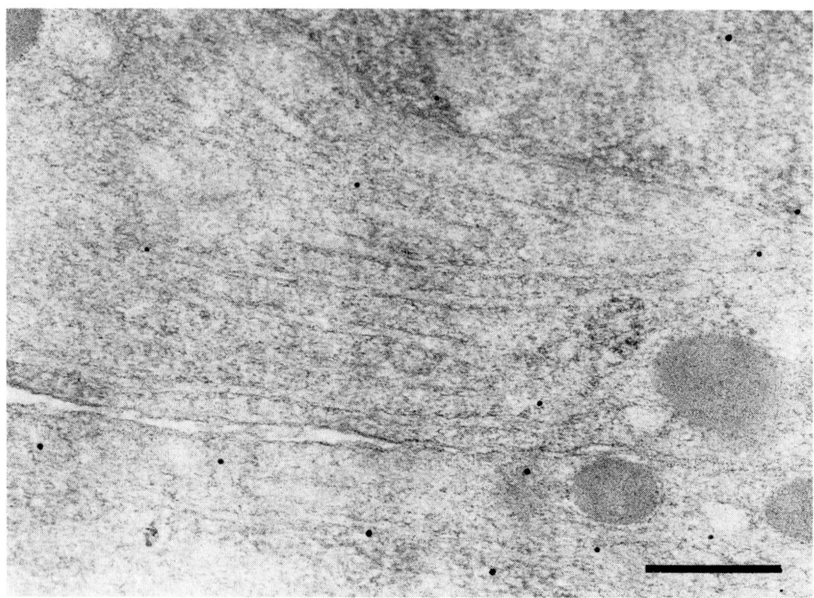

FIGURE 17. Ultrastructural localization of prolactin mRNA in rat anterior pituitary gland using a 30-mer oligonucleotide labeled by tailing 3′ end with digoxigenin-16-dUTP on ultrathin frozen sections. Digoxigenin was revealed using a direct immunocytological reaction (monoclonal antibody against digoxigenin conjugated with 10-nm colloidal gold particles) (as described in Protocol 25). Gold particles were detected in lactotrophs characterized by their cytological aspect: abundant parallel endoplasmic reticulum cysternae and polymorphic secretory granules about 600 to 700 nm in diameter. Gold particles were present mainly in the cytoplasm (endoplasmic reticulum and cytoplasmic matrix) and in the nucleus. The sections were stained with uranyl acetate and embedded in epoxy resin. Bar = 0.5 μm.

uptake of ANF. Three ANF-receptors have been cloned: ANF-receptor A, ANF-receptor B, and ANF-receptor C. A combination of an immunocytological technique, to detect the hormone and thus characterize the cell population, and an *in situ* hybridization technique, to show the presence of ANF-receptor, remains the only possibility of demonstrating the presence of an ANF-receptor type within a particular cell type, as shown in Figure 18.

C. COMPARISON OF TWO METHODS: RADIOACTIVE VS. NONRADIOACTIVE

The choice of label (radioactive or nonradioactive) does not modify the result of the hybridization technique, unlike the choice of probe (cDNA or oligonucleotide). As previously shown growth hormone mRNA was detected in the same anterior pituitary cell type, the somatotrophs.

1. Advantages of Radioactive Hybridization

This technique has four advantages: sensitivity, absence of endogenous activity, efficiency, and standardized method of quantification.[33,38,39]

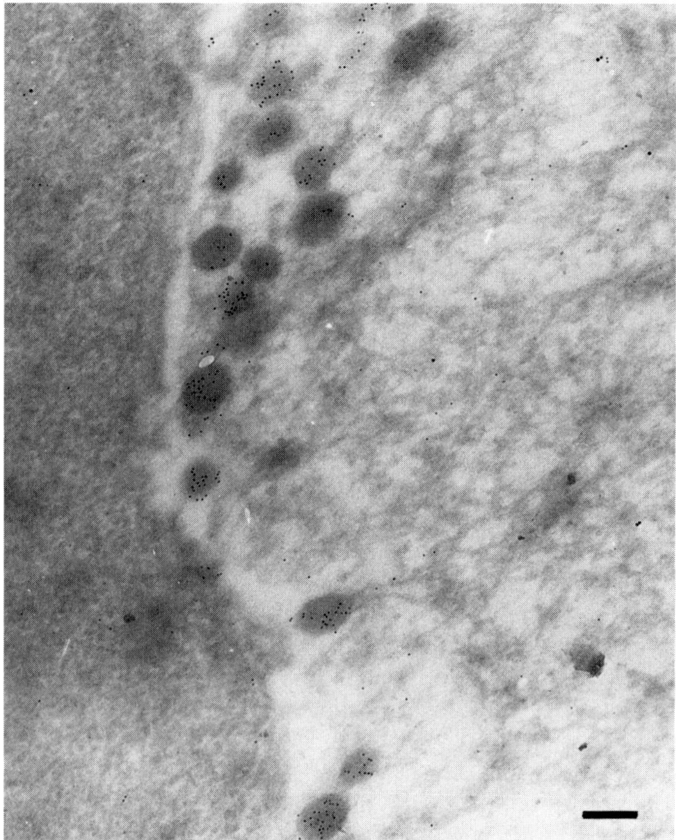

FIGURE 18. Simultaneous ultrastructural localization of atrial natriuretic factor-receptor (ANF-R) mRNA and prolactin (PRL) in rat anterior pituitary gland. Probe was labeled with digoxigenin. Hybrids (ANF-R) and antigen (PRL) were revealed, respectively, using direct immunocytological method (anti-digoxigenin serum conjugated with 15-nm colloidal gold particles) for hybrids detection, and an indirect immunocytological reaction for antigen revelation (rabbit anti-PRL serum, then anti-rabbit serum conjugated with 5-nm colloidal gold particles). The 15-nm gold particles, which represent ANF-R hybrids, were localized in cytoplasmic matrix, while 5-nm gold particles, which represent the PRL-like immunoreactivity, were present on secretory granules. This cell, which stores PRL in secretory granules, is a lactotroph. The ANF-R were expressed by lactotrophs. The sections were stained with uranyl acetate and embedded in methylcellulose. Bar = 0.5 μm.

The sensitivity of *in situ* hybridization with an isotopic labeled probe is the main advantage of the method at the electron microscopic level, as it is, at the light microscopic level. The specific activity of the probe, which is an indication of the number of labeled nucleotides incorporated into the probe, is always higher with isotopic nucleotides than with antigenic nucleotides. Moreover, incorporation of numerous antigenic nucleotides decreases the formation of specific hybrids.

The absence of endogenous activity, such as that of biotin in liver, is not a problem in autoradiography, but positive and negative chemographic effects have been described,[70] which correspond, respectively, to the appearance and the disappearance of silver grains. Some cells, such as eosinophiles and macrophages, as well as some tissues, can fix ^{35}S-labeled probes.

The efficiency is the percentage of disintegrations recorded. Williams[34] gives overall autoradiographic efficiencies for various isotopes, using different developer-emulsion combinations. The sensitivity depends on this factor.

The quantification of autoradiograms uses standardized methods: (1) the probability circle method,[71-75] or (2) the cross-fire method.[76-79]

2. Advantages of Nonradioactive Hybridization

This technique has four advantages: rapidity, resolution, multiple detections, and quantification.

Hybrids labeled with antigen are revealed by an immunocytological method whose duration should not exceed a few hours, while exposure time for autoradiographic detection is always a long period, taking weeks or even months. Moreover ultrathin frozen sections are always difficult to preserve in good condition over long periods.

The resolution depends on the size of the immunoglobulin and gold particles, i.e., 10 nm for immunoglobulin and 5 to 15 nm for gold particles. For the direct method, the resolution could be about 15 nm, and for the indirect method, 25 nm, while for autoradiographic detection the resolution depends on the isotope used.[33,34,39] For example, the resolution for ^{35}S was expressed as the distance from a point radioactive source within which half of the silver grains fall: the half-radius value was 390 nm.[34,39]

Multiple detections of nucleic acids, or nucleic acid and protein, using, respectively, two different labeled nucleotides and a nucleic acid probe as well as another detection method for protein are all possible (see Protocols 28 to 30 for the detection of two different nucleic acids and Protocols 32 to 35 for the detection of nucleic acid and protein). The simultaneous detection of nucleic acid (by autoradiography) and protein (by immunocytology) has been performed by Tong et al.[4] by a combination of a pre-embedding hybridization method and immunocytology on epoxy resin sections in order to reveal prolactin mRNA and prolactin stored in secretory granules of lactotrophs.

The quantification of nonradioactive hybrids revealed with gold particles can be made easier by using a 3′ end labeled oligonucleotidic probe, since each gold particle corresponds to a particular nucleic acid. On the other hand the intensity of the signal is low.

IX. QUANTIFICATION

The aim of quantification is to show, using radioactive and nonradioactive probes:

1. That the signal observed is different from background and nonspecific binding of the probe, and
2. The cellular compartment which contains the nucleic acid studied.

The primary interest of this quantification is to determine background, which can be done simply by comparing the signal over a section to noise over the grid membrane. However, to determine nonspecific binding of the probe, it is necessary to know the cells where the nucleic acid studied is present, and to compare the signal over these cells to the signal over another cell population in the same section. For instance, in pituitary glands the background of *in situ* hybridization used to detect growth hormone mRNA with a radioactive probe was 1.37 silver grains per 100 μm^2 and, over the gonadotrophs, corticotrophs, and thyrotrophs, background plus nonspecific binding came to 3.2 silver grains per 100 μm^2. If a nonradioactive probe is used, background plus nonspecific binding represented about 10% of the signal.

The evaluation of the cellular compartments where the nucleic acid is present depends on the resolution of the *in situ* hybridization method used. If the probe is labeled with a radioactive (^{35}S) nucleotide, the resolution is about 15 times lower than with a nonradioactive nucleotide. For instance if the localization of mRNA within a cell is known, it can be present in the nucleus, in the cytoplasmic matrix, or the endoplasmic reticulum, but only occasionally in other cell compartments, so the analysis is easier and less extended. After autoradiographic revelation of growth hormone mRNA hybrids and intron I hybrids in somatotrophs, their localizations were analyzed by the probability circle method,[80] and their distribution in both the cytoplasm and the nucleus was studied in terms of grain density.[81] Only those somatotrophs sectioned through their nucleus were analyzed. The number of cells studied was between 25 to 116. The number of silver grains in the cytoplasm and the nucleus was calculated, and the percentage of silver grains in the cytoplasm and the nucleus were obtained. The relative concentrations of grains found in the cytoplasm and the nucleus were calculated by dividing the percentage of silver grains by the percentage of the area occupied by each compartment (36.4% for the cytoplasmic matrix and the endoplasmic reticulum; 30.4% for the nucleus and the nuclear membrane)[82] which as previously noted is termed the grain density. In this way it was shown that after *in situ* hybridization with the growth hormone cDNA probe, grain density was twice as high in the cytoplasm as in the nucleus (Figure 19). In contrast, grain density appeared twice as high in the nucleus as in the cytoplasmic matrix after *in situ* hybridization with the growth hormone intron I probe (Figure 19).[9] The effects of a neuropeptide (growth hormone-releasing factor or somatostatin) on the ultrastructural distribution of growth hormone mRNA and growth hormone intron I were shown in the same way.[9]

The quantification of *in situ* hybridization with a nonradioactive probe revealed by gold particles is easier since the resolution is better. In each

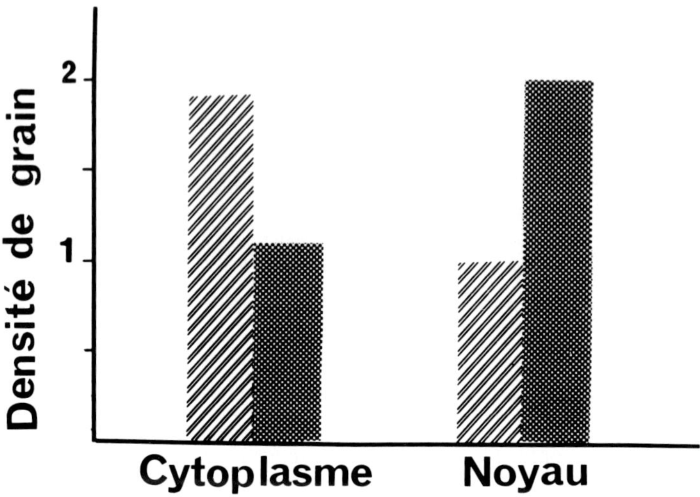

FIGURE 19. Quantification of silver grains obtained from ultrastructural *in situ* hybridization with ^{35}S-labeled growth hormone cDNA on ultrathin frozen sections of pituitary gland. Grain density is the ratio of percentage of silver grains in one compartment of somatotroph to the percentage of the area occupied by that compartment. Labeling associated with the growth hormone cDNA probe was found predominantly in the cytoplasm (cross-hatching) and, with growth hormone intron I probe, in the nucleus (dotted line).[9]

cellular compartment the direct count of gold particles can be divided by the total number of gold particles to obtain the percentage of gold particles in each cellular compartment. Then, as previously noted, the density of gold particles can be compared and analyzed. In other respects, the quantification of chemical revelation of nonradioactive probes, as enzymatic markers, remains difficult at the ultrastructural level.

Quantification and statistical analysis constitute the only way to demonstrate nonconventional localization of nucleic acids.

X. CONCLUSION

The preservation of nucleic acids in frozen tissue is an established fact, like the advantages of *in situ* hybridization on sections. This being the case, the localization of nucleic acids by *in situ* hybridization on ultrathin frozen sections is possible, as well as the use of cDNA or oligonucleotidic probes labeled with radioactive or nonradioactive nucleotides, and the signal can be quantified in similar conditions to those which are found in other methods.

There is no difference between the use of a long probe (cDNA) and that of an oligonucleotidic probe. Labeling is denser with cDNA, but it is more difficult to quantify, and in any case it is always possible to amplify the signal by a mixture of oligonucleotidic probes.

As far as the choice of label is concerned, nonradioactive techniques seem to be the most advantageous. The loss of sensitivity is compensated for by the lower risk involved in the short exposure period and the autoradiographic revelation, which shows ultrastructural damage.

The main disadvantage of *in situ* hybridization on ultrathin frozen sections is the poor preservation of cellular ultrastructures.[66] Its main advantage remains its sensibility.[66] Other positive features are (1) the ability to store frozen specimens as well as ultrathin sections, (2) the rapidity of obtaining the ultrathin sections, and thus the *in situ* hybridization results, and (3) the possibility of simultaneous detections.

ACKNOWLEDGMENTS

I would like to thank the following people for their help in the preparation of this manuscript: J. Doherty (for reading), R. Jambou, H. Mertani, A. Ouhtit (for preparing the protocols), and V. Dumont, M. L. Fluhr, G. Gouru (for their technical assistance).

APPENDIX

The composition of solutions are

- Denhardt solution ($\times$ 50 concentrated)
  Bovine serum albumine, 1%
  Ficoll 400, 1%
  Polyvinylpyrrolidone, 1%
  This solution is stored at −20°C
- Dextran sulfate, 50%
  Dextran sulfate, 50 mg
  Water, 65 μl
  Mix and spin until the solution is clear
  This solution can be stored in aliquot at −20°C
- Phosphate buffer ($\times$ 5 concentrated)
  Na_2HPO_4, 0.5 *M*
  NaH_2PO_4, 0.5 *M*
  pH 7.4 (by addition of the second solution to the first)
  This solution is stored after sterilization at room temperature
- SSC buffer ($\times$ 20 concentrated)
  NaCl, 3 *M*
  NaCitrate, 0.3 *M*
  pH 7
  This solution is stored after sterilization at room temperature
- SSPC buffer ($\times$ 20 concentrated)
  NaCl, 3 *M*
  NaCitrate, 0.3 *M*
  NaH_2PO_4, 0.4 *M*
  pH 7
  This solution is stored after sterilization at room temperature

REFERENCES

1. **Morel, G., Dubois, P. M., and Gossard, F.,** Détection ultrastructurale des ARN messagers codant pour l'hormone de croissance dans l'antéhypophyse du rat par hybridation *in situ, C. R. Acad. Sci. Paris (III),* 302, 479, 1986.
2. **Trembleau, A., Calas, A., and Fevre-Montange, M.,** Ultrastructural localisation of oxytocin mRNA in the rat hypothalamus by *in situ* hybridization using a synthetic oligonucleotide, *Mol. Brain Res.,* 8, 37, 1990.
3. **Guitteny, A. F. and Bloch, B.,** Ultrastructural detection of the vasopressine messenger RNA in the normal and Brattleboro rat, *Histochemistry,* 92, 277, 1989.
4. **Tong, Y., Zhao, H., Simard, J., Labrie, F., and Pelletier, G.,** Electron microscopic autoradiographic localization of prolactin mRNA in rat pituitary, *J. Histochem. Cytochem.,* 37, 567, 1989.

5. **Le Guellec, D., Frappart, L., and Desprez, P. Y.,** Ultrastructural localization of mRNA encoding for EGF receptor in human breast cancer line BT20 by *in situ* hybridization, *J. Histochem. Cytochem.*, 39, 1, 1991.
6. **Puvion-Dutilleul, F. and Puvion, E.,** Ultrastructural localization of viral RNA and DNA by *in situ* hybridization of biotinylated DNA probe on sections of Herpes simplex virus type 1 infected cells, *J. Electron Microsc. Tech.*, 18, 336, 1991.
7. **Binder, M., Tourmente, S., Roth, J., Rebaud, M., and Gehring, W. J.,** *In situ* hybridization at the electron microscopic level: localization of transcripts on ultrathin sections of Lowicryl K4M-embedded tissue using biotinylated probes and protein A-gold complexes, *J. Cell Biol.*, 102, 1646, 1986.
8. **Singer, R. H., Langevin, G. L., and Lawrence, J. B.,** Ultrastructural visualization of cytoskeletal in RNAs and their associated proteins using double-label *in situ* hybridization, *J. Cell Biol.*, 108, 2343, 1989.
9. **Morel, G., Dihl, F., and Gossard, F.,** Ultrastructural distribution of GH mRNA and GH intron 1 sequences in rat pituitary gland: effects of GH-releasing factor and somatostatin, *Mol. Cell. Endocrinol.*, 65, 81, 1989.
10. **Beauvillain, J. C., Mitchell, V., Tramu, G., and Mazzuca, M.,** GABA axon terminals in synaptic contact with enkephalin neurons in hypothalamus of guinea-pig. Demonstration by double immunocytochemistry, *Brain Res.*, 443, 315, 1988.
11. **Kosaka, T., Nagutsu, I., Wu, J. Y., and Hama, K.,** Use of high concentrations of glutaraldehyde for immunocytochemistry of transmitter-synthesizing enzymes in the central nervous systems, *Neuroscience,* 4, 975, 1986.
12. **Stephenson, J. L.,** Ice crystal growth during the rapid freezing of tissue, *J. Biophys. Biochem. Cytol.*, 2, 45, 1956.
13. **Stephenson, J. L.,** Ice crystal formation in biological materials during rapid freezing, *Ann. N. Y. Acad. Sci.*, 85, 535, 1960.
14. **Pryde, J. A. and Jones, G. O.,** Properties of vitreous water, *Nature (London),* 170, 685, 1964.
15. **Brüggeller, P. and Mayer, E.,** Complete vitrification in pure liquid water and dilute aqueous solution, *Nature (London),* 288, 569, 1980.
16. **Dubochet, J. and McDowall, A. W.,** Vitrification of pure water for electron microscopy, *J. Microscopy,* 124, RP3, 1981.
17. **Mayer, E. and Brüggeller, P.,** Vitrification of pure liquid water by high pressure jet freezing, *Nature (London),* 298, 715, 1982.
18. **Nash, T.,** Chemical constitution and physical properties of compounds able to protect living cells against damage due to freezing and thawing, in *Cryobiology,* Meryman, H. T., Ed., Academic Press, New York, 1966, 179.
19. **Meryman, H. T.,** Cryoprotective agents, *Cryobiology,* 8, 173, 1971.
20. **Skaer, H. le B.,** Chemical cryoprotection for structural studies, *J. Microscopy,* 125, 137, 1982.
21. **Morel, G.,** Coupes semi-fines et ultrafines de tissus congelés: congélation, cryo-ultramicrotomie, contraste, in *Techniques en Microscopie Electronique: Cryométhodes, Immunocytologie, Autoradiographie, Hybridation In Situ,* Morel, G., Ed., Editions INSERM, Paris, 1991, 31.
22. **Tokuyasu, K. T.,** Use of poly(vinylpyrrolidone) and poly(vinyl alcohol) for cryoultramicrotomy, *Histochem. J.*, 21, 163, 1989.
23. **Tokuyasu, K. T.,** A technique for ultracryotomy of cell suspensions and tissues, *J. Cell Biol.*, 57, 551, 1973.
24. **Griffiths, G., Simons, K., Warren, G., and Tokuyasu, K. T.,** Immunoelectron microscopy using thin frozen sections: application to studies of the intracellular transport of Semliki Forest Virus spike glycoprotein, *Methods Enzymol.*, 96, 466, 1983.
25. **Boivin, G., Morel, G., Meunier, P. J., and Dubois, P. M.,** Ultrastructural aspects after cryoultramicrotomy of bone tissue and sutural cartilage in neonatal mice calvaria, *Biol. Cell,* 49, 227, 1983.

26. **Morel, G., Biovin, G., David, L., Dubois, P. M., and Meunier, P. J.,** Immunocytochemical evidence for endogenous calcitonin and parathyroid hormone in osteoblasts from the calvaria of neonatal mice. Absence of endogenous estradiol and estradiol receptors, *Cell Tissue Res.*, 240, 89, 1985.
27. **Stierhof, Y. D., Schwarz, H., and Frank, H.,** Transverse sectioning of plastic-embedded immunolabeled cryosections: morphology and permeability to protein A-colloidal gold complexes, *J. Ultrastruct. Mol. Struct. Res.*, 97, 187, 1986.
28. **Keller, G. A., Tokuyasu, K. T., Dutton, A. H., and Singer, S. J.,** An improved procedure for immunoelectron microscopy: ultrathin plastic embedding of immunolabeled ultrathin frozen sections, *Proc. Natl. Acad. Sci. U.S.A.*, 81, 5744, 1984.
29. **Boller, K.,** Coloration et enrobage des coupes congelées ultrafines, in *Techniques en Microscopie Electronique: Cryométhodes, Immunocytologie, Autoradiographie, Hybridation In Situ,* Morel, G., Ed., Editions INSERM, Paris, 1991, 55.
30. **Griffith, G., McDowall, A., Back, R., and Dubochet, J.,** On the preparation of cryosections for immunocytochemistry, *J. Ultrastruct. Res.*, 89, 65, 1984.
31. **Tokuyasu, K. T.,** A study of positive staining of ultrathin frozen sections, *J. Ultrastruct. Res.*, 63, 287, 1978.
32. **Tokuyasu, K. T.,** Immunochemistry on ultrathin frozen sections, *Histochem. J.*, 12, 381, 1980.
33. **Cau, P.,** *Microscopie Quantitative, Stereologie, Autoradiographie et Immunocytochimie Quantitatives,* Cau, P., Ed., Editions INSERM, Paris, 1990, 262.
34. **Williams, M. A.,** Preparation of electron microscope autoradiographs, in *Autoradiography and Immunocytochemistry,* Glauert, A. M., Ed., North-Holland, Amsterdam, 1977, 77.
35. **Seeburg, P. H., Shine, J., Martial, J. A., Baxter, J. D., and Godman, H. M.,** Nucleotide sequence and amplification in bacteria of structural gene for rat growth hormone, *Nature (London),* 270, 486, 1977.
36. **Page, G. S., Smith, S., and Godman, H. M.,** DNA sequence of the rat growth hormone gene: location of the 5′ terminus of the growth hormone mRNA and identification of an internal transposon-like element, *Nucleic Acid Res.*, 9, 2087, 1981.
37. **Miller, W. L. and Eberhardt, N. L.,** Structure and evolution of the growth hormone gene family, *Endocr. Rev.*, 4, 97, 1983.
38. **Rogers, A. W.,** *Techniques of Autoradiography,* Elsevier/North-Holland, Amsterdam, 1979, 429.
39. **Morel, G.,** Electron microscopic autoradiographic techniques, in *Electron Microscopy in Biology, A Practical Approach,* Harris, J. D., Ed., IRL Press, Oxford, 1991, 83.
40. **Lacassagne, A., Lattès, J., and Lavedan, J.,** Etude expérimentale des effets biologiques du polonium introduit dans l'organisme, *J. Radiol. Electro,* 9, 1, 1925.
41. **Kopriwa, B. M.,** A semi-automatic instrument for the radioautographic coating technique, *J. Histochem. Cytochem.*, 14, 923, 1966.
42. **Bouteille, M.,** The "LIGOP" method for routine ultrastructural autoradiography. A combination of single grid coating, gold latensification and phenidon development, *J. Microsc. Biol. Cell.*, 27, 121, 1976.
43. **Bouteille, M., Fakan, S., and Burglen, J.,** Efficiency, resolution, grain size, grain shape and background noise in routine electron microscope autoradiography, *J. Microscopie Biol. Cell.*, 27, 171, 1976.
44. **Chabot, J.-G., Dihl, F., and Morel, G.,** Autoradiographie à l'échelle ultrastructurale sur coupes de tissu inclus en résine époxy ou congelé: mise en évidence de molécules captées par leurs cellules cibles, in *Techniques en Microscopie Electronique: Cryométhodes, Immunocytologie, Autoradiographie, Hybridation In Situ,* Morel, G., Ed., Editions INSERM, Paris, 1991, 361.
45. **Bachmann, L. and Salpeter, M. M.,** Autoradiography with electron microscope, *Lab. Invest.*, 14, 1041, 1965.
46. **James, T. H.,** The site of reaction in direct photographic development. II. Kinetics of development initiated by gold nuclei, *J. Colloid Sci.*, 3, 447, 1948.

47. **Lettré, H. and Paweletz, N.,** Probleme der elektronmikroskopischen autoradiographie, *Naturwissenschaften,* 53, 268, 1966.
48. **Caro, L. G. and Van Tubergen, R. P.,** High resolution autoradiography. I. Methods, *J. Cell Biol.,* 15, 173, 1962.
49. **Bouteille, M., Fakan, S., and Burglen, J.,** Efficiency, resolution, grain size, grain shape and background noise in routine electron microscope autoradiography, *J. Microsc. Biol. Cell.,* 27, 171, 1976.
50. **Moss, J. and Lane, M. D.,** The biotin-dependent enzymes, *Adv. Enzymol.,* 35, 321, 1971.
51. **Avrameas, S. and Uriel, J.,** Méthode de marquage d'antigène et d'anticorps avec des enzymes et son application en immunodiffusion, *C. R. Acad. Sci., Paris,* 262, 2543, 1966.
52. **Nakane, P. K.,** Application of peroxidase labelled antibodies to the intracellular localization of hormones, *Acta Endocrinol (Copenhagen),* 153, 190, 1971.
53. **Stenberger, L. A.,** *Immunocytochemistry,* John Wiley & Sons, New York, 1979, 353.
54. **Roth, J.,** The colloidal gold marker system for light and electron microscopic cytochemistry, in *Techniques in Immunocytochemistry,* Vol. 2, Bullock, G. R. and Petrusz, P., Eds., Academic Press, London, 1983, 217.
55. **Bernadac, A., Bolla, J. M., Lazdunski, C., Inouye, M., and Pages, J. M.,** Precise localization of an overproduced periplasmic protein in *Eschericia coli:* use of double immuno-gold labelling, *Biol. Cell,* 61, 141, 1987.
56. **Bernadac, A., Bolla, J. M., and Pages, J. M.,** Double marquage immunologique aprés cryo-ultramicrotomie: localisation d'une proteine hybride surproduite chez *Escherichia coli,* in *Techniques en Microscopie Electronique: Cryométhodes, Immunocytologie, Autoradiographie, Hybridation In Situ,* Morel, G., Ed., Editions INSERM, Paris, 1991, 249.
57. **Thorpe, J. R.,** A novel methodology for double protein A-gold immunolabeling utilizing the monovalent fragment of protein A, *J. Histochem. Cytochem.,* 40, 435, 1992.
58. **Bendayan, M. and Stephens, H.,** Double labelling cytochemistry applying the protein A-gold technique, in *Immunolabelling for Electron Microscopy,* Polak, J. M. and Vardell, I. M., Eds., Elsevier, Amsterdam, 1984, 143.
59. **Wood, G. S. and Warnke, R.,** Suppression of endogenous avidin-binding activity in tissues and its relevance to biotin-avidin detection systems, *J. Histochem. Cytochem.,* 29, 1196, 1981.
60. **Strass, W.,** Inhibition of peroxidase by methanol or methanol nitroferricyanide for use in peroxidase procedures, *J. Histochem. Cytochem.,* 19, 682, 1971.
61. **Moriarty, G. C. and Halmi, N. S.,** Electron microscopy study of the adrenocorticotropin producing cell with the use of unlabeled antibody and the soluble peroxidase-anti-peroxidase complex, *J. Histochem. Cytochem.,* 20, 590, 1972.
62. **Hemming, F. J., Mesguich, P., Morel, G., and Dubois, P. M.,** Cryoultramicrotomy versus plastic embedding: comparative immunocytochemistry of rat anterior pituitary cells, *J. Microsc.,* 131, 26, 1983.
63. **Barinaga, M., Yamonoto, G., Rivier, C., Vale, W., Evans, R., and Rosenfeld, M. G.,** Transcriptional regulation of growth hormone gene expression by growth hormone-releasing factor, *Nature, London* 306, 84, 1983.
64. **Barinaga, M., Bilezikjan, L. M., Vale, W., Rosenfeld, M. G., and Evans, R.,** Independent effects of growth hormone-releasing factor on growth hormone release and gene transcription, *Nature, London* 314, 279, 1985.
65. **Brazeau, P., Vale, W., Burgus, R., Ling, N., Butcher, M., Rivier, J., and Guillemin, R.,** Hypothalamic polypeptide that inhibits the secretion of immunoreactive growth hormone, *Science,* 179, 77, 1973.
66. **Le Guellec, D., Trembleau, A., Pechoux, C., and Morel, G.,** Ultrastructural nonradioactive *in situ* hybridization of GH mRNA in rat pituitary gland: pre-embedding vs ultra-thin frozen sections vs post-embedding, *J. Histochem. Cytochem.,* 40, 979, 1992.

67. **Cooke, N. E., Coit, D., Welner, R. I., Baxter, J. D., and Martial, J. A.,** Structure of cloned DNA complementary to rat prolactin messenger RNA, *J. Biol. Chem.*, 255, 6502, 1980.
68. **Morel, G.,** Hybridation *in situ* sur coupes de tissus congelés: sonde radioactive, sonde non-radioactive, double marquage, in *Techniques en Microscopie Electronique: Cryométhodes, Immunocytologie, Autoradiographie, Hybridation In Situ,* Morel, G., Ed., Editions INSERM, Paris, 1991, 519.
69. **Morel, G., Chabot, J.-G., Belles-Isles, M., and Heisler, S.,** Synthesis and internalization of atrial natriuretic factor in anterior pituitary cells, *Mol. Cell. Endocrinol.*, 55, 219, 1988.
70. **Mulvaney, B. D.,** Chemography of lysosome-like structures in olfactory epithelium, *J. Cell Biol.*, 51, 568, 1971.
71. **Williams, M. A.,** *Quantitative Methods in Biology,* North-Holland, Amsterdam, 1977, 85.
72. **Williams, M. A.,** in *Advances in Optical and Electron Microscopy,* Vol. 3, Barer, R. and Cosslett, V. E., Eds., Academic Press, London, 1969, 219.
73. **Salpeter, M. M.,** H^3 Proline incorporation into cartilage: electron microscope autoradiographic observations, *J. Morphol.*, 124, 387, 1968.
74. **Nadler, N. J.,** The interpretation of grain count in electron microscope autoradiography, *J. Cell Biol.*, 49, 877, 1971.
75. **Salpeter, M. M. and McHenry, F. A.,** Electron microscope autoradiography. Analysis of autoradiograms, in *Advanced Techniques in Biological Electron Microscopy,* Koehler, J. K., Ed., Springer-Verlag, New York, 1973, 113.
76. **Blackett, N. M. and Parry, D. M.,** A new method for analysis electron microscope autoradiographs using hypothetical grain distributions, *J. Cell Biol.*, 57, 9, 1973.
77. **Parry, D. M.,** Practical approaches to the statistical analysis of electron microscope autoradiographs, *J. Microsc. Biol. Cell.*, 27, 185, 1976.
78. **Parry, D. M. and Blackett, N. M.,** Analysis of electron microscope autoradiographs using the hypothetical grain analysis method, *J. Microsc.*, 106, 117, 1976.
79. **Blackett, N. M. and Parry, D. M.,** A simplified method of ''hypothetical grain'' analysis of electron microscope autoradiographs, *J. Histochem. Cytochem.*, 25, 206, 1977.
80. **Morel, G., Leroux, P., and Pelletier, G.,** Ultrastructural autoradiographic localization of somatostatin-28 (SS-28) in the rat pituitary gland, *Endocrinology,* 116, 1615, 1985.

Chapter 7

ULTRASTRUCTURAL *IN SITU* HYBRIDIZATION OF mRNA ON LOWICRYL SECTIONS USING QUICK-FREEZING, FREEZE-SUBSTITUTION, AND LOW TEMPERATURE EMBEDDING

Lucien Frappart and Dominique Le Guellec

TABLE OF CONTENTS

0-8493-4414-X/93/$0.00 + $.50

I. INTRODUCTION

Detection of nucleic acid sequences at the ultrastructural level has allowed us to better understand the expression of genes in different tissues such as the localization of viral nucleic acids,[1-4] nuclear and mitochondrial rRNA[5-8] or mRNA encoding neuropeptide,[9-11] extracellular matrix,[12] or intracellular molecules.[13-17] By this way, Singer and Lawrence[18] have demonstrated that mRNA encoding actin was associated with actin filaments, and Guitteny and Bloch[9] have observed that a gene mutation involved different sites of mRNA. Other results ask the question of the role of the nucleolus in the moving of mRNA[12,19] and the role of intron sequences in the cytoplasm.[20]

These studies have been carried out following different methods, pre-embedding,[11] ultrathin frozen section,[10] or post-embedding[12,21,22] methods that have not the same sensitivity, the same ultrastructure preservation, and the same detection system. Recently, we have compared these three methods[23] in terms of specificity, sensitivity, resolution, and cytological preservation. The pre-embedding method gave the best ultrastructural preservation with low resolution with the enzymatic detection system (gold particle could not be used without pretreatment) and an intermediate sensitivity. The frozen section method gave the best sensitivity but the lowest ultrastructure preservation. The post-embedding method resolution was as high as with the frozen section method, ultrastructure conservation was intermediate, and sensitivity was low. We have chosen to develop this last method[24] which seems to be a good compromise between sensitivity and ultrastructure preservation. The resin used to embed the tissue must conserve the cell ultrastructure and the accessibility of mRNA. That is why the hydrophilic Lowicryl resins have been chosen by several authors.[1,21] We have tested *in situ* hybridization on Epon sections without results. *In situ* hybridization carried out after tissue embedding in LR White or other methacrylate resins has been described.[25] During the classical embedding process in Lowicryl resin, after dehydration by progressive lowering of temperature, the critical part is the high temperatures during the early dehydration steps. The only way to circumvent this inconvenient step is offered by rapid freezing and freeze substitution followed by low temperature embedding. In this chapter, we are describing an *in situ* hybridization method to localize at the electron microscope level the mRNA coding the receptor of epidermal growth factor (EGF-R) in BT20 human carcinoma cell line using ^{35}S cDNA probes, and we are comparing the results obtained after the two embedding methods.

II. FIXATION

Fixation is used only with the progressive lowering temperature (PLT) method.

A. FIXATIVE

Commercially available solutions of formaldehyde are unsuitable for electron microscopy because of their content of methanol. Formaldehyde solutions are therefore prepared in the laboratory from powdered paraformaldehyde (trioxy-methylene) just before use (see Protocols 1, 2, and 3).

PROTOCOL 1 — FIXATIVE PREPARATION (20% PARAFORMALDEHYDE)

The fixative should be prepared and stored in a clean, glass-stoppered bottle and is stable for 1 or 2 days if kept at 4°C in the dark in a refrigerator.

1. Dissolve 2 g of paraformaldehyde in 10-ml double-distilled water by heating to 80°C with continuous stirring.
2. Add 1 *N* sodium hydroxyde until the solution clears.

The resulting solution of 20% paraformaldehyde is allowed to cool before mixing with the other components.

The pH of the fixative may drop slightly with time and should be tested before use.

Note: it is essential to work in a fume cupboard.

PROTOCOL 2 — PHOSPHATE BUFFER 0.2 *M* pH 7.4

Na H_2PO_4, 2 H_2O	5, 92 g
K_2H PO_4, 12 H_2O	28, 21 g
Distilled water	to make 1 ℓ

PROTOCOL 3 — PHOSPHATE BUFFER 0.175 *M* pH 7.4

Na H_2PO_4, 2 H_2O	5, 18 g
K_2H PO_4, 12 H_2O	24, 66 g
Distilled water	to make 1 ℓ

We currently use two types of fixative (1) paraformaldehyde 4% in phosphate buffer 0.2 *M*, pH 7.4 and (2) paraformaldehyde 2% in phosphate buffer 0.2 *M*, pH 7.4.

Fixatives containing both paraformaldehyde and glutaraldehyde give better preservation of a wide variety of tissues than either aldehyde alone. As a consequence they are very widely used as primary fixatives at the present time. Formaldehyde penetrates tissues much more rapidly than glutaraldehyde and it is thought that the formaldehyde temporarily stabilizes structures which are subsequently fixed more permanently by the glutaraldehyde. In our experiment the addition of 0.1% of glutaraldehyde in the fixative inhibits the hybridization.

B. MATERIAL FIXATION

Cells growing on petri dishes are harvested after trypsination and washed in culture medium using centrifugation at 800 rpm for 10 min. The fixative is run slowly down the side of the centrifuge tube so as to keep the pellet intact; the volume of fixative added should be at least ten times the volume of the pellet. The cells are fixed for 1 h at 4°C. After the fixation the cells are then centrifuged to a pellet again before the next step in the preparative procedure: the fixative is carefully removed from the pellet and replaced with the washing solution: phosphate buffer solution, 0.175 *M*, pH 7.4, 2 × 5 min, and dehydrated as indicated in Section III.B.

III. EMBEDDING PROCESS IN LOWICRYL

A. CHARACTERISTICS OF THE RESINS

Lowicryls are a highly cross-linked acrylate- and methacrylate-based embedding media which have been designed for use over a wide range of embedding conditions. Both resins are photopolymerized by long wavelength (360 nm) ultraviolet light. Since the initiation of the photopolymerization is largely independent of temperature, blocks may be polymerized at the same temperatures that are used for infiltration.

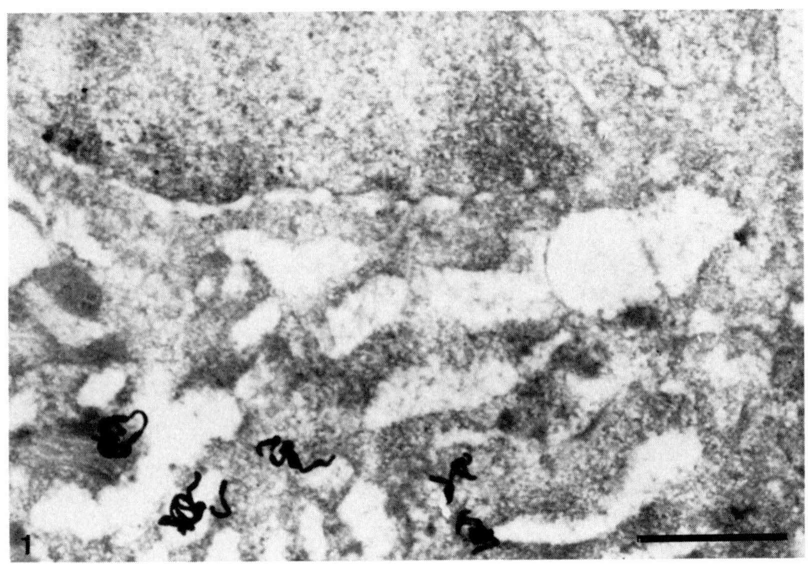

FIGURE 1. *In situ* hybridization using ^{35}S-labeled EGF-R cDNA on BT20 human mammary carcinoma cell line dehydrated by progressive lowering of temperature and embedded in Lowicryl K4M. Labeling is observed principally over the cytoplasm and the endoplasmic reticulum. Bar = 1 μm; magnification ×13,000.

1. Lowicryl K4M

This resin has been formulated to provide low viscosity at low temperature: K4M is a polar (hydrophilic) resin usable down to −35°C. The hydrophilic properties of K4M provide two distinct advantages. During dehydration and infiltration the specimens may be kept in a partially hydrated state. Second, K4M is particularly useful for immunolabeling of sections using specific antisera or lectins. This type of resin is assumed to maintain the intramolecular and intermolecular hydrogen bonds, replacing water by the resin solvent solution.

The composition of the resin mixture is as follows.[26] (Table 1)

Lowicryl K4M: hydrophilic, polar resin

Weight percent	Monomer	Side chain (R)
48.4	Hydroxypropyl methacrylate	Ch_2-CH_2-CH_2OH
23.7	Hydroxyethyl acrylate	CH_2-CH_2-OH
9.0	*n*-Hexyl methacrylate	CH_2-$(CH_2)4$-CH_3
13.9	Crosslinker	

TABLE 1
A Schematic Formula Showing the Basic Monomers and the Reaction Pathways for Polymerization

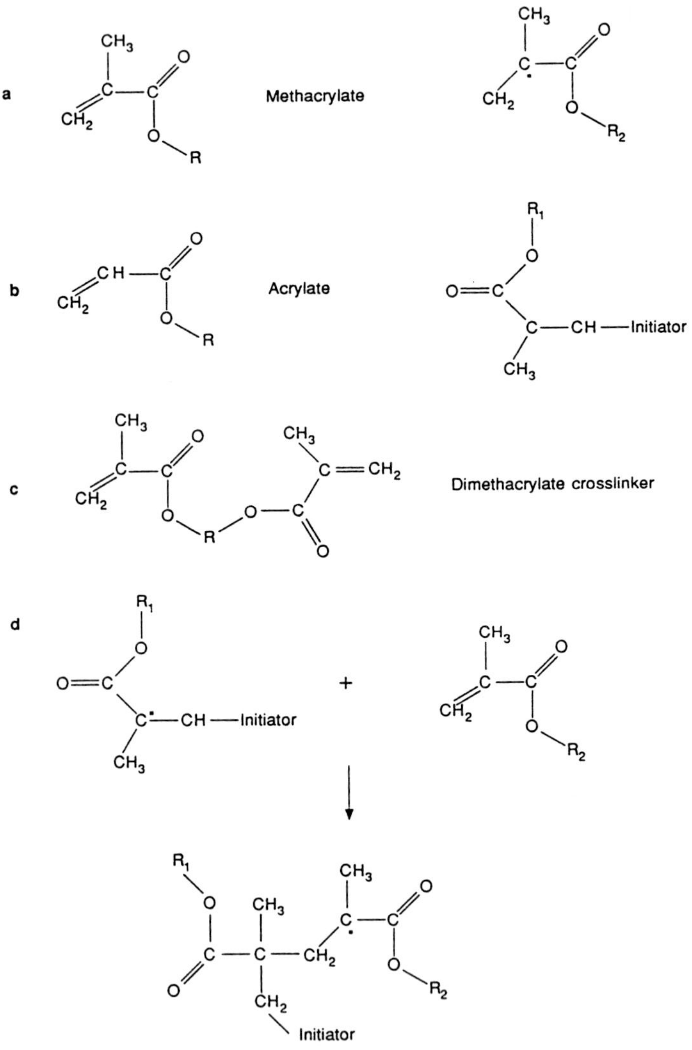

a — Methacrylate

b — Acrylate

c — Dimethacrylate crosslinker

d

Note: in a and b, R represents the position of the side group in the monomer. For c, R represents a dialcohol linker between the two methacrylate groups. In d, a methacrylate monomer has been activated with an initiator to form an unpaired electron at C2. This radical now attacks another monomer to begin the growth of the polymer chain. The crosslinker used is triethylene glycol dimethacrylate (TEGDMA). The amount of crosslinker used here will yield blocks of medium hardness. The relative amount of crosslinker can be varied within a certain range.

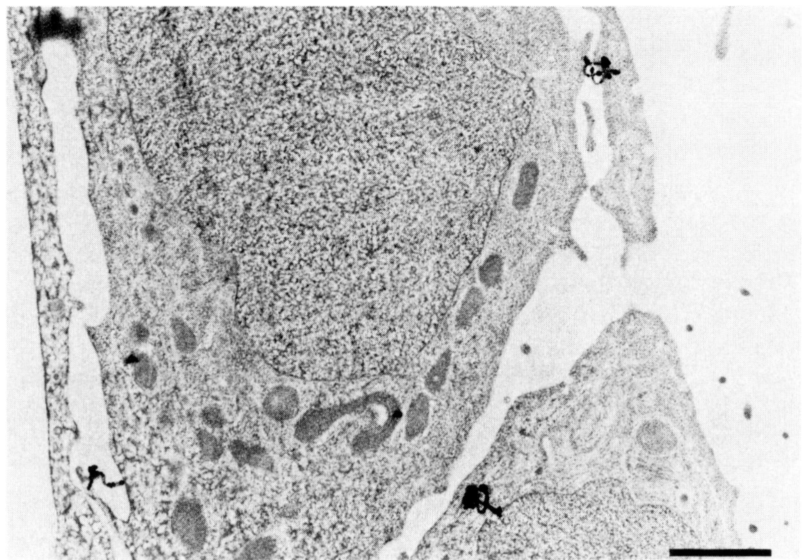

FIGURE 2. *In situ* hybridization using ^{35}S-labeled EGF-R cDNA on BT20 human mammary carcinoma cell line, after quick-freezing and Lowicryl K11M embedding. Better ultrastructural preservation is obtained with this procedure. Labeling is observed principally over the cytoplasm and the microvilli. Bar = 1 μm; magnification ×9,100.

The commercially available kit contains the resin as three components to be mixed in the following proportions:

Monomers B	13.7 g
Crosslinker A	2.7 g
Photo-initiator C	0.1 g

2. Lowicryl K11M

Lowicryl K11M has similar properties as K4M but can be applied at 20 to 30°C lower temperature. K11M has a much lower viscosity compared with K4M. The composition of the resin mixture is as follows.[27]

Lowicryl K11M: hydrophilic, polar resin

Weight percent	Monomer
12.2	n-Butyl methacrylate
10.2	2-Methoxy ethyl methacrylate
11.2	2-Ethoxy ethyl methacrylate
20.6	2-Hydroxy ethyl acrylate
41.0	2-Hydroxy propyl methacrylate
4.8	1,3 Butanediol dimethacrylate
0.5	Benzoin methyl ether or benzoin ethyl ether, or benzyl dimethyl acetal

The commercial kit contains the resin as three components to be mixed in the following proportions:

Monomers I	95.2 weight%	19 g
Crosslinker H	4.8 weight%	1 g
Photo-initiator C		0.1 g

3. Preparation of the Resins

Methacrylates may cause skin reactions in sensitive individuals. The following basic rules should be observed when handling Lowicryl:

- Work in a good fume hood.
- Wear safety clothing and safety glasses with side protection.
- Use a barrier hand cream and proper gloves (latex gloves are only partially adequate; their protective duration is a matter of minutes; 4H gloves (obtainable from Safety 4 A/S, Lyngby, Denmark) are a good choice, but can become brittle at low temperatures, and also have a large seam).
- Avoid any direct contact with liquid or vapors by manipulating liquid mixtures or components with tweezers or other mechanical means.

As the resins are of very low viscosity they require little mixing. Indeed, excessive stirring results in the incorporation of too much oxygen into the mixture and this may interfere with polymerization. The monomer and crosslinker should be weighed out into a preweighed vial and then mixed gently for 3 to 5 min by one of the following methods: (1) bubble a stream of dry nitrogen gas through the mixture via a Pasteur pipette or (2) mix gently with a glass rod.

When the monomer and crosslinker have been mixed, the initiator is added and mixing is continued until it has been completely dissolved in the resin.

B. DEHYDRATION AND IMPREGNATION

Cells and tissue samples are prepared for Lowicryl resin embedding according to two methods:

1. The progressive lowering temperature (PLT) technique, using Lowicryl K4M resin
2. The cryosubstitution or freeze substitution, using Lowicryl K11M resin

1. PLT Technique

a. Dehydration at Low Temperature (see Protocol 4)

PROTOCOL 4 — DEHYDRATION AT LOW TEMPERATURE

Ethanol (v% in H_2O)	Temperature (°C)	Time (min)
30	0	30
50	−20	60
70	−20	60
95	−20	60
100	−20	60
100	−20	60

b. Infiltration (see Protocol 5)

PROTOCOL 5 — TYPICAL INFILTRATION SCHEDULE

Resin: ethanol (v/v)	Temperature (°C)	Time (min)
1:2	−20	60
1:1	−20	60
2:1	−20	60
Pure resin	−20	Overnight
Pure resin	−20	1 Day

c. Polymerization (see Protocol 6)

The samples are next put in Beem capsules which should be completely filled with fresh precooled resin. The capsules should be filled to the top, to minimize dead air space over the resin. To minimize the condensation of water and the crystallization of ice on the sample, vials, capsules, and all apparatus should be precooled. Samples may be polymerized in Beem capsules. A suitable capsule holder is required so that the capsules receive UV irradiation for all sides.

The polymerization chamber — A polymerization chamber can be constructed which will fit in a deep chest-type freezer or in a cold room. To provide diffuse illumination, a right-angle reflector is suspended below the UV lamp. All six inner surfaces, as well as the reflection, should either be constructed of a UV-reflective material or should be lined with aluminum foil. The capsule holder is placed 30 to 40 cm below the fluorescent lamp. The light source must be 360-nm long wavelength UV (two 15-watt fluorescent tubes). Ventilation from the top and bottom will provide air circulation.

PROTOCOL 6 — ULTRAVIOLET POLYMERIZATION

1. Polymerize K4M for at least 24 h under UV light at −20°C
2. Remove the capsules from the cold
3. Continue "curing" under UV for 2 to 3 d at room temperature

2. The Cryosubstitution or Freeze-Substitution without Pretreatment at Ambient Pressure

Cryosubstitution is a method which does not require any chemical fixation of tissue and so avoids its damaging effects on sensitive and labile sites. It uses a physical fixation in which the rapid freezing of the specimen immobilizes all the cellular components. After freezing, the ice in the specimen is replaced by an organic solvent which is miscible with the embedding resin (acetone or methanol).

a. Method of Freezing and Cell Preparation

The cells on plastic coverslips (13 mm in diameter) Thermanox were plunged into liquid propane using a KF 80 Apparatus (Reichert-Jung, Vien, Austria) after elimination of the culture medium. The freezing must achieve very fast cooling rates so as to minimize ice-crystal formation. The largest depth of well-frozen material when measured from the freezing edge is 5 to 6 μm with this method. The samples were then transferred to CS Auto Substitution Apparatus (Reichert-Jung) where they are cryosubstituted.

b. Substitution Media

The ice is removed by substitution with an organic solvent: 100% methanol (freezing point: −94°C). Methanol can substitute ice even in the presence of 10% water.

PROTOCOL 7 — SUBSTITUTION SCHEDULE

	Temperature (°C)	Duration (h)
100% Methanol	−90	48
The temperature is raised to	−80	10
100% Methanol	−80	48
The temperature is raised to	−60	20
100% Methanol	−60	48
The temperature is raised to	−50	10

c. *Infiltration and Embedding (see Protocols 7 and 8)*

PROTOCOL 8 — INFILTRATION SCHEDULE

K11M: methanol (v/v)	Temperature (°C)	Time (h)
1:2	−50	1
1:1	−50	1
2:1	−50	1
Pure resin	−50	1
Pure resin	−50	Overnight

d. *Polymerization (see Protocol 9)*

PROTOCOL 9 — POLYMERIZATION SCHEDULE

1. The samples are put in fresh pure Lowicryl K11M, 8 h at −50°C.
2. Polymerization by indirect UV light (360-nm long wavelength) for 3 d at −50°C.
3. The temperature is then raised to room temperature (10°C/h).

IV. SECTIONING AND STAINING

A. THE LOWICRYL RESINS

When they are of the correct hardness, they are easily sectioned with either glass or diamond knives (sectioning speeds of 2 to 5 mm/s are recommended; thickness of the sections: 80 nm). K4M and K11M are hydrophilic resins. Precautions should be taken to ensure that the block face does not become wet during sectioning. This can be reached with a level of fluid in the trough which is slightly below normal.

B. GRIDS

Copper grids are not adequate for *in situ* hybridization studies because the formamide present in the hybridization buffer is highly reactive with the metal. Thin sections are mounted on nickel grids, G 200 Hex, preliminary degreased in acetone.

C. STAINING OF SECTIONS

Since K4M and K11M are hydrophilic resins, the sections should be incubated on drop of the stains for short periods of time. Prolonged staining may cause contamination and distortion of the sections. Sections may be stained with either saturated aqueous or alcoholic solutions of uranyl acetate. Both Reynolds lead citrate and Millonig's lead acetate give good results.

V. PREPARATION AND LABELING OF THE PROBE

The probe was a human 1.3-kb *Eco* RI cDNA corresponding to epidermal growth factor receptor (EGF-R) transmembrane and protein kinase domains. The entire recombinant plasmid (vector + insert) can be used when the probe is labeled by nick translation, but not when it is labeled by the "random priming" method (see Protocol 10), because in this case, there is no amplification of the labeling by the vector that can increase the aspecific signal. Then, it is preferable to purify the insert.

PROTOCOL 10 — PROBE LABELING BY RANDOM PRIMING

Stock solutions (stored at −20°C)

- Linearize cDNA probe diluted in water or TE 1× (10 m*M* Tris-HCl pH 8, 1 m*M* EDTA) (25 ng/ml)
- Buffer labeling 10× (Boehringer): 70 m*M* Tris-HCl; 70 m*M* $MgCl_2$; 300 m*M* NaCl; 100 m*M* DTT; 1 m*M* EDTA; pH 7.5
- 0.5 m*M* dCTP (Boehringer)
- 0.5 m*M* dGTP (Boehringer)
- 0.5 m*M* dTTP (Boehringer)
- Klenow fragment of DNA polymerase I 25 U/µl
- $^{35}S\alpha$-dATP (800 Ci/mmol) (Amersham) (stored at −80°C)
- 0.2 m*M* EDTA pH 8
- 5 mg/ml t-RNA in water
- 10 m*M* DTT
- Water bath at 37°C

Method

1. Denaturate 25 ng of probe at 95°C for 5 to 10 min (depending on the size of the cDNA)
2. Cool on ice
3. Add the following to an Eppendorf vial on ice and make up to final volume 20 µl

    - 2 µl buffer labeling
    - 1 µl dCTP
    - 1 µl dGTP
    - 1 µl dTTP
    - 1 µl DTT
    - 50 µCi ^{35}S α-dATP
    - 1 µl Klenow enzyme

4. Mix gently and touch spin
5. Incubate at least for 30 min at 37°C (a higher specific activity is obtained with a 1 h and 30 min incubation)
6. Stop the reaction by adding 2 μl 0.2 *M* EDTA
7. Add 10 μg t-RNA

The labeled probe is purified through Sephadex G50 and/or by ethanol precipitation (see Protocol 11).

PROTOCOL 11 — PROBE PURIFICATION THROUGH SEPHADEX G50

Preparation of Sephadex G50[28]

1. Add about 5 g of Sephadex in 100 ml of 1 × TE
2. Autoclave for 15 min on liquid cycle
3. Allow to cool at room temperature and replace the supernatant with an equal volume of 1 × TE
4. Store the solution at +4°C

Method

1. Block up the bottom of a 2-ml disposable syringe with sterile glass wool
2. Insert the syringe into a centrifuge tube
3. Refill with Sephadex G50 without making bubbles
4. Centrifuge at 1600 *g* for 4 min
5. Add the labeled probe in a total volume of 250 μl (use 1 × TE)
6. Insert a decapped Eppendorf tube in the bottom of centrifuge tube
7. Centrifuge at 1600 *g* for 4 min

The purified labeled probe is collected in the Eppendorf tube (see Protocol 12).

PROTOCOL 12 — ETHANOL PRECIPITATION AND MEASUREMENT OF THE SPECIFIC ACTIVITY

1. Add in the labeled probe solution 1/10^{e} 3 *M* sodium acetate pH 5 and 2 volumes of cold ethanol 100%
2. Mix well and store at −20°C for at least 2 h or at −80°C for at least 30 min (if the size of DNA is small or when it is present in small amounts, the period of storage should be extended)
3. Centrifuge at 12,000 *g* for 30 min at 4°C
4. Discard the supernatant
5. Rinse the pellet (often it is not visible) with cold 70% ethanol
6. Centrifuge at 12,000 *g* for 10 min at 4°C

7. Discard the supernatant
8. Dry the pellet
9. Dissolve the labeled DNA in the desired volume of 1 × TE
10. Specific activity: measure the amount of radioactivity in order to calculate the specific activity (dpm/μg). With ^{35}S, it should be at least 2.10^8 dpm/μg

VI. HYBRIDIZATION

The hybridization process consists in depositing the denaturated probe on a tissue section that makes hydrogen bonds with the interesting mRNA. The stringence of hybridization has to be rigorous to hybridize only homologous mRNA without aspecific hybridization. Two parameters are important to have a good hybridization, the temperature and the cation monovalent concentration. The melting temperature (Tm) is the temperature at which 50% of two ADN strands are denatured. The Tm depends on the hybrid size, the sodium concentration, and the percentage of G-C bases. The Tm increases by 16.6°C every time that the sodium concentration increases 10 times. For a hybrid whose size is more than 200 base pairs, the Tm is as follows:

$$Tm = 16.6 \log (Na^+) + 0.41 (\% G - C) + 81.5$$

The nearer the Tm the hybridization temperature is, the more specific the hybridization. Usually the hybridization temperature is 20 to 25°C below the Tm. In fact this temperature is too high and not consistent with a good structure preservation of the tissues. That is why formamide is added in the hybridization buffer to diminish the hybridization temperature. For every percent of formamide the Tm decreases by 0.65°C. In these conditions the hybridization temperature with formamide is only 5°C below the Tm. With 50% formamide, the hybridization temperature is usually between 37 and 42°C (see Protocols 13, 14, and 15).

PROTOCOL 13 — PREPARATION OF STOCK SOLUTIONS

1. Prepare deionized formamide.

   a. Add into a 10-ml tube 1/10° of mixed bed resin (200 to 400 mesh) (Biorad) and 9/10° of formamide (Merck)
   b. Stir gently for 30 min in darkness
   c. After decantation or filtration, keep the supernatant
   d. Deionized formamide can be stored at −20°C

The role of formamide is to reduce the hybridization temperature.

2. Prepare SSC 20× (saline sodium citrate buffer): NaCl (175.3 g); sodium citrate (88.2 g)

   a. Dissolve in 800 ml of distilled water
   b. Adjust pH to 7 with 1 *N* NaOH
   c. Make up final volume to 1 ℓ
   d. Sterilize by autoclaving
   e. Conserve the solution at room temperature

The hybridization between two sequences of nucleic acids is directly dependent on ionic force. This solution stabilizes the pH of the hybridization buffer.

3. Prepare 50% dextran sulfate (Sigma)
   It is preferable to prepare this solution just before using.

   a. To prepare 100 ml of a 50% solution, weigh 50 mg of dextran sulfate (PM = 500,000) in an Eppendorf tube
   b. Add 65 μl of autoclaved distilled water
   c. Dissolve the powder completely with a Pasteur pipette tip
   d. Centrifuge a few seconds at 12,000 rpm to eliminate bubbles

The dextran sulfate accelerates the speed of the hybridization process.

4. Prepare 100× Denhardt's solution
   BSA (Sigma), 2 g
   Polyvinylpyrolydone (Sigma), 2 g
   Ficoll 400 (Sigma), 2 g

   a. Adjust the volume to 100 ml with autoclaved distilled water
   b. After dissolution, store aliquots at −20°C

The role of this solution is to diminish the background by elimination of electrostatic reactions. Besides, the presence of high-molecular-weight molecules favor the hybridization.

5. Denaturated salmon sperm DNA

   a. Dissolve the DNA (Boehringer) in autoclaved distilled water to a concentration of 25 mg/ml. The dissolution can be helped by stirring on a magnetic stirrer for several hours.
   b. Denature the DNA for 10 min by heating at 95°C
   c. Cool in ice
   e. Store at −20°C in small aliquots

Before every use heat the DNA at 95°C for 5 min and chill it quickly in ice.

6. Prepare 1 *M* dithiotreitol (DTT)

Dissolve 1.5 g in 10 ml of 0.01 *M* sterilized sodium acetate (pH 7.2). Store at −20°C in small aliquots. Avoid freezing after freezing.

PROTOCOL 14 — PREPARATION OF HYBRIDIZATION BUFFER

50% Dextran sulfate	200 μl
Deionized formamide	500 μl
20× SSC	200 μl
100× Denhardt's solution	10 μl
Denaturated salmon sperm DNA (25 mg/ml)	10 μl
1 *M* DTT	10 μl
Autoclaved distilled water	70 μl

The viscosity of the dextran solution is very important with respect to the difficulty to pipette. Therefore, it is preferable to prepare this solution first and to add the other solutions directly in the same tube.

PROTOCOL 15 — HYBRIDIZATION PROCESS

1. Denaturate the labeled probe by boiling for 5 min
2. Cool immediately in ice
3. Centrifuge a few seconds at 10,000 rpm
4. Add the hybridization buffer to have a probe concentration of 50 ng/ml
5. Mix gently
6. Deposit a drop of 5 μl on a ceramic plate for dilution (Pelanne Instruments)
7. Float one grid per drop
8. Humidify with the hybridization buffer and close the plate
9. Incubate 4 h at 37°C

The hybridization temperature is determined from the Tm. It has been observed that no more hybridization is obtained after 5 h of hybridization. A long time of incubation could involve a loss of ultrastructure preservation, but in our experiments, we have not observed morphology changes after an overnight incubation.

VII. WASHING

The aim of washing is to eliminate the aspecific hybridization. As for the hybridization step, the two parameters of washing are the temperature and the ionic force. During washing the temperature is increased and the sodium concentration reduced (see Protocol 16). Two notions are important in the washing step: the stringence that allows eliminating the aspecific hybridization and the time that allows eliminating the unhybridized probe. These two parameters are dependent on the probe and have to be determined experimentally.

PROTOCOL 16 — WASHING CONDITIONS

Reagents

- Formamide (Merck)
- 20× SSC

Method

Float the grids successively on:

a. 100 μl of 50% formamide, 2× SSC for 15 min
b. 100 μl of 2× SSC for 15 min
c. 100 μl of 1× SSC for 15 min
d. 100 μl of 0.5× SSC for 30 min
e. 100 μl of 0.5× SSC for 30 min at 42°C

VIII. AUTORADIOGRAPHY: THE RING METHOD

This method was introduced by Caro et al.[29] It allows one to emulsify directly ultrathin sections previously collected on grids.

A. MATERIAL

1. Grids (Ni, 200 mesh) — The ultrathin sections are collected on grids coated with a film of Formvar or collodion to provide a smooth surface for coating. The ribbons are collected onto the bright side of the grids, thereby making it easier to visualize them in a dark room. The grids are then covered with carbon.
2. Ilford L4 emulsion — It is used because of the quality of its grains whose diameter is 0.18 μm (supplied in a 50-ml flask; delivery time 5 to 6 weeks; to be stocked at 4°C; well away from any source of radioactivity).
3. Platinum or nickel/chrome ring, 10/30 mm, mounted on a mandrin
4. Hygrometer
5. Dark room with one or several red lamps with a Kodak Wratten II filter — The distance between the lamp and the working area should be at least 1 m under indirect lighting.
6. Porcelain spoon
7. Double-sided sellotape and glass slides

8. Kodak Microdol X
9. Actigel
10. Black sellotape, aluminum foil
11. Borel tube
12. Two waterbaths — The first preset at 40°C, the other at 25°C, taking care to cover or remove the control lamp
13. A Balzer's type grid support
14. Black plastic Kartell box
15. Sodium thiosulfate: $Na_2S_2O_3$-$5H_2O$

B. METHOD

1. Emulsion Preparation in Dark Room

Ilford L4 emulsion is used for best reproductibility, grain size (0.18 μm), and then resolution. The emulsion L4 is diluted to 40%, in bidistilled water, then placed in a waterbath at 40°C, and slowly stirred for at least 1 h. The emulsion is then cooled down to room temperature. Room temperature should be 28°C and hygrometry 80 to 90%. The emulsion is debubbled before hand by dipping several clean glass slides into it.

2. Grid Emulsification in Dark Room

The grids are placed on top of metal poles of the Balzer's supports, shiny side up. A 10/30 mm platinum ring is dipped in the emulsion. As the emulsion film gels vertically, a gray zone develops at the top and exhibits a metallic sheen when observed in the reflecting safe light (in TEM, this zone appears as a monolayer of crystal with almost 100% reproductibility). The excess emulsion gathers in the lower 1/3. The convection movements of the emulsion gradually subside, then stop. One should then apply the emulsion (upper 1/3 of the ring) by a quick movement on the grid.

Controls should be prepared: collodion-coated, emulsified grids examined immediately on TEM. A monolayer film is obtained, if the conditions have been respected. It is sometimes necessary to bring into play several parameters: dilution of the emulsion, emulsion temperature, room temperature, and degree of hygrometry to obtain an emulsion of good quality.

After a few minutes the grids are then taken with a tweezer (Dumont number 7), then stuck by their edges on a double-sided sellotape previously placed on a glass slide. Three or four slides, each with two or three grids, are prepared per sample so as to develop them at different times.

The slides are stored in plastic boxes containing a desiccant (or Actigel), the boxes thoroughly sealed with black tape, wrapped in aluminum paper, and stored at 4°C away from any source of radioactivity.

3. Development in Dark Room (see Protocol 17)

We use Kodak Microdol X, prepared the day before with commercial powder. It should be kept away from light and contain no air bubbles. The fixative is prepared at the time of use. The dishes, Microdol, distilled water, fixative, and slide box containing the grids should be placed in a room at a temperature of 17°C. The bath temperature should be carefully checked and the duration of the different baths be carefully timed.

PROTOCOL 17 — DEVELOPMENT

	Temperature (°C)	**Time (min)**
Microdol	17	4
Distilled water	17	1
$Na_2S_2O_3$-$5H_2O$ (30%)	17	10
Distilled water	17	30

4. Grid Contrast

If the number of grids allow it, they can be divided into two batches. The first will be used for classical contrast (see Protocol 18).

PROTOCOL 18 — CLASSICAL CONTRAST

	Time (min)
Saturated uranyl acetate in alcohol 100%	15
Alcohol 100%	3
Lead citrate	3
Distilled water	2 × 3

The second method is treated as follows, so as to eliminate, at least partially, the gelatin, which is the source of many contrast artifacts (see Protocol 19).

PROTOCOL 19 — TO ELIMINATE THE GELATIN

	Time (min)	**Temperature (°C)**
Distilled water	30	37
Acetic acid 0.25 *N*	15	37
Distilled water	2 × 30	20

Then carry out a classical contrast.

Because of the risks of contrast artifacts, it is nevertheless necessary to space out the different staining times so as to let the grids dry properly: first day — development and second day — contrast with lead citrate.

IX. COMMENTS

In this model, we have obtained a better ultrastructural preservation with quick freezing and Lowicryl K11M embedding than with Lowicryl K4M embedding. Organelles such as mitochondria and endoplasmic reticulum are more preserved; the localization of the labeling is easier. However, only one reduced thickness (4 to 5 μm) is quickly frozen and the implementation of this technique requires a lot of material. The use of this technique is required when the cells or tissues are fragile like tumoral cells. Lowicryl K4M represents a morphological compromise and a large number of cells or tissue fragment can be embedded in this type of resin.[23,24] In terms of sensitivity, we have not observed a difference in the intensity of the labeling between the two methods. In all cases the sensitivity is low compared with the other methods of ISH at the ultrastructural level because of the difficulty of the probe to penetrate inside the Lowicryl section. We have demonstrated (data not shown) that radiolabeled probes penetrate a few μm inside a 50-μm Lowicryl section. We have compared ^{35}S- and ^{3}H-labeled probes. ^{3}H gave the best resolution, but the exposure time was very long (3 to 6 months). ^{35}S probe seems to be a good compromise between resolution and exposure time. However, if radiolabeled probes are more sensitive, the best resolution is obtained with biotinylated probe.[12,23]

ACKNOWLEDGMENT

This research was supported by the Ligue contre le Cancer du Rhone.

REFERENCES

1. **Puvion-Dutilleul, F. and Puvion, E.,** Ultrastructural localization of defined sequences of viral RNA and DNA by *in situ* hybridization of biotinylated DNA probes on sections of Herpes simplex virus type 1 infected cells, *J. Electron Microsc. Tech.*, 18, 336, 1991.
2. **Puvion-Dutilleul, F., Bachellerie, J. P., and Puvion, E.,** Nucleolar organization of HeLa cells as studied by *in situ* hybridization, *Chromosoma*, 100, 395, 1991.
3. **Puvion-Dutilleul, F. and Puvion, E.,** Sites of transcription of adenovirus type 5 genomes in relation to early viral DNA replication in infected HeLa cells. A high resolution *in situ* hybridization and autoradiographical study, *Biol. Cell*, 71, 135, 1991.
4. **Troxler, M., Pasamontes, L., Egger, D., and Bienz, K.,** *In situ* hybridization for light and electron microscopy: a comparaison of methods for the localization of viral RNA using biotinylated DNA and RNA probes, *J. Virol. Methods*, 30, 1, 1990.
5. **Escaig-Haye, F., Grigoriev, V., Peranzi, G., Lestienne, P., and Fournier, J. G.,** Analysis of human mitochondrial transcripts using electron microscopic in situ hybridization, *J. Cell Sci.*, 100, 851, 1991.

6. **Géraud, M. L., Herzog, M., and Soyer-Gobillard, M. D.,** Nucleolar localization of rRNA coding sequences in *Prorocentrum micans* Ehr. (dinomastigote, kingdom protoctist) by *in situ* hybridization, *BioSystems,* 26, 61, 1991.
7. **Wachtler, F., Mosgöller, W., and Scwarzacher, H. G.,** Electron microscopic *in situ* hybridization and autoradiography: localization and transcription of rDNA in human lymphocyte nucleoli, *Exp. Cell Res.,* 187, 346, 1991.
8. **Tourmente, S., Savre-Train, I., Berthier, F., and Renaud, M.,** Expression of six mitochondrial genes during Drosophila oogenesis: analysis by *in situ* hybridization, *Cell Differ. Dev.,* 31, 137, 1990.
9. **Guitteny, A. F. and Bloch, B.,** Ultrastructural detection of the vasopressin messenger RNA in the normal and Brattleboro rat, *Histochemistry,* 92, 277, 1989.
10. **Morel, G., Chabot, J. G., Gossard, F., and Heisler, S.,** Is atrial natriuretic peptide synthesized and internalized by gonadotrophs?, *Endocrinology,* 124, 1703, 1989.
11. **Trembleau, A., Calas, A., and Fevre-Montange, M.,** Ultrastructural localization of oxytocin mRNA in the rat hypothalamus by *in situ* hybridization using a synthetic oligonucleotide, *Mol. Brain Res.,* 8, 37, 1990.
12. **Le Guellec, D., Frappart, L., and Wilhems, R.,** Ultrastructural localization of fibronectin mRNA in chick embryo by *in situ* hybridization using ^{35}S or biotin labeled cDNA probes, *Biol. Cell.,* 70, 159, 1990.
13. **Pomeroy, M. E., Lawrence, J. B., Singer, R. H., and Billings-Gagliardi,** Distribution of myosin heavy chain mRNA in embryonic muscle tissue visualized by ultrastructural *in situ* hybridization, *Dev. Biol.,* 143, 58, 1991.
14. **Silva, F. G., Lawrence, J. B., and Singer, R. H.,** Progress toward ultrastructural identification of individual mRNAs in thin section: myosin heavy chain in developing myotubes, in *Techniques in Immunocytochemistry,* Vol. 4, Bullock, G. and Petrusz, P., Eds., Academic Press, London, 1989, 77.
15. **Tong, Y., Zhao, H., Simard, J., Labri, F., and Pelletier, G.,** Electron microscopic autoradiographic localization of prolactin mRNA in a rat pituitary, *J. Histochem. Cytochem.,* 37, 567, 1989.
16. **Wenderroth, M. A. and Eisenberg, B. R.,** Ultrastructural distribution of myosin heavy chain mRNA in cardiac tissue: a comparison of frozen and LR White embedment, *J. Histochem. Cytochem.,* 39, 1025, 1991.
17. **Singer, R. H., Lawrence, J. B., Silva, F., Langerin, G. L., Pomeroy, M., and Billings-Gagliardi,** Strategies for ultrastructural visualization of biotinated probes hybridized to messenger RNA *in situ,* in *Current Topics in Microbiology and Immunology,* Vol. 143, Springer-Verlag, Berlin, 1989, 55.
18. **Singer, R. H., Langevin, G. H., and Lawrence, J. B.,** Ultrastructural visualization of cytoskeletal messenger RNAs and their associated proteins using double label *in situ* hybridization, *J. Cell Biol.,* 108, 2343, 1989.
19. **Cochrane, A. W., Perkins, A., and Rosen, C. A.,** Identification of sequences important in the nucleolar localization of human immunodeficiency virus rev: relevance of nucleolar localization to function, *J. Virol.,* 64, 881, 1990.
20. **Morel, G., Dihl, F., and Gossard, F.,** Ultrastructural distribution of GH mRNA and GH intron I sequences in rat pituitary gland: effects of GH releasing factor and somatostatin, *Mol. Cell Endocrinol.,* 65, 81, 1989.
21. **Binder, M., Tourmente, S., Roth, J., Renaud, M., and Gehring, W. J.,** *In situ* hybridization at the electron microscope level: localization of transcripts on ultrathin sections of lowicryl K4M-embedded tissue using biotinylated probes and protein A-gold complex, *J. Cell Biol.,* 102, 1646, 1986.
22. **Puvion-Dutilleul, F. and Puvion, E.,** Analysis by *in situ* hybridization and autoradiography of sites of replication and storage of single and double-stranded adenovirus type 5 DNA in lytically infected HeLa cells, *J. Struct. Biol.,* 103, 280, 1990.

23. **Le Guellec, D., Trembleau, A., Pechoux, C., Gossard, F., and Morel, G.,** Ultrastructural non radioactive *in situ* hybridization of GH mRNA in rat pituitary gland: pre-embedding vs ultrathin frozen sections vs post-embedding, *J. Histochem. Cytochem.*, in press.
24. **Le Guellec, D., Frappart, L., and Desprez, P. Y.,** Ultrastructural localization of mRNA encoding for the EGF receptor in human breast cell cancer line BT 20 by *in situ* hybridization, *J. Histochem. Cytochem.*, 39, 1, 1991.
25. **Brangeon, J., Nato, A., and Forchioni, A.,** Ultrastructural detection of ribulose-1,5-biphosphate carboxylase protein and its subunit mRNAs in wilde type and holoenzyme-deficient Nicotiana using immuno-gold and *in situ* hybridization techniques, *Planta*, 177, 151, 1989.
26. **Carlemalm, E., Garavito, R. M., Villiger, W.,** Resin development for electron microscopy: an analysis of embedding at low temperature, *J. Microsc.*, 126, 123, 1982.
27. **Acetarin, J. O., Carlemalm, E., and Villiger, W.,** Developments of new Lowicryl resins for embedding biological specimens at even lower temperatures, *J. Microsc.*, 143, 81, 1986.
28. **Maniatis, T., Fritsch, E. F., and Sambrook, J.,** *Molecular cloning. A Laboratory Manual,* Cold Spring Harbor Laboratory, Cold Spring Harbor, NY, 1982.
29. **Caro, L. G. and Van Tubergen, R. P.,** High-resolution autoradiography. I. Methods, *J. Cell Biol.*, 15, 173, 1962.

Chapter 8

ELECTRON MICROSCOPIC *IN SITU* HYBRIDIZATION OF rRNA ON LOWICRYL SECTIONS

Jean-Guy Fournier and Françoise Escaig-Haye

TABLE OF CONTENTS

0-8493-4414-X/93/$0.00 + $.50

I. INTRODUCTION

The analysis of gene transcription can be carried out at the cellular level using the *in situ* hybridization technique and light microscopy. This approach is a powerful tool now widely used to determine in which type of cell either cellular or viral RNA is present and to localize at the single cell level the specific regions in which RNA is preferentially found.[1-6] However, it does not permit observation of the exact position of the nucleic acid molecules in relation to the fine structure of the cell and its organelles. This analysis is only possible if one is able to observe molecule detection with a high cell resolution. Several attempts have been made, particularly during the past few years, to adapt an *in situ* hybridization protocol to electron microscopy. The basic methods, such as pre-embedding,[7-12] post-embedding,[13-15] and ultracryomicrotomy[16] have been used, and also whole mounted cells deposited on grids.[17-19] With the introduction of nonisotopically labeled probes, especially those which have incorporated biotinylated nucleotides,[20,21] this approach offers a very rapid and effective method for the detection of cellular[22-26] and viral[27-31] nucleic acid sequences providing the highest possible resolution, since the detection of biotin occurs at the exact site of hybridization.

One demonstration of the necessity of visualizing the transcripts integrated in their subcellular environment is the analysis of genome expression from a cytoplasmic organelle such as mitochondria. In this case, the ultrastructural detection of nucleic acid sequences is the only means of examining the molecular aspect of the relationship between structure and function, since the morphological organization of the mitochondria is visible only under the electron microscope. Another example is a cellular organelle localized in the nucleus, the nucleolus representing the morphological expression of the nuclear ribosomal genes. For a better understanding of the nucleolus morphofunctional aspect, it is essential to detect the products of the ribosomal gene expression in relation with the different substructures of the organelle.[32]

In this chapter, we will describe a post-embedding methodology using the Lowicryl hydrosoluble resins and biotinylated probes coding for ribosomal

RNA of either cellular or mitochondrial origin and subsequently identified by a colloidal gold immunocytochemistry marker technology to reveal the site of the specific hybrid formation.

II. FIXATION

A. FIXATIVES

The objective in fixation is double when studying cells by a method of *in situ* hybridization and constitutes a central question for the methodology. Fixatives should preserve tissue and cell morphological integrity and conserve endogenous nucleic acid molecules, thus permitting probe access to their sequences. In practice, it is difficult to fully meet these criteria because the best conditions of fixation generally affect the efficiency of hybridization reaction. Thus, a compromise depending on several parameters (e.g., type of cell and probe and system of visualization) should be found to optimize the protocol of fixation. A wide variety of fixatives is used for *in situ* hybridization at the light microscope level, but at the present time it seems that aldehyde-based fixatives, such as paraformaldehyde and glutaraldehyde,[2,21,33] have the preference of the majority of authors because they permit the best cell RNA retention particularly when protease digestion is required. However, it is worth noting that glutaraldehyde at high concentrations should only be used when the explored cellular RNA is present in very high abundance. Indeed, this dialdehyde fixes the RNA-associated proteins too extensively, thus inducing carbon bridges between the molecules and partially too extensively, thus inducing carbon bridges between the molecules and partially denying the probe access to the RNA sequences. It now appears well established that the most suitable base fixative for general purposes is constituted by paraformaldehyde, either alone or combined with a low concentration of glutaraldehyde. It is clear that these fixatives are particularly well suited to electron microscopic *in situ* hybridization studies.

We currently use two types of fixative: paraformaldehyde 4% and glutaraldehyde 0.1% (fixative A) and paraformaldehyde 2% and glutaraldehyde 0.2% (fixative B).

PROTOCOL 1 — FIXATIVE PREPARATION

The mixture of aldehydes is diluted in phosphate buffer:

- Stock solution A: NaH_2PO_4 0.4 *M*
- Stock solution B: Na_2HPO_4,12 H_2O 0.4 *M*
- Stock solution C: 16 ml A + 84 ml B pH 7.40

Solution C is diluted 1/2 in distilled water before being used for fixation.

Dissolve 2 g of paraformaldehyde in 25-ml distilled water by heating to 65°C and by adding 1 *N* sodium hydroxide until the solution clears. Once the solution is cool, the fixative is prepared as follows:

	Fixative A	**Fixative B**
8% Paraformaldehyde in water	5 ml	2.5 ml
0.2 *M* Phosphate buffer pH 7.40	5 ml	5 ml
10% Glutaraldehyde[a]	0.1 ml	0.2 ml
Distilled water	—	2.3 ml

[a] The glutaraldehyde solution is prepared extemporarily from 25% stock glutaraldehyde diluted in distilled water.

B. MATERIAL FIXATION

1. Cells

Monolayers of cells growing on supports are harvested after trypsinization and washed in culture medium, while cells cultivated in suspension are directly washed in medium using centrifugations at 1200 rpm for 20 min.

Remove the medium and then add the aldehyde fixative solution prepared extemporarily. Pour the fixative slowly on the side of the centrifuge tube so as not to destroy the pellet.

Fix cells for 2 h at +4°C. After 1 h, the pellet is gently detached from the bottom of the tube using a Pasteur pipette. The fixative is changed and the pellet is kept floating for 1 h.

The cell pellet is next rinsed in buffer solution and cut in several small fragments before the dehydration step.

2. Tissue Samples

Human (biopsies and post-mortem human material) and animal tissues samples can be fixed by immersion in the fixative. Small (about 1 mm^3) pieces of organ should be used. With animal tissues, fixation can also be obtained by vascular perfusion, placing a cannula into the aorta via the left ventricle.

After the fixation step, tissue samples are rinsed 2 × 5 min in phosphate buffer 0.1 *M*. They can be stored in a buffer complemented of 0.3 *M* saccharose for several weeks before embedding.

III. EMBEDDING PROCEDURE IN LOWICRYL

A. CHARACTERISTICS OF THE RESINS

Lowicryls are acrylate-methacrylate mixtures of low viscosity that polymerize at temperatures from −35 to −80°C under indirect UV light. They were developed originally for embedding tissues at low temperatures to improve the preservation of supramolecular and molecular organization and antigenicity.[34]

TABLE 1
Lowicryl Resins

Resin	Polar	Hydrophilic	Lowest usable temperature
K4M	+	+	−35°C
HM20	−	−	−40°C
K11M	+	+	−60°C
HM23	−	−	−80°C

The general schematic formula is as follows:

CH_2=C(CH^3)-CO-O-R, methacrylate
Ch_2=CH-CO-O-R, acrylate
CH_2 =C(CH^3)-CO-O-R-O-CO-(CH^3)C=CH^2, dimethylacrylate cross-linker

Among resins available (see Table 1), K4M represents the most popular for immunocytochemistry.[35-37] We currently use this type of resin for our *in situ* hybridization studies.[24,30,31,38,39] After several experiments we have observed that the HM20 resin improves the morphology, with an hybridization efficiency similar to K4M resin. However, a high background was sometimes observed. We suggest using this resin when the morphological quality of the material is crucial. Other acrylic resins such as LR Gold[23] or LR White[11,29,40] have also been used with success for post-embedding electron microscopic *in situ* hybridization. To date, no study has been carried out to compare the performance of these different resins.

1. Lowicryl K4M Resin

K4M is a polar hydrophilic resin which is miscible with all polar dehydration agents including ethylene glycol. This type of resin is assumed to maintain the intramolecular and intermolecular hydrogen bonds, replacing water by the resin-solvent solution.

The composition of the monomer is as follows:

Monomer	Wt %	Side chain (R)
Hydroxypropyl methacrylate	48.4 g	–CH^2–CH^2–CH^2–OH
Hydroxyethyl acrylate	23.7 g	–CH^2–CH^2–OH
n-Hexyl methacrylate	9.0 g	–CH^2–(CH^2)4–CH^3
Crosslinker		
Triethylene glycol dimethacrylate	13.9 g	–(CH^2–CH^2–O)^3–

2. Lowicryl HM20 Resin

HM20 is a nonpolar hydrophobic resin which is miscible with most polar dehydrating agents, but not with ethylene glycol. It can be used at lower temperatures than Lowicryl (see Table 1).

The composition is as follows:

Monomer	Wt %	Side chain (R)
Ethyl methacrylate	68.5 g	$-CH^2-CH^3$
n-Hexyl methacrylate	16.6 g	$-CH^2-(CH^2)4-CH^3$
Crosslinker		
Triethylene glycol dimethacrylate	14.9 g	$-(CH^2-CH^2-O)^3-$

Lowicryl resins are commercially available in kits containing three components A, B, and C.

	K4M	HM20
Crosslinker A	2.70 g	2.98 g
Monomer B	17.30 g	17.02 g
Initiator C	0.10 g	0.10 g

The initiator compound necessary to start the reaction by activating the methacrylate monomer is generally dibenzoyl peroxide.

PROTOCOL 2 — LOWICRYL RESIN PREPARATION

- Weigh out in a brown glass container the crosslinker and the monomer. Gently mix the compounds with a glass rod.
- Add the initiator and mix until complete dissolution in the resin. Since the resin components have a low viscosity, the mixture is stirred gently to avoid air bubbles and subsequent oxygen dissolution in the resin that can interfere during the polymerization. Make sure that the mixture is thoroughly homogeneous before use.

Safety — Vapors of Lowicryl resins are highly toxic, and care must be taken when handling mixtures. Use a fume hood and wear gloves for subsequent steps involving use of the resins, since contact with the skin may cause eczema.

B. DEHYDRATION AND INFILTRATION

Cells and tissue samples are prepared for Lowicryl resin embedding according to the progressive lowering temperatures (PLT) technique. The protocol provided for the K4M resin is also used in our laboratory for the HM20 resin (see Protocols 3 and 4).

PROTOCOL 3 — DEHYDRATION

30% Ethanol at 0°C, 30 min
50% Ethanol at −20°C, 60 min
70% Ethanol at −20°C, 60 min
90% Ethanol at −20°C, 60 min
100% Ethanol at −20°C, 60 min

PROTOCOL 4 — INFILTRATION

100% Ethanol/K4M (1/1) at −20°C, 60 min
100% Ethanol/K4M (1/2) at −20°C, 60 min
100% Lowicryl at −20°C, 60 min
100% Lowicryl at −20°C, overnight

The samples are next put in gelatin or Beem capsules, which should be completely filled with the resin and capped in order to avoid trapping air bubbles for reasons explained above.

C. POLYMERIZATION

The polymerization reaction is carried out at low temperature (−35°C) under indirect UV irradiation.

There are several systems to perform the polymerization. An apparatus is provided by Balzer Co. which offers security and easy handling.

However, it is possible to construct a polymerization chamber at low cost with high efficiency. Whatever the system chosen, the important thing is that it should permit indirect UV-irradiation on all sides of the capsules filled with resin and containing the samples.[41]

Equipment needed:

- Plexiglass plaque 10 × 20 × 0.5 cm with holes
- Cross-piece support and grippers
- UV lamps of 15 W emitting 360 nm wavelength
- Aluminum foil
- Freezing chamber to −35°C

PROTOCOL 5 — POLYMERIZATION

The capsules are held in a 0.5-cm thick plexiglass plaque drilled with several holes of a diameter slightly smaller than that of the capsules.

The plaque is maintained by a gripper to the cross-piece support and is placed 30 cm below the UV lamps covered with aluminum foil. The system is set up in a freezer with a chamber of 50 × 60 × 80 cm, the walls of which are covered with aluminum foil, thus permitting the reflection of UV light by the capsules.

The polymerization is continued for at least 48 h at −35°C.

The system is next removed from the cold and the capsules are kept under UV irradiation for an additional 2 to 3 d at room temperature.

If the resin is not of the correct hardness, UV irradiation may be continued at room temperature.

Comment: The temperatures and times we use for embedding in Lowicryl are slightly different than those generally recommended. This indicates that the resins are sufficiently versatile to allow embedding at different temperatures and under different conditions, which should be adapted by each investigator. However, they have some limitations, including bubble formation in the block and evolution of heat during polymerization. The fact that the results are nonreproducible remains a critical aspect of the use of these acrylic resins.

IV. SECTIONING

A. GENERAL CONSIDERATIONS

The Lowicryl resins can be cut in similar ways than others resins, using either glass or diamond knives. Though it is recommended to use a speed with ultramicrotome higher than for Epon (2 to 5 mm/s vs. 1 to 1.5 mm/s), we currently cut the Lowicryl resins with the speed used for Epon.

When hydrophilic polar resins such as K4M are sectioned, they should be kept dry during the manipulation. This can be reached with a level of water in the trough slightly lower than normal.

Generally we cut sections with a thickness of about 80 nm. It should be noted that the surface of Lowicryl sections are rougher than the surface of Epoxy sections because the resins are in way less hard, and the diamond knife scrapes in cutting the surface. Consequently, supramolecular structures of thickness estimated at about 3 nm emerge from the resin surface, facilitating the detection of the molecules without the necessity of removing part of the resin after etching treatment as for Epoxy resins. This constitutes with the conservation of native structure of the molecules discussed above, another advantage that may explain the high performance of Lowicryl resins for post-embedding detection.

Comment: The difficulty for sectioning can be sometimes linked with the fact that the Lowicryl blocks contain humidity. Store the blocks in a dry atmosphere keeping them *in vacuo* with desiccant.

B. GRIDS

Copper grids are not adequate for *in situ* hybridization studies because the formamide present in the hybridization buffer is highly reactive with the metal. Thus, as usually employed in immunocytochemistry techniques, nickel or gold grids are appropriate. In our laboratory, thin sections are mounted on naked-gold grids of 600 mesh, preliminary degreased in acetone for at least 1 h. The grids can be stored several days in dry atmosphere before use.

V. PREPARATION OF THE PROBE

The ribosomal probe of nuclear origin corresponds to the *E. coli* fragment of mouse genomic DNA encompassing the sequences of 18S, 5.8S, and 28S. The 6.7-kb fragment is inserted into pBR-322.[24,38] (Figures 1 to 4).

The mitochondrial probe contains genes coding for the 16S and 18S ribosomal sequences. After restriction of enzyme digestion of the human

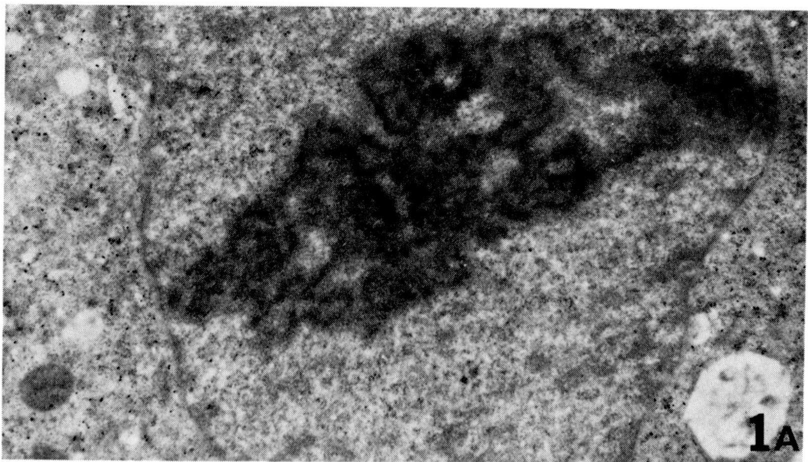

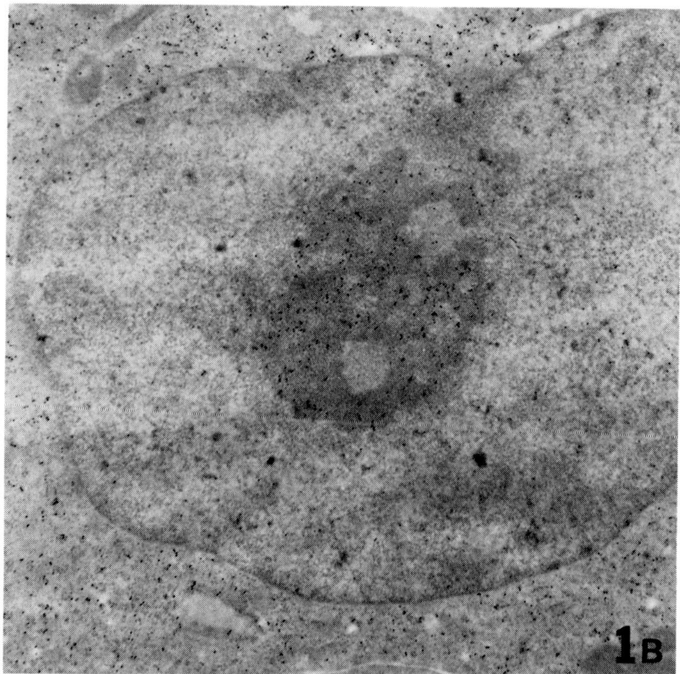

FIGURE 1. Immunogold detection of nuclear rRNA sequences in Vero cells. (A) Direct procedure of biotinylated probe visualization with streptavidine conjugated with 10-nm gold particles. The labeling is localized principally on the dense fibrillar component of the nucleolus and on the cytoplasm. (B) Indirect procedure with anti-rabbit anti-biotin as a first antibody and goat anti-rabbit conjugated with gold particles of 10 nm as a secondary antibody. The intensity of labeling is higher than in (A). (A) and (B) ×13,000.

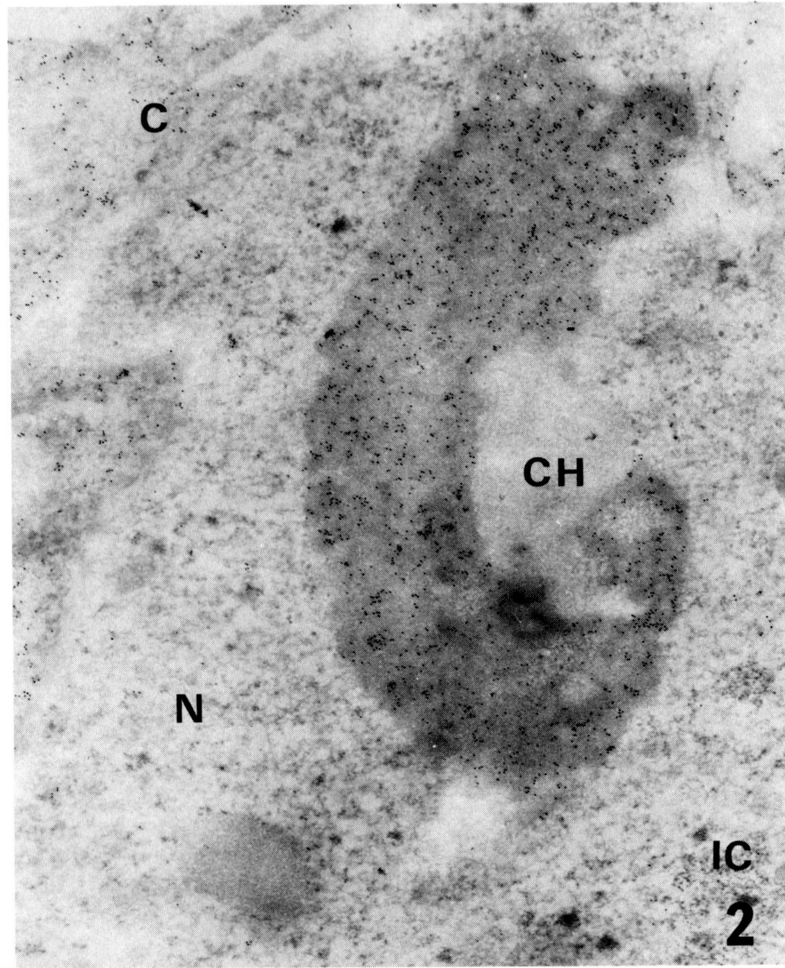

FIGURE 2. Immunogold detection of nuclear rRNA sequences in L mouse cells. Intense labeling is observed on the nucleolus after indirect procedure. The nucleoplasm (N) does not react with the ribosomal probe. CH (chromatin), IC (interchromatin grains), and C (cytoplasm) $\times$15,000.

mitochondrial genome, the 1.5-kb XbaI 1-2 DNA fragment has been cloned in pUC 19[38] (Figure 5).

As mentioned by Singer et al.,[43] double-strand DNA constituted by the entire recombinant plasmid (vector + insert) represents the best choice of probe. Indeed, it has been demonstrated that the vector contributes to form a network, amplifying the number of reporter molecule at the hybridization site.

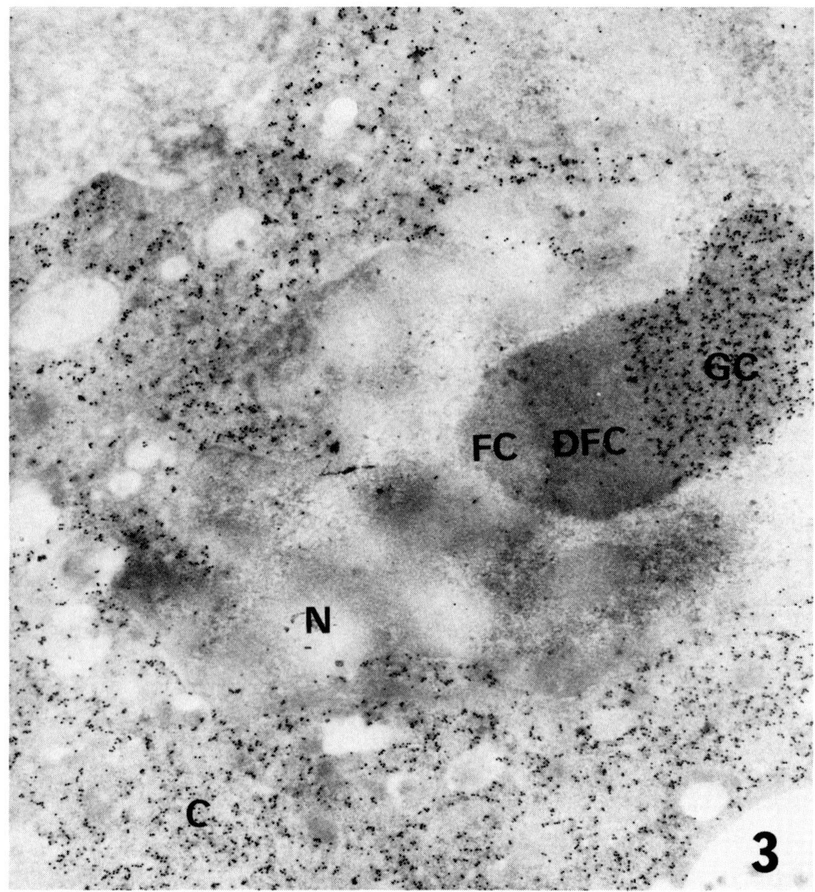

FIGURE 3. Immunogold detection of nuclear rRNA sequences in L mouse cells treated with Actinomycin D for 2 h. On the segregated nucleolus, the labeling is exclusively associated with the granular component (GC). Fibrillar center (FC), dense fibrillar component (DFC), nucleoplasm (N). The cytoplasm (C) is intensively labeled on ribosome-rich areas. ×21,000.

PROTOCOL 6 — PROBE LABELING (NICK TRANSLATION)

Reagents used and stored at −20°C

- Recombinant plasmid probe diluted in water (1 μg/μl)
- Nick translation buffer ×10
  Tris-HCl 50 m*M*; $MgCl_2$ 10 m*M*; DTT 100 m*M*; 50 μg/ml BSA; pH 7.2
- dNTP (Pharmacia) stock 100 m*M*
- DNase I (Boehringer) 5 μg/ml diluted in distilled water with 50% glycerol.

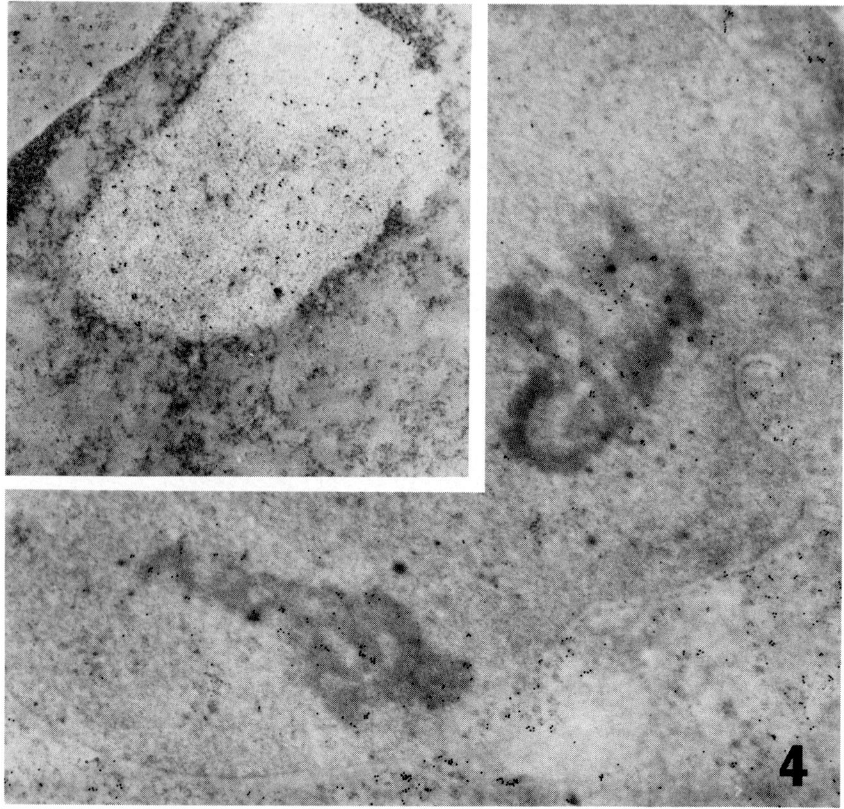

FIGURE 4. Immunogold detection of nuclear rRNA sequences in human lymphoid cells (CEM). The intensity of nucleolar labeling is similar with (insert) or without protease pretreatment. ×14,000.

- DNase polymerase *E. coli* (Boehringer)
- Bio-11-dUTP (Enzo Diagnostics)
- EDTA 500 m*M*; pH 8
- Ethanol 100%
- Sonicated salmon sperm DNA (Sigma) 10 mg/ml in distilled water
- *E. coli* RNA (Sigma) 20 mg/ml in distilled water
- Waterbath placed in cold room and adjusted to 15°C

Method

1. Add in Eppendorf tube the following in order
   27 µl sterile distilled water
   10 µl Bio-11-dUTP
   5 µl buffer

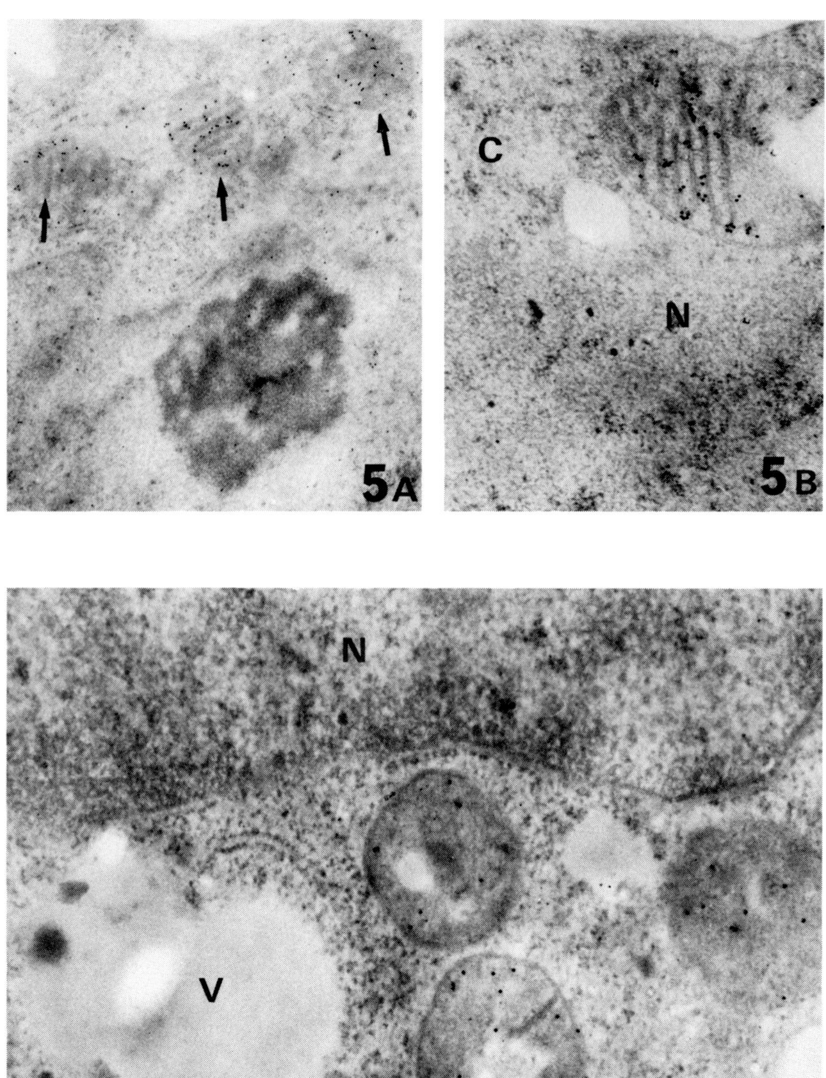

FIGURE 5. Immunogold detection of mitochondrial rRNA sequences in CEM cells (A) and (B) and in human blood lymphocytes (C). The hybridization gold signal is localized preferentially over mitochondria (arrows) close to the inner membrane particularly the cristae (B). N (nucleus), C (cytoplasm), V (vacuole). (A) ×13,000; (B) ×26,000; (C) ×26,000.

5 μl dNTP (CTP, GTP, ATP) diluted 1:100 in distilled water
1 μl probe
5 μl DNase I

2. Mix and touch spin
3. Incubate for 2 to 3 min at 37°C
4. Stop the reaction in ice and add 2 μl DNA polymerase
5. Gently agitate the mixed reagents **(Do not use vortex)**
6. Incubate for 75 min at 15°C
7. Stop the reaction by addition of 3 μl of EDTA

The uncorporated nucleotides are eliminated after precipitation in ethanol (see Protocol 7).

PROTOCOL 7 — ETHANOL PRECIPITATION OF LABELED PROBE

1. Add in the nick translated probe tube

   - 5 μl Sonicated salmon sperm DNA 10 mg/ml diluted in water
   - 5 μl *E. coli* RNA (Sigma) 20 mg/ml diluted in water
   - 7 μl Sodium acetate

2. Mix well and touch spin
3. Add 140-μl cold ethanol (100%)
4. Incubate at −20°C overnight
5. Spin for 30 min
6. Discard the supernatant
7. Rinse the pellet four times with 70% ethanol without resuspension
8. Keep dry 2 to 3 h at room temperature.

Comment: To improve the quality of a radioactive DNA probe it can be made a phenolchloroform extraction before ethanol precipitation. In the case of a DNA probe labeled with biotin, this step should be avoided because biotin is lipophilic and a fraction of DNA could be lost into the organic phase decreasing the amount of recovered probe. The size of DNA is monitored by gel electrophoresis and depends of the concentration of DNase I. The optimal concentration should be determined experimentally because the sensitivity of the enzyme varies with the DNA preparations. Generally, for double stranded DNA, the most efficient size of fragments appears to be about 300 nucleotides in length. After drying, the pellet is directly resuspended in hybridization buffer.

VI. PREPARATION OF HYBRIDIZATION BUFFER

The reactivity of the probe with the cellular RNA takes place in a suitable buffer which allows a correct association between the two nucleic acid molecules. Basically the conditions of hybridization are similar to those performed at the light microscope level.[33]

Reagents

Formamide (Fluka puris) (see Protocol 8) — Denaturing agent of nucleic acid duplexes allowing specific hybridization at temperatures lower than that required.

20× SSC (see Protocol 9) — Buffer for maintaining the pH close to the neutrality.

Dextran sulfate (Pharmacia) (see Protocol 10) — This polymer is assumed to accelerate the rate of hybridization reaction.[44]

Sonicated salmon sperm DNA (Sigma) — Used to sature the nonspecific sites of adsorption and reduce the background binding of the probe.

Biotin-labeled DNA probe — Probe concentration usually used is about 1 μg/ml both at the light and electron microscope level. We and others have observed that the optimal concentration to sature the target sites is in the range of 10 to 20 μg/ml.[28–30,38]

PROTOCOL 8 — DEIONIZATION OF FORMAMIDE

Formamide is used after deionization in MB 1 Amberlit monobed (Bio-rad).

1. Treat 100 ml of formamide with 3 g of resin by gentle stirring for 30 min in darkness
2. Eliminate the resin particles by filtration
3. Aliquote in Eppendorf tubes
4. Store at −20°C

Comment: Generally, the formamide is kept no more than 1 month. This is an important point because the quality of this reagent is decisive for the success of the hybridization reaction.

PROTOCOL 9 — 20× SSC HYBRIDIZATION BUFFER

NaCl 173.3 g (3 *M*)
Na citrate 88.2 g (0.3 *M*)

1. Dissolve in 800 ml of water and adjust pH to 6.2[a]
2. Bring up the volume to 1 liter
3. Autoclave to sterilize

[a] Adjustment of the pH to slightly lower than 7 is necessary to reach this value after subsequent dilutions, which influence the protonic activity of this buffer.

PROTOCOL 10 — PREPARATION OF DEXTRAN SULFATE SOLUTION

To prepare rigorously 50% (weight/volume) dextran sulfate, it is necessary to consider the volume occupied by the dextran sulfate powder. The ratio should be in that case 1.3 instead of 2, for obtaining 50 mg of product in a final volume of 100 μl.

1. Weigh out directly in an Eppendorf tube 50 mg of dextran sulfate
2. Add 65 μl of distilled water
3. Mix with a plugged-up end Pasteur pipette until complete dissolution of powder (2 to 3 min).
4. Centrifuge for 30 s to eliminate air bubbles from the liquid

Comment: The advantage of this protocol is obtaining easily and rapidly the dextran sulfate solution, thus allowing its preparation extemporaneously. However, the quality of this reagent is crucial and the storage at −20°C can modify its efficiency. The solution is very viscous and should be taken slowly with a tip cut 0.5 cm from its extremity to facilitate the rising of the liquid.

PROTOCOL 11 — PREPARATION OF THE HYBRIDIZATION MIXTURE

1. Add in the Eppendorf tube containing the dried probe pellet

   - 25 μl formamide deionized
   - 10 μl 20× SSC; pH 6.2
   - 10 μl Dextran sulfate 50%
   - 5 μl sonicated salmon sperm DNA 10 mg/ml

2. Make sure the solution is homogeneous in mixing with a Pasteur pipette
3. Centrifuge 15 s
4. Take the volume necessary for experiments (2 to 3 μl per grids) in an Eppendorf tube, the cap of which is perforated with a needle
5. Denature the probe by plunging the tube in boiling water for 3 min and then chill on ice.
6. The remaining probe hybridization solution is stored at −20°C for only 4 to 5 d

VII. THIN SECTION HYBRIDIZATION PROCEDURES

A. PRETREATMENT OF SECTIONS

Before hybridization, sections can be treated with different enzymatic solutions following the type of information desired (see Protocols 12 and 13)

1. Proteases: To increase the sensitivity of the method by improving the probe access to the RNA sequences. In general, this treatment, efficient

at the light microscope level, gives poor results at the electron microscope level[13,29,38,45] (Figure 4).

2. Nucleases: To control the specificity of the RNA-DNA hybridization reaction by eliminating the RNA sequences at the surface of the ultrathin sections.[25,27,28,38]

PROTOCOL 12 — PROTEASE PRETREATMENT

Proteinase K 20 mg/ml in water
Buffer: 20 m*M* Tris-HCl pH 7.4; 2 m*M* $CaCl^2$

1. Incubate the section with proteinase K diluted 1/100 in the buffer for 15 min
2. Rinse in distilled water

PROTOCOL 13 — NUCLEASE PRETREATMENT

RNase A (Sigma) 10 mg/ml in water
Buffer: 2× SSC
Incubate the sections with the RNAse A diluted 1/10 in the buffer for 1 h.

B. INCUBATION OF SECTIONS WITH PROBES

The purpose of this step is to establish a base specific association, according to the Watson-Crick criteria, between the nucleotide sequences of the DNA probe and the complementary cellular RNA sequences localized at the surface of the Lowicryl section (see Protocol 14).

PROTOCOL 14 — HYBRIDIZATION OF THIN SECTIONS

1. Deposit 2 to 3 μl of denaturated hybridization solution in one well of the multiwell porcelain plaque (Pelanne Instrument)
2. Cover the hybridization solution drop with the grid section side down onto the liquid
3. Repeat following the number of grids used and deposit 100 μl of hybridization buffer in 2 other wells to prevent drying
4. The plaque is then covered with the cap and sealed with parafilm
5. Incubate overnight at 37°C

1. Optimal Incubation Time

By analyzing the time course of hybridization, it is possible to specify the optimal incubation time corresponding to the hybridization of all accessible cellular RNA sequences and survey the state of cell morphology during incubation in the hybridization solution. Indeed, this solution contains formamide known to have a detrimental effect on the cellular components. To date, this analysis has only been carried out at the light microscope level and then extrapolated to the electron microscopic level.[13,16] We have examined the kinetic of DNA-RNA annealing directly at the ultrastructural level in

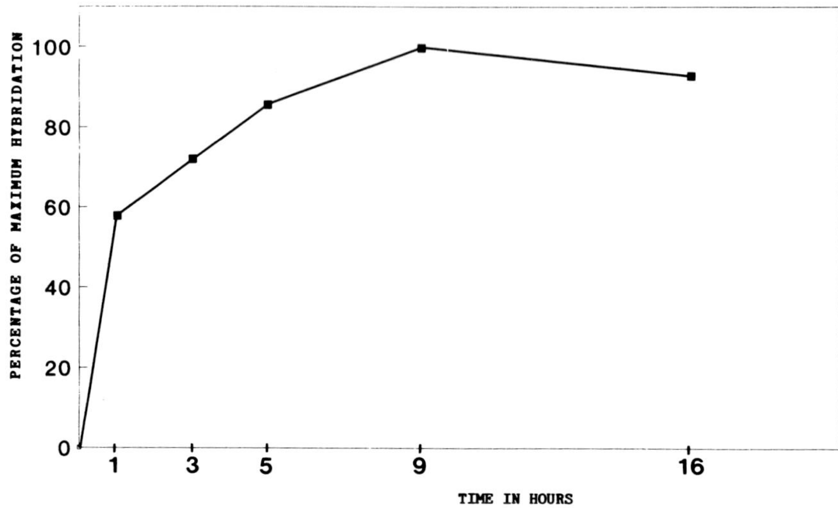

FIGURE 6. Kinetics of hybridization reaction on thin Lowicryl-embedded cell sections incubated with a mitoribosomal probe.

detecting the mitochondria ribosomal RNA sequences (Figure 5). The localization of the hybridization signal in a well-defined cellular organelle greatly facilitates this analysis.[38] As shown in Figure 6 the optimal hybridization reaction time is reached after 9 h incubation, while 57% of the maximum signal is obtained after 1 h. As for the loss of the cell morphology, it is especially visible during the first hour and subsequently remains unchanged. For practical reasons incubation overnight constitutes a good compromise.

C. WASHINGS

To eliminate the unhybridized probe, sections are transferred in several washing solutions. The protocol (see Protocol 15) described is sufficiently stringent to retain only the probe on the specific cellular RNA sequences

PROTOCOL 15 — WASHING CONDITIONS

Reagents:

- Formamide (Fluka) low grade
- 8× SSC prepared from 20× SSC

Float the grids successively on:

- 100 μl of 100% nondeionized formamide diluted 1/1 in 8× SSC twice for 5 min
- 100 μl of 4× SSC twice for 5 min

- 100 μl of 0.2× SSC twice for 5 min
- 100 μl of distilled water once for 2 min

All steps are performed at room temperature.

D. POSTTREATMENT OF SECTIONS

To assess the validity of the hybridization reaction, it is possible to verify if the signal observed on the cells is the result of the RNA-DNA hybrid formation, using RNAse H enzyme (see Protocol 16). The particularity of this enzyme is to digest only RNA sequences which are established hydrogen bonds with DNA sequences.[5] In the present case, the digestion of the cellular RNA molecules hybridized with the labeled DNA probe gives rise to the elimination of the latter and the reduction of the labeling. The best result is obtained when digestion is performed before immunocytochemistry. A post-immunocytochemistry treatment slightly reduces the labeling, probably because of the difficulty in eliminating the probe carrying the gold marker system.[38]

PROTOCOL 16 — POSTHYBRIDIZATION RNase H TREATMENT

Reagents
RNase H. Stock 1U/μl (Boehringer)
Buffer

- 100 m*M* KCl
- 20 m*M* Tris-HCl; pH 7.5
- 1.5 m*M* $MgCl_2$
- 50 μg/ml BSA
- 1 m*M* DTT
- 0.7 m*M* EDTA
- 13 m*M* HEPES

1. Incubate the sections in solution of RNase H at 16 U diluted in 1 ml of the buffer for 1 h at 37°C
2. Rinse in buffer
3. Rinse in distilled water

E. CONTROL PROBE

In parallel with hybridization performed with the specific probe, similar thin sections must be hybridized in the same conditions with a control probe to assess the nonspecific affinity of Lowicryl sections for nucleic acid probe. In general, the control probe used corresponds to the plasmid vector or the pBR322 pasmid labeled in a similar way than that of the specific probe (see Table 2).

TABLE 2
Density of Labeling

Probes	pMtrRNA	pBR322	Buffer	RNaseH
Mitochondria	52.26 + 2.70	1.93 + 0.37	2.96 + 0.47	7.60 + 1.30
Nucleus	2.20 + 0.35	0.91 + 0.19	0.74 + 0.15	1.18 + 0.24
Cytoplasm	0.71 + 0.12	1.10 + 0.17	0.50 + 0.05	0.40 + 0.02

VIII. IMMUNOCYTOLOGICAL VISUALIZATION OF HYBRIDIZED PROBE

A. GENERAL CONSIDERATIONS

The probe hybridized to cellular RNA is visualized by a colloidal gold immunocytology marker following a direct (see Protocol 17) or indirect procedure (see Protocol 18). As is generally observed, the indirect staining leads to greater sensitivity (Figure 1). The double advantage of using colloidal gold as a marker is that it gives the best resolution and allows quantitative studies. It is known that the size of the gold particles influences the resolution and intensity of labeling.[46] We believe that for current experiments, the most appropriate size corresponds to 10 nm gold particles.

PROTOCOL 17 — DIRECT PROCEDURE

Phosphate-buffered solution

NaCl (MW 58.44)	80 g
KCl (MW 74.56)	2 g
$NaHPO_4$, 12 H_2O	28 g
Na_2HPO_4	11 g
KH_2PO_4	2 g
Distilled water	1000 ml
BSA (Sigma fraction V)	

- The buffer is used with 1% BSA PBS/BSA 1%
  or with 0.05% Tween-20 PBS/Tween 0.05%
- Streptavidin conjugated with gold particles of 10 nm (Janssen Life Sciences)

Method

All the steps are done at room temperature.

1. Incubate grids on blocking solution PBS/BSA 1% for 15 min
2. Incubate grids on gold-conjugated streptavidin diluted 1/50 in PBS/BSA 1% for 45 min
3. Rinse the grids in PBS/Tween 0.05% and distilled water for 10 min each

PROTOCOL 18 — INDIRECT PROCEDURE

Reagents

- Rabbit anti-biotin (ENZO Diagnostics)
- Goat anti-rabbit antibody (GAR) complexed with 10-nm gold particles (Jansen Life Sciences)

Method

1. Incubate sections on blocking solution PBS/BSA 1% for 15 min
2. Incubate sections on anti-biotin solution diluted 1% in PBS/BSA 1% for 60 min
3. Rinse grids in PBS/Tween 0.05 twice for 10 min
4. Incubate on GAR solution diluted 1/50 in PBS/BSA 1% for 60 min
5. Rinse in PBS/Tween 0.05% and distilled water for 10 min each

During all steps of the procedure the sections should be kept wet.

B. CONTROLS

Controls should be performed in addition to those mentioned above to evaluate the background noise introduced by the system of visualization (see Table 2).

Immunocytochemistry is carried out on sections after incubation with the hybridization buffer alone. The labeling obtained estimates the level of background noise introduced by the first antibody to be in contact with the section (antibiotin for indirect procedure; streptavidin for direct procedure).

IX. STAINING

Since the osmification of the material is not possible before embedding because it prevents the polymerization, sections offer a very low contrast under the electron beam and thus should be stained (see Protocol 19).

PROTOCOL 19 — STAINING

We routinely used a saturated aqueous solution of uranyl acetate.

1. Keep the sections stained for 15 min
2. Wash in distilled water for twice 5 min
3. Keep the sections dry for 30 min before observation with the electron microscope. To increase the contrast of the image, it is recommended accelerating the electrons with a voltage of 60 kV

X. QUANTITATION

The quantitative analysis of gold particles can be applied for several purposes:

1. Technical controls. To evaluate the specificity of the labeling in comparing its intensity in different cell compartments and after hybridization with unrelated probe or after RNase treatments.
2. Biological analysis. To compare the intensity of the hybridization signal between different physiological situations and different gene expressions.

We have chosen for evaluating the density of labeling an analysis system Mode Biocom 200 to measure the surface area of different cell compartments, such as mitochondria, nucleus, and cytoplasm, the value of which is determined by subtracting the nucleus and mitochondria surfaces from the total surface of the negative electron micrographs. In order to recognize the structures, the micrographs were taken at minimal magnification of × 11,500. The limits of the cell nucleus and mitochondria are drawn on the screen and each area is yielded automatically. For each area, the number of gold particles is counted manually on the corresponding negative electron micrographs. The labeling density is expressed as the mean number of gold particles per μm^2 standard deviation of different samples, and statistical analysis is performed utilizing the Student *t* test determination.

For example, Table 2 shows the labeling density over different cell compartments after hybridization of human lymphoid cells with a probe coding for human mitochondrial rRNA (Figure 5) and compared with controls.

XI. COMMENTS

The methodological approach that we have described here, which permits the identification of RNA molecules with a high resolution in their subcellular environment, offers an appropriate solution for investigating the relationship between gene expression and fine-level cellular organization.

Its principal limitation is related to the low sensitivity which requires the analysis of intracellular nucleic acid sequences present in relatively high abundance. To assess correctly the specificity of the hybridization gold signal, well-controlled experiments should be performed with eventually the introduction of quantitative analysis.

With the use of other reporter molecules introduced in probes which can be identified by immunocytology, one can envisage in the near future the possibility to detect simultaneously two different types of RNA molecules.[30] The molecular analysis at the submicroscopical level can open up a new chapter in the field of molecular cytology and histology.

REFERENCES

1. **Fournier, J. G., Kessous, A., Tiollais, P., Simard, R., Brechot, C., and Bouteille, M.,** Hepatitis B virus genome expression detected in *in situ* hybridization in a human hepatoma cell line, *Biol. Cell,* 44, 197, 1982.
2. **Fournier, J. G., Rozenblatt, S., and Bouteille, M.,** Localization of measles virus acid nucleic sequences in infected cells by *in situ* hybridization, *Biol. Cell,* 49, 287, 1983.
3. **Lawrence, J. B. and Singer, R. H.,** Intracellular localization of messenger RNAs for cytoskeletal proteins, *Cell,* 45, 497, 1986.
4. **Meyer, J. L., Fournier, J. G., and Bouteille, M.,** Expression of integrated hepatitis B virus DNA in PLC/PRF/5, Hep 3B, and L6EC3 cell lines detected by *in situ* hybridization, *Med. Biol.,* 64, 367, 1986.
5. **Lawrence, J. B., Singer, R. H., and Marselle, L. M.,** Highly localized tracks of specific transcripts within interphase nuclei visualized by *in situ* hybridization, *Cell,* 57, 493, 1989.
6. **Fournier, J. G., Prevot, S., Audoin, J., Letourneau, A., and Diebold, J.,** Fine analysis of HIV-1 RNA detection by *in situ* hybridization, in *Modern Pathology of AIDS and Other Retroviral Infections,* Racz, P., Haase, H., and Gluckmann, J. C., Eds., S. Karger, Basel, 1990, 62.
7. **Trembleau, A., Calas, A., and Fevre-Montange, M.,** Ultrastructural localization of oxytocin mRNA in the rat hypothalamus by *in situ* hybridization using a synthetic oligonucleotide, *Mol. Brain Res.,* 8, 37, 1990.
8. **Guitteny, A. F. and Bloch, B.,** Ultrastructural detection of the vasopressine messenger RNA in the normal and Brattleboro rat, *Histochemistry,* 92, 277, 1989.
9. **Wolber, R. A., Beals, T. F., and Massab, H. F.,** Ultrastructural localization of Herpes simplex virus RNA by *in situ* hybridization, *J. Histochem. Cytochem.,* 37, 97, 1989.
10. **Tong, Y., Zhao, H., Simard, J., Labri, F., and Pelletier, G.,** Electron microscopic autoradiographic localization of prolactin mRNA in rat pituitary, *J. Histochem. Cytochem.,* 37, 567, 1989.
11. **Brangeon, J., Nato, N., and Forchioni, A.,** Ultrastructural detection of ribulose-1,5-biphosohate carboxylase protein and its subunit mRNAs in wilde type and holoenzyme-deficient Nicotiana using immuno-gold and *in situ* hybridization techniques, *Planta,* 177, 151, 1989.
12. **Pomeroy, M. E., Lawrence, J. B., Singer, R. H., and Billings-Gagliardi, S.,** Distribution of myosin heavy chain mRNA in embryonic muscle tissue visualized by ultrastructural in situ hybridization, *Dev. Biol.,* 143, 58, 1991.
13. **Binder, M., Tourmente, S., Roth, J., Rebaud, M., and Gehring, W. J.,** *In situ* hybridization at the electron microscope level: localization of transcripts on ultrathin sections of Lowicryl K4M embedded tissue using biotinylated probes and protein A-gold complexes, *J. Cell Biol.,* 102, 1646, 1986.
14. **Webster, H. E., Lamperth, L., Favilla, J. T., Lemke, G., Tesin, D., and Manuelidis, L.,** Use of a biotinylated probe and *in situ* hybridization for light and electron microscopic localization of Po mRNA in myelin-forming Schwan Cell, *Histochemistry,* 86, 441, 1987.
15. **Leguellec, D., Frappart, L., and Desprez, P. Y.,** Ultrastructural localization of mRNA encoding for the EGF receptor in human breast cell cancer line BT 20 by *in situ* hybridization, *J. Histochem. Cytochem.,* 39, 1, 1991.
16. **Morel, G., Dihl, F., and Gossard, F.,** Ultrastructural distribution of GH mRNA and GH intron I sequences in rat pituitary gland: effects of GH releasing factor and somatostatin, *Mol. Cell. Endocrinol.,* 65, 81, 1989.
17. **Hutchinson, N., Langer-Safer, P., Ward, D., and Hamkalo, B.,** *In situ* hybridization at the electron microscope level: hybrid detection by autoradiography and colloidal gold, *J. Cell Biol.,* 95, 609, 1982.
18. **Radic, M. Z., Lundgren, G., and Hamkalo, B. A.,** Curvature of mouse satellite DNA and condensation of heterochromatine, *Cell,* 50, 1101, 1987.

19. **Singer, R. H., Langevin, G. L., and Lawrence, J. B.,** Ultrastructural visualization of cytoskeletal messenger RNAs and their associated protein using double label *in situ* hybridization, *J. Cell Biol.*, 108, 2343, 1989.
20. **Langer, P. R., Waldrop, A. A., and Ward, D. C.,** Enzymatic synthesis of biotin-labeled polynucleotides novel nucleic acid affinity probes, *Proc. Natl. Acad. Sci. U.S.A.*, 78, 6633, 1981.
21. **Singer, R. H. and Ward, D. C.,** Actin gene expression visualized in chicken muscle tissue culture by using *in situ* hybridization with biotinated nucleotide analogue, *Proc. Natl. Acad. Sci. U.S.A.*, 79, 7331, 1982.
22. **Harris, N. and Croy, R.,** Localization of mRNA for pea legumin: *in situ* hybridization using a biotinylated cDNA probe, *Protoplasma*, 130, 57, 1986.
23. **Mc Fadden, G. I., Cornish, E. C., Boning, I., and Clarke, A. E.,** A simple fixation and embedding method for use *in situ* hybridization histochemistry of plant tissues, *Histochem. J.*, 20, 575, 1988.
24. **Escaig-Haye, F., Grigoriev, V., and Fournier, J. G.,** Detection ultrastructurale d'ARN ribosomal par hybridation *in situ* à l'aide d'une sonde biotinylée sur coupes ultrafines de cellules animales en culture, *C. R. Acad. Sci., Paris, Ser. D*, 309, 429, 1989.
25. **Thiry, M. and Thiry-Blaise, L.,** *In situ* hybridization at the electron microscope. Improved method for precise localization of ribosomal DNA and RNA, *Eur. J. Cell Biol.*, 50, 235, 1989.
26. **Silva, F. G., Lawrence, J. B., and Singer, R. H.,** Progress towards ultrastructural identification of individual mRNAs in thin section: myosin heavy chain in developing myotubes, in *Techniques in Immunochemistry*, Vol. 4, Bullock, G. and Petrusz, P., Eds., Academic Press, London, 1989, 77.
27. **Puvion-Dutilleul, F. and Puvion, E.,** Replicating single-strand adenovirus type 5 DNA molecules accumulate within well-delimited intranuclear area of lytically infected Hela cells, *Eur. J. Cell Biol.*, 52, 379, 1990.
28. **Puvion-Dutilleul, F. and Puvion, E.,** Ultrastructural localization of defined sequences of viral RNA and DNA by *in situ* hybridization of biotinylated DNA probes on sections of herpes simplex virus type 1 infected cells, *J. Electron Microsc. Tech.*, 18, 336, 1991.
29. **Troxler, M., Pasamontes, L., Egger, D., and Biez, K.,** *In situ* hybridization for light and electron microscopy: a comparison of methods for the localization of viral RNA using biotinylated DNA and RNA probes, *J. Virol. Methods*, 30, 1, 1990.
30. **Fournier, J. G., Escaig-Haye, F., and Grigoriev, V.,** Visualization of ribonucleic acid sequences after *in situ* hybridization with non-isotopic probes and immunogold electron microscopy, *Boll. Soc. Ital. Microsc. Elett.*, 2, 104, 1991.
31. **Fournier, J. G., Escaig-Haye, F., and Grigoriev, V.,** Ultrastructural detection of cellular and viral RNA with biotinylated DNA probes, *Bull. Assoc. Anat.*, 75, 13, 1991.
32. **Jordan, E. G.,** Interpreting nucleolar structure: where are the transcribing genes?, *J. Cell Sci.*, 98, 437, 1991.
33. **Haase, A. T.,** Analysis of viral infections by *in situ* hybridization, in *In Situ Hybridization, Applications to Neurobiology*, Valentino, K. L., Ebrwine, J. H., and Barchas, J. D., Eds., Oxford University Press, New York, 1987, 197.
34. **Carlemalm, E., Villiger, W., Hobot, J. A., Acetarin, J. D., and Kellenberger, E.,** Low temperature embedding with Lowicryl resins: two new formulations and some applications, *J. Microsc.*, 140, 55, 1985.
35. **Bendayan, M.,** Protein A-gold and protein G-gold post-embedding immunoelectron microscopy, in *Colloidal Gold: Principles, Methods, and Applications*, Vol. 1, Hayat, M. A., Eds., Academic Press, San Diego, 1989, 33.
36. **Grigoriev, V., Kadochnicov, J., Rudneva, I. A., Karamanov, E., Klimenko, S., Escaig, F., and Fournier, J. G.,** Detection by immuno-gold techniques of HIV antigens in Lowicryl ultrathin sections of infected cells, *Path. Res. Pract.*, 184, 494, 1989.

37. **Grigoriev, V., Escaig-Haye, F., Sharova, N. K., Rudneva, I. A., Simonov, V., Kazennova, E. K., Kush, A. A., Klimenko, S. M., Buckrinskaya, A., and Fournier, J. G.,** Localization by immunogold labelling of HIV-1 structural proteins on Lowicryl embedded HIV-1 infected cell ultrathin sections, *J. Submicrosc. Cytol. Pathol.,* 24, 163, 1992.
38. **Escaig-Haye, F., Grigoriev, V., Peranzi, G., Lestienne, P., and Fournier, J. G.,** Analysis of human mitochondrial transcripts using electron microscopic *in situ* hybridization, *J. Cell Sci.,* 100, 851, 1991.
39. **Escaig-Haye, F., Grigoriev, V., Sharova, I. A., Rudneva, V., Buckrinskaya, A., and Fournier, J. G.,** Ultrastructural localization of HIV-1 RNA and core proteins: Simultaneous visualization using double immunogold labelling after *in situ* hybridization and immunocytochemistry, *J. Submicrosc. Cytol. Pathol.,* 24, in press.
40. **Wenderroth, M. A. and Eisenberg, B. R.,** Ultrastructural distribution of myosin heavy chain mRNA in cardiac tissue: a comparison of frozen and LR White embedment, *J. Histochem. Cytochem.,* 39, 1025, 1991.
41. **Völker, W., Frick, B., and Robenek, H.,** A simple device for low temperature polymerization of Lowicryl K4M resin, *J. Microsc.,* 138, 91, 1985.
42. **Nelson, I., Ponsot, G., Marsac, C., Degoul, F., Vayssiere, J. L., Obermaïer, B., Romero, N., Fardeau, M., and Lestienne, P.,** General heteroplasmic mitochondrial DNA markers in Kearns-Sayre syndrome, *Nucleic Acids Res.,* 17, 8117, 1989.
43. **Singer, R. H., Lawrence, J. B., and Rashtchian, R. N.,** Towards a rapid and sensitive *in situ* hybridization methodology using isotopic and non-isotopic probes, in *In Situ Hybridization: Application to the Central Nervous System,* Valentino, K., Eberwin, J., and Barchas, J., Eds., Oxford University Press, New York, 1987, 71.
44. **Meinkoth, J. and Wahl, G.,** Hybridization of nucleic acids immobilized on solid support, *Anal. Biochem.,* 138, 267, 1984.
45. **Puvion-Dutilleul, F., Bachellerie, J. P., and Puvion, E.,** Nucleolar organization of Hela cells as studied by *in situ* hybridization, *Chromosoma,* 100, 395, 1990.
46. **De Mey, J.,** The preparation of immunoglobulin gold conjugates (IGS reagents) and their use as markers for light and electron microscopy, in *Immunocytochemistry,* Cuello, A. C., Ed., John Wiley & Sons, New York, 1983, 347.

Chapter 9

PROCEDURES OF *IN SITU* NUCLEIC ACID HYBRIDIZATION TO DETECT VIRAL DNA AND RNA IN CELLS BY ELECTRON MICROSCOPY

Francine Puvion-Dutilleul

TABLE OF CONTENTS

0-8493-4414-X/93/$0.00 + $.50

I. INTRODUCTION

The technique of *in situ* nucleic acid hybridization allows the specific detection of defined nucleic acid sequences in biological material and, therefore, it promises to be very useful for the identification of viral nucleic acids in clinical tissue specimens, especially when virions and their antigens are not detectable. Until recently, the study of both active and latent viral infections could be performed only at the optical level, and it revealed only which cells contained the viral nucleic acid within the large population of normal cells of tissue sections and cell cultures.

The extension of *in situ* nucleic acid hybridization to the electron microscope (EM) level permits the identification of nucleoprotein structures which contain specific viral nucleic sequences, provided that the specificity and sensitivity of the detection is accompanied by good ultrastructural preservation. Electron microscopy *in situ* nucleic acid hybridization is ideal for the analysis of productive infections in which the number of viral genomes is large and the level of gene expression, as indicated by the amount of viral RNA, is intense. Latent infection in which the amount and expression of viral genomes is greatly diminished is more difficult to analyze because it is more difficult to detect the reactive molecules.

Electron microscope *in situ* nucleic acid hybridization has been used to analyze viral infections in plant and animal cells.[1-13] The subcellular localization of viral nucleic acids was performed either by pre-embedding or post-embedding *in situ* nucleic acid hybridization. The latter procedure requires the use of hydrophilic resins and has the advantage of maintaining well-preserved structural details, especially following Lowicryl K4M embedding.[14-16] The specific nucleic acid sequences which are entirely enclosed within the interior of the ultrathin section are not accessible to the probe and, therefore, are unlabeled. Labeling is restricted to those portions of the nucleic acid sequences which are exposed at the surface of the section. Since the post-embedding procedure preserves morphological detail better and allows parallel immunological detection of proteins and various cytochemical techniques, it is at present the best available method for identifying with certainty the structures which are specifically labeled.

Labeling of the probes with a biotinylated nucleotide,[17] instead of the usual radioactive nucleotides, and subsequent detection of hybrids by biotin-binding proteins (streptavidine or anti-biotin antibody) labeled with colloidal gold particles has markedly improved hybrid detection by increasing the resolution of the *in situ* nucleic acid hybridization and by allowing a more rapid signal detection. The association of biotinylated probes and Lowicryl K4M embedding was first described by Binder et al.,[18] for localizing cellular RNA. It is now extensively employed for studying the variation in nucleic acid expression during various physiological and pathological metabolic activities of the cell.

Because progress in the electron microscopic analysis of infected cells is obtained by the full exploitation of the available techniques, we have used EM *in situ* nucleic acid hybridization to extend our knowledge of the structure-function relationships in herpes simplex virus type 1 (HSV-1) and adenovirus type 5 (Ad5) infected cells.[5-11] These infected cells produce large amounts of viral DNA-binding proteins, known to accumulate in late-infected nuclei,[19,20] which could hide some nucleic acid sequences and/or exhibit interactive sites of binding with the probe. In addition, during the intracellular development of HSV-1 and adenovirus, single-stranded portions occur within the viral genomes which, therefore, might be detected simultaneously with the viral RNA sequences.[5,7-11] These biological characteristics of both HSV-1 and Ad5 infections could lead to false positive or negative results. Therefore, *in situ* hybridization localization of viral nucleic acid sequences in infected cells requires appropriate experimental conditions for eliminating nonspecific binding of the probe and rendering the targets more accessible to the probe. The various procedures for the detection of different types and forms, e.g., single- or double-stranded, of nucleic acids by *in situ* hybridization contain some identical steps, but they vary in other important details. Therefore, there is no universal procedure. This is the reason why we do not use the imperative form throughout this text but, instead, emphasize the variations in procedures and the importance of the controls. Because the biological characteristics of infected cells which may affect the binding of the probe probably depend on the types of viruses, this article describes only the protocols that we have used for the viruses which we have studied. The technique of *in situ* nucleic acid hybridization will be referred to simply as ''hybridization'' throughout the rest of the paper. (We have published technical chapters in French on the subject.[21-23])

II. BIOLOGICAL MATERIAL

A. CELLS AND VIRUSES

Rabbit fibroblasts (line RS537) and HeLa cells are seeded in 50-mm plastic culture dishes (5×10^5 cells per dish) and incubated in Eagle's minimum essential medium (MEM) supplemented with 5% fetal calf serum. When monolayers are near confluence, the cells are incubated with 200 μl of virus inoculum at an input multiplicity of 5 to 10 plaque-forming units per cell. HSV-1 and Ad5 are used for infection of rabbit fibroblasts and HeLa cells, respectively. Adsorption is for 30 min at 37°C. The end of the adsorption period was designated as time zero. Six milliliters of culture medium are then added and cultures are incubated at 37°C for different periods of time (from time 0 to 24 h) before being fixed.

B. FIXATION (SEE PROTOCOL 1)

Infected and noninfected cell monolayers are fixed *in situ* for 1 h at 4°C with either 1.6% glutaraldehyde (glutaraldehyde EM (25% aqueous solution), Taab Lab. Equip. Reading, U.K.) or 4% formaldehyde (Merck, Darmstadt, Germany) in 0.1 *M* Sörensen phosphate buffer, pH 7.3. Both fixatives give good results for the detection of both viral and cellular (ribosomal and U3) RNA, although the hybridization signal is slightly lower following glutaraldehyde fixation.[9-11,24,25] In experiments in which denaturation of the DNA of the specimen is required, glutaraldehyde fixation results in a very slight labeling. We, therefore, use formaldehyde for localizing viral and ribosomal DNA.[5-9,24]

The cells are scraped from the container during fixation and are centrifuged at about 4°C. The pellets are rinsed for 2 h in ice-cold phosphate buffer and then dehydrated prior to Lowicryl K4M embedding.

PROTOCOL 1 — FIXATION OF CELL CULTURE MONOLAYERS

1. Remove the culture medium
2. Fix monolayers for 1 h at 4°C
   a. For DNA detection, use 4% formaldehyde (freshly prepared from paraformaldehyde) in 0.1 *M* phosphate buffer, pH 7.3
   b. For RNA detection, use either 4% formaldehyde or 1.6% glutaraldehyde in the phosphate buffer
3. During the 1-h fixation
   a. Scrape the cells from the substratum with a piece of silicone
   b. Centrifuge for 15 min at 5000 *g*
4. Rinse the pellets with ice-cooled phosphate buffer for 2 h

C. LOWICRYL EMBEDDING PROCEDURE (SEE PROTOCOL 2)

Embedding in Lowicryl K4M is performed according to Bendayan's technique which is a slight modification of the technique of Carlemalm et al.[14-16]

After fixation and washing (see Protocol 1), the resulting pellets are dehydrated in increasing concentrations of methanol. Methanol dehydration provides better structural preservation of viral DNA, which is only loosely associated with protein, especially in herpes virus nucleoids.[26] Methanol 30% and 50% are successively used for 5 min at 4°C. Methanol 70% and 90% are then used for 5 min and 30 min, respectively, at about −20°C. Then, pellets are successively plunged in mixtures of methanol and Lowicryl K4M medium (Chemische Werke Lowi, Waldkraiburg, Germany) (Lowicryl K4M medium = 4 g crosslinker + 26 g monomer + 0.15 g initiator) as follows: (1/1) methanol 90%/Lowicryl for 1 h; (1/2) methanol 90%/Lowicryl for 1 h, Lowicryl for 1 h. Finally, pellets in Lowicryl are maintained overnight at −20°C in a freezer prior to embedding in fresh Lowicryl K4M medium. Polymerization is performed for 5 d at about −30°C under long wavelength UV light (Philips fluorescence tubes TL 6W).

PROTOCOL 2 — LOWICRYL K4M EMBEDDING

1. Last rinse of pellets with ice-cooled phosphate buffer
2. 30% methanol for 5 min at 4°C
3. 50% methanol for 5 min at 4°C
4. 70% methanol for 5 min at ≈ −20°C
5. 90% methanol for 30 min at ≈ −20°C
6. 90% methanol and Lowicryl K4M[a] (1:1) for 60 min at ≈ −20°C
7. 90% methanol and Lowicryl K4M (1:2) for 60 min at ≈ −20°C
8. Lowicryl K4M for 1 h at ≈ −20°C
9. Lowicryl K4M overnight at ≈ −20°C
10. Embed in fresh Lowicryl K4M mixture (gelatin capsules)
11. Polymerize under UV light for 5 days at ≈ −30°C[b]

[a] Lowicryl K4M mixture consists of 4 g crosslinker, 26 g monomer, 0.15 g initiator.
[b] Four Philips fluorescence tubes TL6W/05 are arranged in a sandwich fashion at 30 cm distance from the capsules.

D. SECTIONING

We use a LKB ultramicrotome Nova with a diamond knife. For the hybridization procedure, ultrathin sections are mounted on carbon-Formvar-coated gold grids (mesh 200) and stored in complete darkness at room temperature until use (up to 1 year).

III. PROBES AND HYBRIDIZATION SOLUTIONS

We use commercial probes for localizing viral nucleic acids by EM hybridization.[5-11] For the detection of cellular nucleic acids, however, we use DNA probes (ribosomal DNA and U3 DNA probes) biotinylated by a coworker, which give intense hybridization signals.[24,25]

A. PROBES

The viral HSV and Ad5 probes, biotinylated by nick-translation, are purchased from Enzo (Enzo Biochem. Inc., New York, NY, U.S.). We divide them into 16-μl aliquots which can be stored for more than 3 years at −20°C without diminution of their activity. They are used at a final concentration of 10 μg/ml and 3 to 4 μg/ml, respectively.

The HSV probe is a mixture of two clones of HSV (1 and 2) DNA sequences and consists of double-stranded DNA with a fragment size ranging between 200 and 2000 bp. This HSV probe is similar to that used by Wolber et al.[13] for localizing the thymidine kinase mRNA in HSV-2 infected cells by means of a pre-embedding hybridization method. In lot number 7DBA3 used in our experiments on HSV-1 infection, the insert size of HSV-1 DNA cloned in pBR322 is 8.0 kb and corresponds to fragment F of the genomic map of HSV-1 as determined by restriction enzymes.[27,28] The fragment includes the U_L 48 gene coding for the α trans-inducing factor (αTIF),[29-31] a late protein which is located in the tegument of the viruses and which is required for the induction of immediate-early genes at the beginning of the infectious cycle.

The Ad5 probe (lot number 5JBG4) was prepared for the entire Ad5 genome and consists of double-stranded genomic Ad5 DNA with a fragment size ranging between 50 and 500 bp.

B. HYBRIDIZATION SOLUTIONS

They are prepared in a 1.5-ml Eppendorf tube and stored at 4°C for several months without diminution of their activity. They consist of:

- 50 μl deionized formamide, pH 6.8 to 7.2 (stock solution stored at 4°C in darkness)
- 20 μl dextran sulfate at 50% in distilled water (stored at 4°C)
- 10 μl 20× SSC buffer (1× SSC = 0.15 *M* NaCl, 0.015 *M* sodium citrate) (stored at 4°C)
- 16 μl biotinylated double-stranded viral DNA probe (stored at −20°C)
- 4 μl salmon sperm DNA acting as competitor. The stock solution (Enzo), at 10 mg/ml in water, is distributed in aliquots of 4 μl and stored at −20°C until use in the hybridization solution.

The final concentrations of the different products is as follows: 50% deionized formamide, 10% dextran sulfate, 2× SSC buffer, 400 μg/ml competitor DNA, and either 10 μg/ml HSV probe or 3 to 4 μg/ml Ad5 probe.

IV. PRINCIPAL STEPS FOR *IN SITU* NUCLEIC ACID HYBRIDIZATION

A. ENZYMATIC DIGESTIONS (SEE PROTOCOLS 3 TO 5)

Prior to hybridization, protease and nuclease enzymatic treatments are often required in order to increase the accessibility of the nucleic acids in the

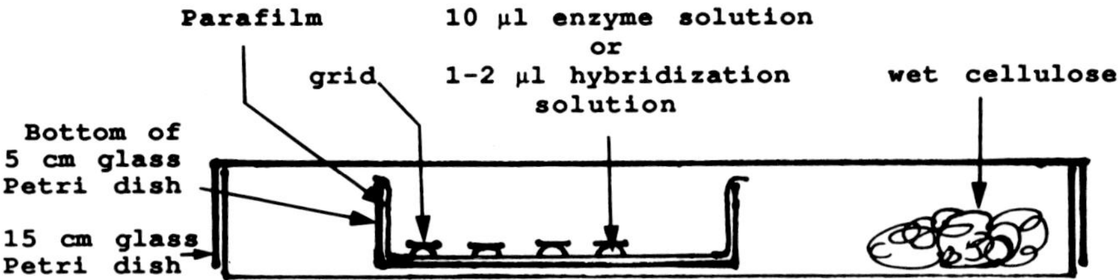

FIGURE 1. Moist chamber used for both enzymatic digestions and hybridization.

sections to the probe and also to increase the specificity of the hybridization signal. The composition of the enzymatic solutions and the required equipment are described below in Section VIII.

All enzymatic treatments are performed at 37°C by floating the sections, previously dried onto grids, on 10 µl drops of reactive solutions distributed on the surface of a sheet of parafilm in a moist chamber (Figure 1). After incubation in the presence of enzymes, the grids are floated successively on 3 drops of distillated water for 1 to 2 min each, rinsed in a jet of distilled water, and finally air-dried (Figure 2). Enzymatic digestions of sections can be performed as much as several weeks prior to the hybridization procedure.

Removal of the proteins of a section by protease treatment (15 min) prior to hybridization dissolves the DNA-binding proteins and renders the nucleic acid molecules more accessible to the probe (see Protocol 3 and Figure 2). It also suppresses a possible interaction between the proteins and the probe. Appropriate control experiments for each experimental model are mandatory to ensure that the probe specifically hybridizes to nucleic acids rather than being bound to proteins. In fact, Galdieri and Monesi demonstrated that a nonspecific binding of radioactive RNA probes to basic proteins of the section was restricted to the nuclei of mouse spermatides at a late stage of differentiation.[32]

PROTOCOL 3 — PROTEASE TREATMENT OF LOWICRYL SECTIONS

1. Mount Lowicryl thin sections on 200-mesh gold grids having a carbon-coated Formvar supporting film.
2. Prepare just before use 0.2 mg/ml protease in distilled water
3. Float the grids on a 10 µl drop of protease solution for 15 min at 37°C in a moist chamber
4. Rinse the grids on 3 drops of distilled water
5. Wash the grids in a jet of distilled water
6. Air-dry the grids

Hydrolysis of RNA molecules of sections by a RNase treatment (1 h) (Protocol 4 and Figure 2) entirely suppresses the interaction of the probe with the RNA of the specimen and, therefore, hybridization of the probe is restricted

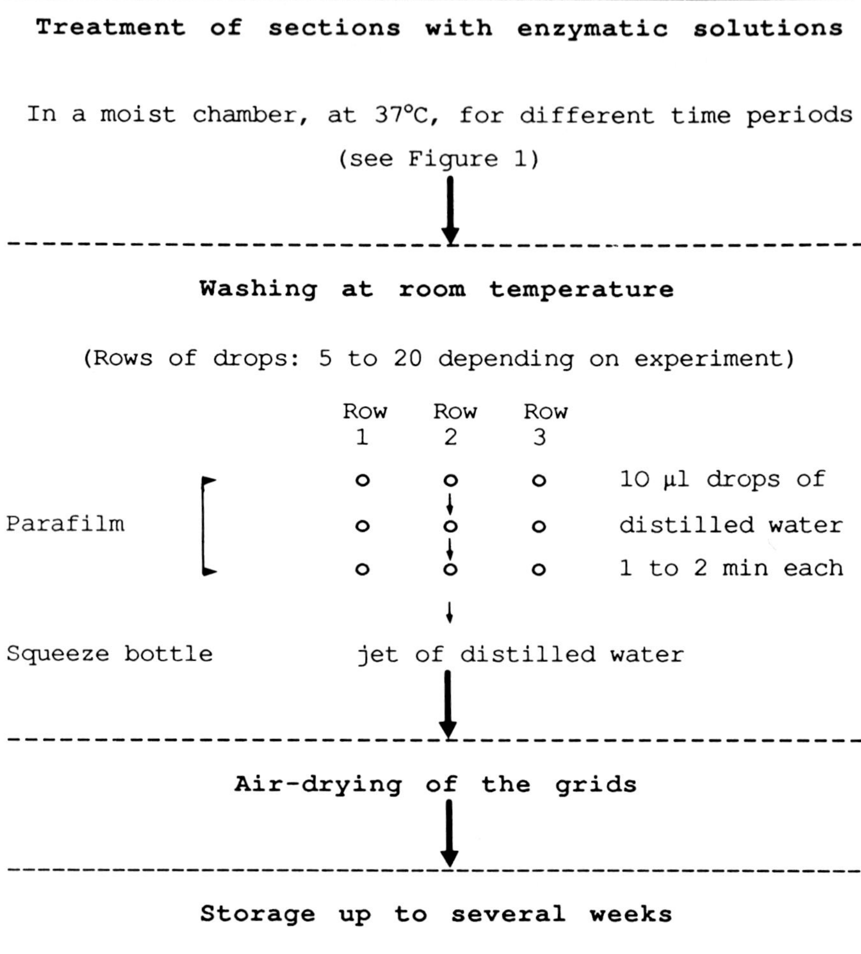

FIGURE 2. Successive steps for enzymatic digestion of sections.

to complementary sequences of DNA. RNase digestion is required, therefore, for specifically localizing DNA in sections by hybridization. On the other hand, RNase digestion also is used prior to the detection of RNA as a negative control of the specificity of the hybridization signal.

PROTOCOL 4 — RNase TREATMENT OF LOWICRYL SECTIONS

1. Mount Lowicryl thin sections on 200-mesh gold grids having a carbon-coated Formvar supporting film.
2. Treat sections with protease (see Protocol 3) (optional)

3. Prepare just before use 1 mg/ml RNase in 10 m*M* Tris-HCl buffer, pH 7.3
4. Float the grids on a 10-μl drop of RNase solution for 1 h at 37°C in a moist chamber
5. Rinse the grids on 3 drops of distilled water
6. Wash the grids in a jet of distilled water
7. Air-dry the grids

Hydrolysis of DNA of sections by a DNase treatment (1 h) (see Protocol 5 and Figure 2) restricts the interaction of the probe to RNA and ensures that the hybridization signal also does not localize to single-strands of DNA. DNase treatment is not used, except as control, for detecting cellular ribosomal and messenger RNA because cellular genes are double-stranded and, therefore, will not bind the probe in the absence of alkali treatment of sections (see below).[24,25] However, DNase treatment is mandatory for localizing viral RNA in HSV-1 and Ad5 infection,[9-11] which are both known to contain portions of viral single-stranded DNA. The DNase solution contains $MgCl_2$, which is mandatory for DNase activity, and RNasin, an inhibitor of RNase, which is effective only in the presence of dithiothreitol.

PROTOCOL 5 — DNase TREATMENT OF LOWICRYL SECTIONS

1. Mount Lowicryl thin sections on 200-mesh gold grids having a carbon-coated Formvar supporting film
2. Treat sections with protease (see Protocol 3) and/or RNase (see Protocol 4) (optional treatments)
3. Rinse all laboratory material with DEPC (see Protocol 10) and air-dry
4. Prepare just before use 1 mg/ml DNase in 10 m*M* Tris-HCl buffer, pH 7.3 containing 5 m*M* $MgCl_2$, 2% RNasin, and 2 m*M* dithiothreitol
5. Float the grids on a 10-μl drop of DNase solution for 1 h at 37°C in a moist chamber
6. Rinse the grids on 3 drops of distilled water
7. Wash the grids in a jet of distilled water
8. Air-dry the grids

Effectiveness of the enzymatic digestions is imperative and must be verified as follows. The effectiveness of the protease treatment for removing viral proteins is demonstrated by the total absence of labeling following immunogold detection of major viral DNA-binding proteins on protease-digested sections.[6,7] The effectiveness of nuclease treatments is demonstrated by the total absence of autoradiographic labeling of nuclease-treated sections of cells previously exposed to tritiated nucleotides prior to Lowicryl embedding.[10] We insist on the importance of the digestions, especially the nuclease digestions, for specifically detecting viral nucleic acid sequences.

B. DENATURATION OF THE DNA OF THE SECTION (SEE PROTOCOL 6)

For localizing viral genomes in sections of HSV-1 and Ad5 infected cells, a treatment of the sections which denatures the DNA is required, in order to separate the double-stranded portions of the viral sequences into single strands which are recognized by the probe. We have tested different protocols in parallel.[5,6] They included 4 min treatments of sections with 0.07 *N* NaOH, a concentration routinely used for denaturing DNA fibers in whole mount chromosomes,[33] 0.5 *N* NaOH or *N* NaOH, a 30 min treatment with 5 *N* HCl used for releasing purine bases from resin-embedded DNA molecules,[34] a 4-min immersion in boiling water, and, finally, a 4 min floating of a piece of silicone bearing the grids on the surface of boiling water. Under our experimental conditions, we verified that heat, HCl, and 0.07 *N* NaOH treatments denature DNA in Lowicryl sections but give a reduced hybridization signal.[6] NaOH denaturation at the higher concentrations (0.5 *N* or *N*) consistently gives the best results for viral DNA detection and also for localizing cellular ribosomal genes.[5-9,24]

Routinely (see Protocol 6 and Figure 3), digested and nondigested grids are placed on a drop of 0.5 *N* NaOH at room temperature for 4 min in order to denature cell and viral DNA present in the sections. This is followed by three changes of distilled water over a total period of about 3 min and a wash in a jet of distilled water. Grids are again air-dried. They are used a few minutes later (up to 15 min).

Structural detail is well preserved during alkali denaturation. Following 0.5 *N* NaOH treatment, all DNA molecules of the section, including both the free and encapsidated viral genomes, are able to react with the probe.[6]

PROTOCOL 6 — DENATURATION OF THE DNA MOLECULES OF THE LOWICRYL SECTIONS

1. Float grids, with or without previous enzymatic treatments (see Protocols 3 to 5), on 10-μl 0.5 *N* NaOH drops for 4 min at room temperature
2. Rinse on 3 drops of distilled water
3. Wash in a jet of distilled water
4. Air-dry the grids[a]

[a] NaOH-treated grids are used for hybridization a few minutes later (up to 15 min).

C. DENATURATION OF THE DNA OF THE PROBE (SEE PROTOCOL 7)

The DNA molecules of the hybridization solution (viral DNA and competitor [carrier] DNA) are double-stranded. Heat treatment is required for their denaturation. The resulting single-stranded DNA molecules are able to form hybrids with the complementary nucleic acid sequences (both single-stranded RNA and DNA sequences) which are accessible at the surfaces of the Lowicryl sections.

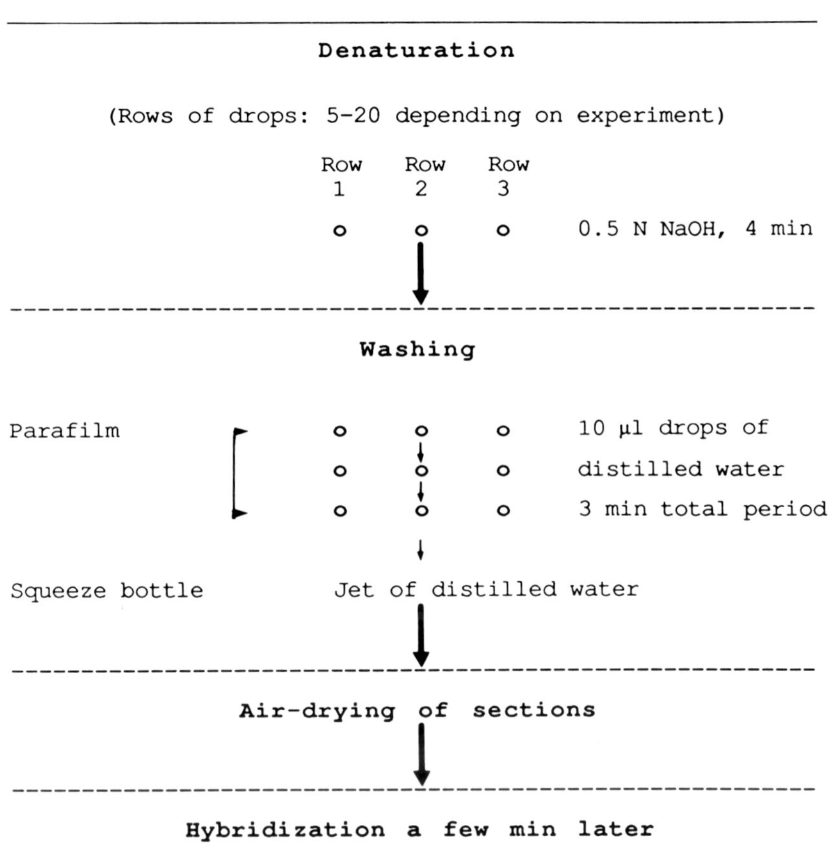

FIGURE 3. NaOH denaturation at room temperature of the DNA of sections for the subsequent detection of total viral DNA (both single-stranded and double-stranded DNA).

Just before use (see Protocol 7 and Figure 4), the Eppendorf tube containing the hybridization solution is plunged for 4 min in boiling water and immediately chilled in melting ice. Subsequently, the solution is separated into aliquots of 1 or 2 µl for hybridization at the surface of a sheet of parafilm. The remaining hybridization solution is kept at 4°C and denatured again before each further hybridization experiment.

PROTOCOL 7 — DENATURATION OF THE DNA OF THE HYBRIDIZATION SOLUTION AND HYBRIDIZATION

1. Prepare the hybridization solution by mixing in an Eppendorf tube:

- 50 µl deionized formamide, pH 6.8 to 7.2
- 20 µl 50% dextran sulfate in distilled water

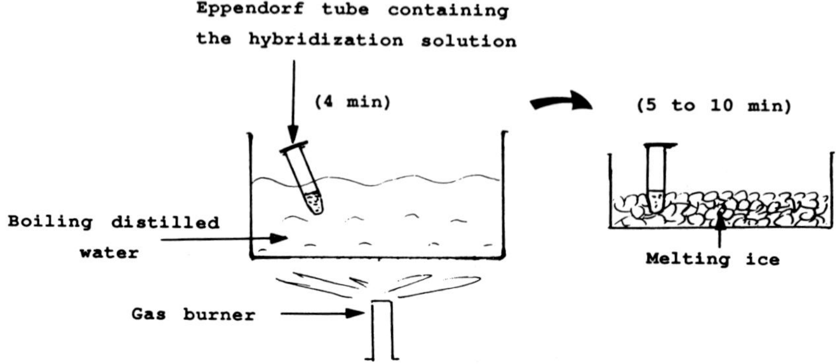

FIGURE 4. Denaturation of the DNA of the hybridization solution.

- 10 μl 20 × SSC buffer
- 16 μl biotinylated DNA probe
- 4 μl salmon sperm DNA

2. Immediately before use, plunge the Eppendorf tube in boiling water for 4 min
3. Immediately chill the tube in melting ice for ≈5 min
4. Separate the hybridization solution into 1- to 2-μl drops at the surface of a parafilm sheet[a]
5. Float grids, with or without enzymatic treatments (see Protocols 3 to 5) and with or without denaturation treatment (see Protocol 6), on the 1- to 2-μl drops, for ≈1 h, at 37°C
6. Rinse the grids on 2 drops of distilled water and then on a drop of PBS (at room temperature)
7. Detect hybrids (see Protocol 8)

[a] Keep the remaining hybridization solution at 4°C and denature it before each further experiment.

D. HYBRID FORMATION (SEE PROTOCOL 7)

Sections with or without prior enzymatic digestion treatment and with or without denaturation treatment are floated on drops of 1 to 2 μl of freshly denatured hybridization solution distributed on the parafilm of the moist chamber (see Protocol 7 and Figures 1, 5) for 1 h at 37°C. No evident evaporation of the microdrops of hybridization solution occurs, due to the viscosity of the dextran sulfate in the solution, even following longer incubation time (24 h) at 37°C; therefore, a larger amount of hybridization solution is not required. Grids are then floated at room temperature on 2 drops of distilled water for 1 min each and then on a drop of phosphate buffered saline (PBS) for 5 min. The grids are immediately processed for the immunocytological detection of biotin.

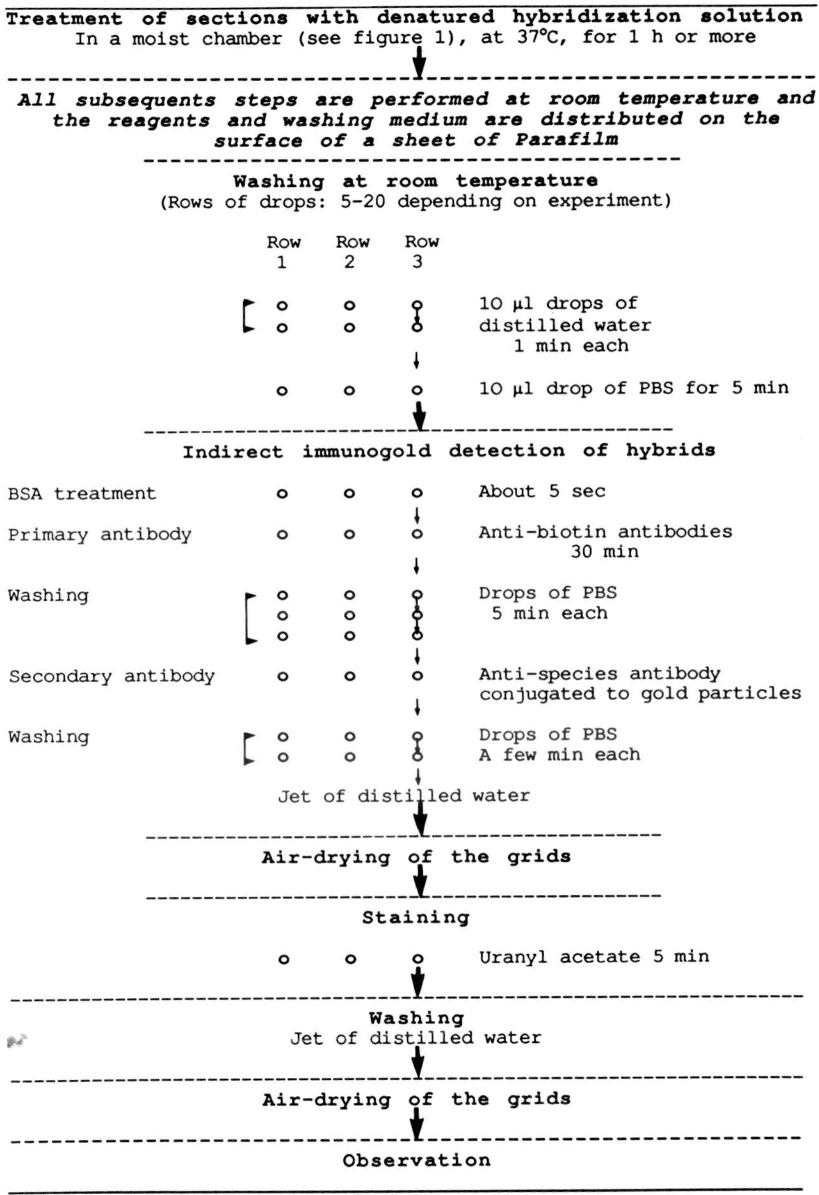

FIGURE 5. Successive steps for formation of hybrids, detection of hybrids, and staining of sections.

E. HYBRID REVELATION (SEE PROTOCOLS 8 AND 9)

The biotin of the hybrids which are formed at the surface of the Lowicryl sections are revealed immediately by indirect or direct immunocytological procedures performed at room temperature (Figure 5). In both cases, wet grids are first placed on a drop of bovine serum albumin (BSA) solution for about 5 sec in order to suppress the nonspecific binding of antibodies to the sections and are then floated on drops of antibodies. Details on the reagents used are given in Section VIII.

Indirect immunocytological detection (see Protocol 8) is performed in two steps as follows. Grids are floated on a drop of antibiotin antibody diluted 1/100 in PBS for 30 min. After washing on 3 drops of PBS (3 × 5 min), the grids are floated on a drop of anti-rabbit IgG conjugated to gold particles (10 nm in diameter) diluted 1/100 in PBS, i.e., $O.D._{520}$ nm = 0.02, for 30 min. Finally, the grids are washed in a jet of distilled water, air-dried, and stained (see below).

PROTOCOL 8 — INDIRECT IMMUNODETECTION OF HYBRIDS AND STAINING

1. Last rinse the hybridized grids with PBS (see Protocol 7)
2. Float the wet grids on 10 μl anti-biotin antibody (1/100 in PBS) for 30 min at room temperature)
3. Rinse the grids on 3 drops of PBS (3 × 5 min)
4. Float the grids on 10-μl anti-rabbit IgG conjugated to gold particles, 10 nm in diameter (1/100 in PBS), for 30 min at room temperature
5. Rinse the grids on 3 drops of PBS (3 × 5 min)
6. Wash the grids in a jet of distilled water
7. Air-dry the grids
8. Stain the grids with 5% aqueous uranyl acetate for 10 min
9. Wash the grids in a jet of distilled water
10. Air-dry the grids

Direct immunocytological detection (see Protocol 9) is performed in one step by floating the grids on a drop of goat antibiotin immunoglobulins conjugated to gold particles (10 nm in diameter) diluted 1/25 in PBS. At present we still have not used this one-step hybrid detection for localizing viral DNA, but we have used it with success for detecting both ribosomal nucleic acids and U3 RNA in normal cells.[25]

PROTOCOL 9 — DIRECT IMMUNODETECTION OF HYBRIDS AND STAINING

1. Last rinse the hybridized grids with PBS (see Protocol 7)
2. Float the wet grids on 10 μl anti-biotin antibody conjugated to gold particles, 10 nm in diameter (1/25 in PBS), for 30 min at room temperature
3. Rinse the grids on 3 drops of PBS (3 × 5 min)
4. Wash the grids in a jet of distilled water
5. Air-dry the grids
6. Stain the grids with 5% aqueous uranyl acetate for 10 min

7. Wash the grids in a jet of distilled water
8. Air-dry the grids

F. STAINING PROCEDURES AND OBSERVATION

To stain the constituents of the cells and viruses (Figure 5), the dried grids are floated on a drop of 5% uranyl acetate in water, extensively washed in a jet of distilled water, and air-dried (see Protocols 8 and 9). Counterstaining with lead citrate is not recommended because a thin, electron-opaque granulosity of the Lowicryl is revealed. The EDTA regressive staining of Bernhard, which preferentially stained ribonucleoprotein structures,[35] can be used although in this case the Lowicryl granulosity is slightly revealed. In addition, the higher staining of RNP structures, including the nucleolus, renders the observation of the gold particles more difficult. We prefer the uranyl acetate staining.

Stained sections are observed in a transmission electron microscope at 80 kV, at magnifications varying from 4000× to 20,000×.

G. CONTROLS OF THE SPECIFICITY OF THE REACTION

Protocols for controls of the specificity of the hybridization detection are summarized in Figure 6.

To demonstrate the specificity of the hybridization reaction, we perform parallel experiments on noninfected cells. In addition, we apply the Ad5 DNA probe to HSV-1 infected cells and, conversely, the HSV DNA probe to both cytomegalovirus and Ad5 infected cells. The total absence of labeling following these procedures confirms that the probes we use do not contain sequences homologous with cell structures and heterologous viruses.

We verify that no labeling occurs when sections are digested with DNase and RNase in the case of hybridization detection of DNA and RNA, respectively.

We also verify that our biological material is devoid of endogenous biotin, given that some cells are known to contain biotin. To do this, all steps of the procedure are carried out except the exposure of the sections to the probe. A total absence of gold labeling demonstrates that no anti-biotin antibodies bind to the section in the absence of hybrid formation.

V. DETECTION OF VIRAL GENOMES

A. CHARACTERISTICS OF THE HSV-1 AND Ad5 SYSTEMS

Since viral RNAs are synthesized extensively during the infectious cycles of HSV-1 and Ad5 viruses, their removal from the sections by RNase treatment prior to hybridization is required.

Among the numerous viral proteins which are synthesized, some DNA-binding proteins are produced in large amounts and accumulate in the nucleus of the infected cells, where they are tightly bound to the viral genomes (ICSP11,12 protein for HSV-1; ICP8 protein for HSV-2; and 72-kDa protein for the adenoviruses).[19,20] As a result, some portions of the viral genomes are

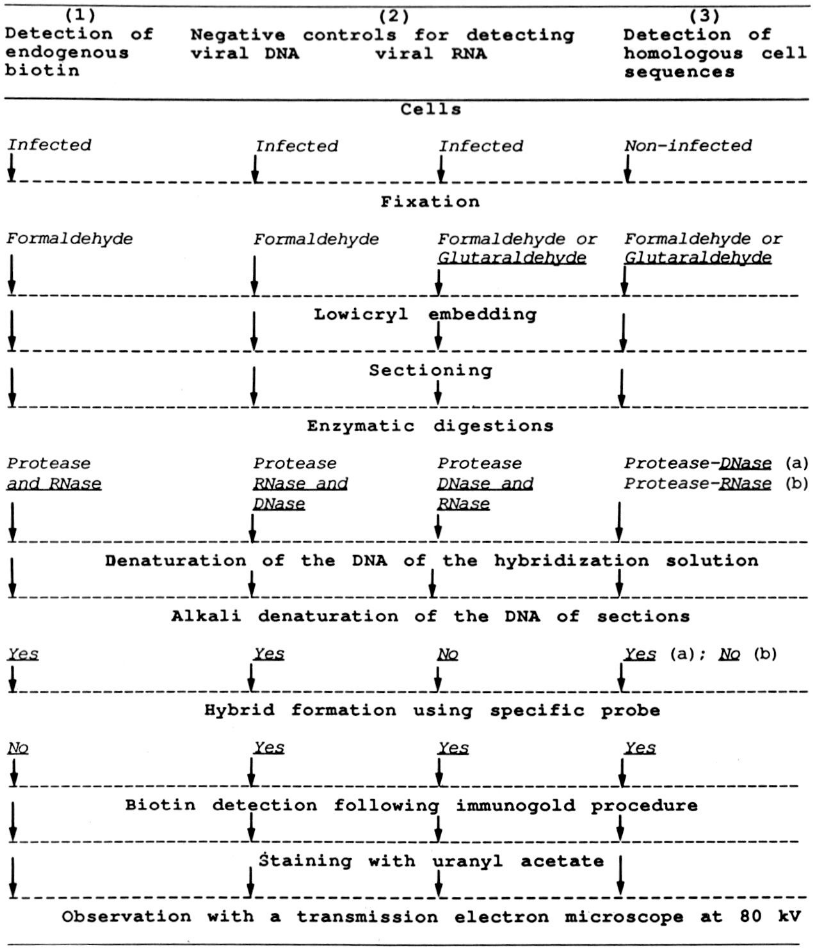

FIGURE 6. Protocols for controls of the specificity of the hybridization reaction: (1) detection of endogenous biotin; (2) negative controls for detecting either viral DNA or viral RNA; (3) detection of homologous cell sequences. Successive steps are indicated by bold letters; common details are indicated by italics, and exceptional procedures are underlined. An additional negative control is not mentioned herein. It is the use of a heterologous probe instead of a homologous probe in order to detect a possible binding of DNA molecules to cell and/or viral components.

hidden by proteins. Therefore, a proteolytic digestion of the sections is required to increase the accessibility of the viral genomes to the probe and, also, to suppress a possible binding of the probe onto interactive sites of the proteins.

Finally, viral genomes are double-stranded, so a denaturation of the DNA of the section prior to hybridization also is essential.

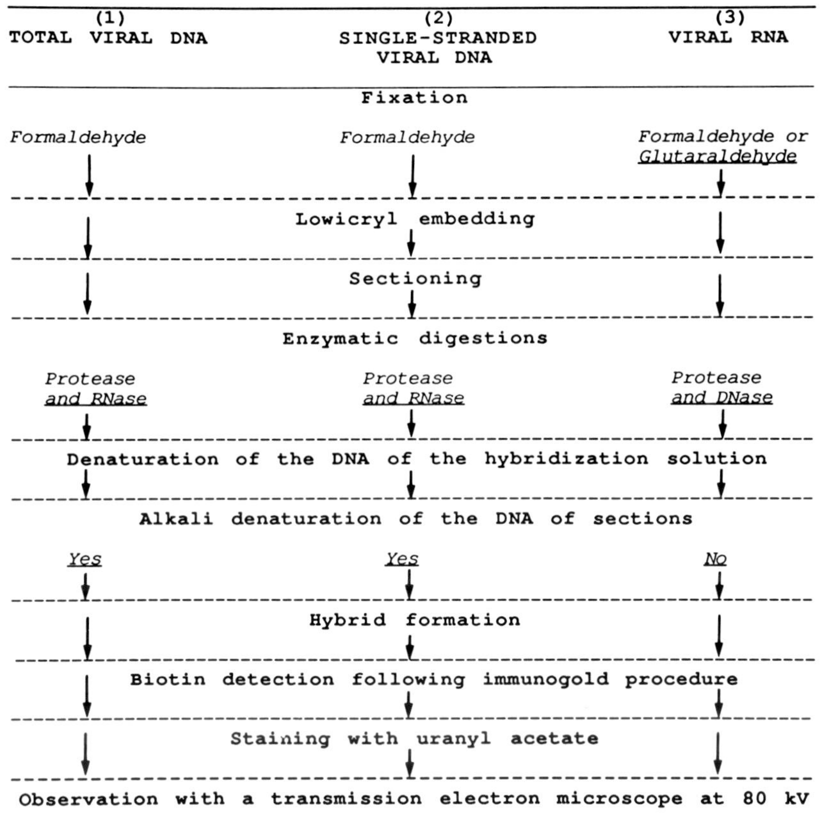

FIGURE 7. Protocols for specific detection by EM hybridization of (1) total viral DNA (both single-stranded and double-stranded viral DNA), (2) single-stranded viral DNA, and (3) viral RNA on Lowicryl sections using a biotin-labeled viral double-stranded DNA probe. The controls are not included here. Successive steps are indicated by bold letters; common details are indicated by italics and exceptional procedures are underlined.

B. METHOD

The successive steps are as follows (Figure 7):

1. *Material* (see Protocols 1 and 2). Lowicryl sections of formaldehyde-fixed cells recovered onto Formvar-carbon-coated gold grids (these can be prepared a long time in advance, i.e., as much as 1 year).
2. *Enzymatic digestions* (see Protocols 3 and 4). Protease (recommended) followed by RNase (required) (both can be performed weeks in advance).
3. *DNA denaturation* (see Protocols 6 and 7). Alkali denaturation (0.5 *N* NaOH, 4 min) of the DNA of the sections and heat denaturation of the DNA of the hybridization solution (4 min in boiling H_2O, then chilling in melting ice) (both are performed just before hybridization).

4. *Hybrid formation* (see Protocol 8). The grids are floated over the freshly denatured hybridization solution for 1 h at 37°C. (We have verified that a longer incubation time up to 24 h does not increase the intensity of the labeling in these systems, and that a higher incubation temperature up to 64°C decreases it.)
5. *Hybrid revelation* (see Protocols 9 and 10). Immunocytological procedure using gold particles as marker. High precision of labeling is obtained even following indirect detection, and is attested by the precise labeling of the nucleoids of the viruses.[6]
6. *Staining of sections* (Protocols 9 and 10). Sections are stained with uranyl acetate. This is followed by observation and eventual quantitative evaluation of gold labeling over the different cell compartments.
7. *Controls.* In parallel, control experiments are performed as indicated above at the end of Section IV.

C. RESULTS AND COMMENTS

Although the development of HSV-1 and Ad5 viruses in permissive cells exhibits common features, several biological characteristics lead to a different intracellular distribution of the viral genomes.

The productive cycle of such viruses takes place in the nucleus of the permissive cells. The beginning of the synthesis of viral genomes appears a few hours after the initiation of the infection and proceeds until cell lysis occurs. It is clearly established that viral genomes in cells infected with either HSV-1 or Ad5 are double-stranded and display two kinds of organization, i.e., either extended or aggregated in viral capsids. However, there are two differences. First, the viral nucleoids of adenovirus (only about 20 nm in diameter) are markedly smaller than those of HSV-1 (about 80 nm in diameter) and, therefore, many adenovirus nucleoids are located within the width of the section, whereas almost all HSV nucleoids are sectioned and accessible to the probe. Second, both nonencapsidated and encapsidated viral genomes in adenovirus infection are restricted to the nucleus of the infected cells, that is, the viruses accumulate in the enlarged infected nucleus.[8] Contrarily, the cytoplasm and extracellular space of HSV-1 infected cells contain viruses which leave the nucleus to attain the extracellular space after the envelopment of the viral DNA-containing capsids to form full capsids.[6] In herpetic infection, therefore, both encapsidated and nonencapsidated viral genomes are present in the nucleus, but the extranuclear regions also contain encapsidated viral genomes. A hybridization protocol is suitable if the localization of the hybridization signal is in agreement with the biological characteristics of the virus and if both encapsidated and nonencapsidated genomes can be detected.

EM hybridization is sufficiently sensitive to detect HSV-1 infection when only a few HSV-1 genomes are present in the cell.[5] At the initial stage of infection, the viral nucleoids of the parental viruses which penetrate the target cells and are released from the nucleocapsids in the cytoplasm become swollen in the juxtanuclear cytoplasm. In this form, they enter the nuclei to attain the regions of host chromatin. Total disaggregation of the viral genomes occurs only in more central regions of the nuclei. Subsequent replication of viral genomes results in the appearance of unusually clear regions within the nucleoplasm.

At 4 h and 7 h after infection, small numbers of gold particles appear within the faintly fibrillar clear regions of the nucleus.[5] Later in the infection, the fibrillar component of the now enlarged intranuclear clear regions is intensely labeled (Figure 8A). Clustered gold particles are present also over the encapsidating and fully encapsidated viral nucleoids whatever their stage of formation and whatever their location in the cell (Figure 8C).[6,9] Therefore, whereas nonencapsidated genomes are exclusively clustered in the clear virus-specific regions of the nucleus and never intermingled with the marginated host chromatin, viral DNA which is inserted into viral capsids can appear within the chromatin. In the cytoplasm and extracellular space, only viral genomes which are packaged within the viral nucleoids of the mature viruses are present and are labeled. In addition, our data reveal that the encapsidated viral genomes are not hidden by the proteins of the viral nucleoids, since they are accessible to the probe even in the absence of protease treatment (Figures 8C, 8D). The accessibility of the nonencapsidated viral genomes, however, is markedly increased following the elimination of the proteins from the sections (Figures 8A, 8B), which indicates that portions of the nonencapsidated HSV-1 genomes are masked by proteins including the major viral DBP.[19]

With respect to adenovirus infection,[7,8] in the earliest stage of virus-induced nuclear transformation, gold particles are located over discrete masses of thin fibrils which, subsequently, develop into labeled, large, pleomorphic, compact, fibrillar masses surrounded by a labeled, irregular, fibrillo-granular network (Figures 9A, 9B). At a still more advanced stage of nuclear transformation by Ad5 infection, in addition to the above labeled structures, all of the fibrillar contents of a new large, centrally located fibrillar inclusion body are heavily labeled. Some, but not all, viruses are decorated by gold particles. The unlabeled viruses are those which are in the interior of the sections. All other components of the cell are unlabeled. The hybridization allows us to determine the distribution of the Ad5 DNA in the numerous components of the virus-induced nuclear structures which gradually appear during the course of the infectious cycle. Our data indicate that many viral DNA sequences which could be recognized by the viral DNA probe are masked by proteins, since the omission of protease digestion markedly lowers the intensity of the labeling, especially over the small masses of thin fibrils at early stages of infection and over the compact, fibrillar masses which appear later in infection.[7,8]

No labeling is obtained in the control experiments.

Our EM hybridization protocol for HSV-1 and Ad5 DNA detection displays high sensitivity and high specificity of labeling. It is able, therefore, to detect viral genomes not only in favorable systems when viral genomes are abundant, but also when only a few viral genomes are present in the cell.

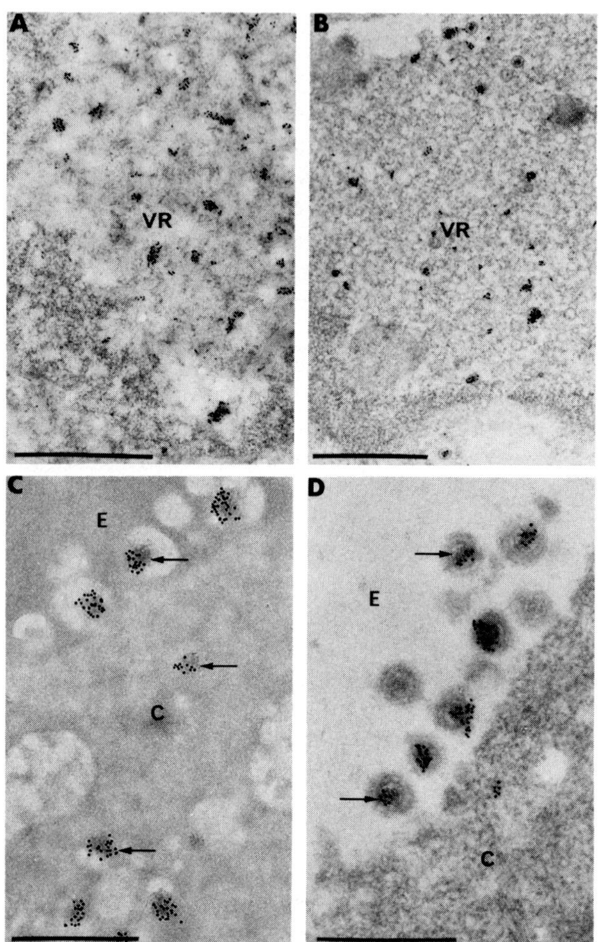

FIGURE 8. Hybridization on 0.5 *N* NaOH-treated Lowicryl sections of 17-h HSV-1 infected cells. Comparison between (A, C) protease-RNase digestion and (B, D) RNase digestion alone. Formaldehyde fixation. Uranyl acetate staining. (A) and (B): Intranuclear gold labeling is similar in distribution after both types of enzymatic digestion, but the labeling of the fibrillo-granular network of the virus-induced clear region (VR) is more extensive in the protease-RNase digested section (A). Bars = 1 μm. ([B] from Puvion-Dutilleul, F., in *Techniques en Microscopie Electronique: Cryométhodes, Immunocytologie, Autoradiographie, Hybridation In Situ,* Morel, G., Ed., Editions INSERM, Paris, 1991, 553. With permission.) (C) and (D): Parts of cytoplasm (C) and extracellular space (E). Clusters of gold particles decorate the nucleoids of virions (arrows). The intensity of the labeling and the number of labeled nucleoids is similar with (C) or without (D) protease digestion. Bars = 0.5 μm. (From Puvion-Dutilleul, F. and Puvion, E., *Eur. J. Cell Biol.,* 49, 99, 1989. With permission.)

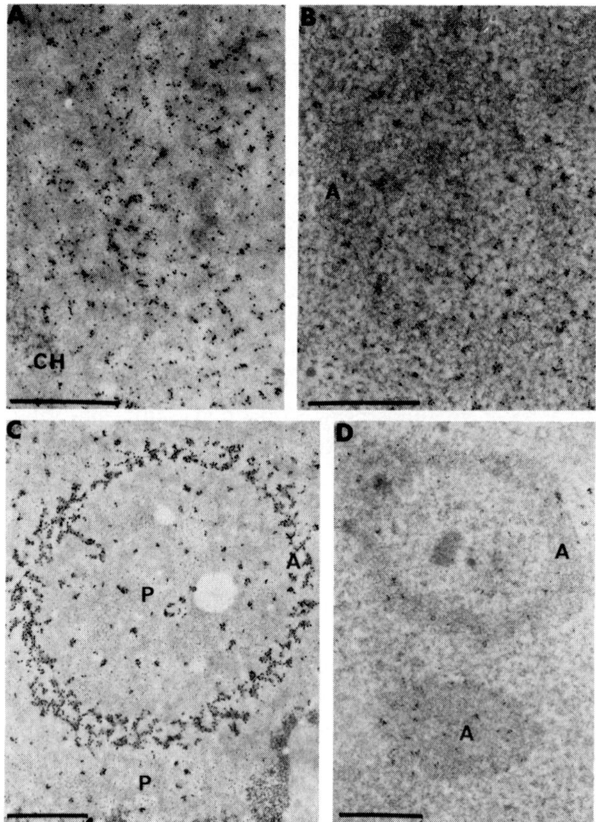

FIGURE 9. Effects of proteins on the accessibility of Ad5 DNA in Lowicryl sections to the Ad5 DNA probe. Seventeen hours after infection. Intermediate stages of virus-induced nuclear transformation. Formaldehyde fixation. Uranyl acetate staining. Bars = 1 μm. (A) and (B): Localization of total viral DNA on 0.5 *N* NaOH-treated sections. (From Puvion-Dutilleul, F. and Puvion, E., *Eur. J. Cell Biol.*, 52, 379, 1990. With permission.) (A): Protease-RNase digestion of sections prior to hybridization. Gold particles are numerous over the entire nucleus except the condensed chromatin (CH). (B): RNase digestion alone. Gold labeling is similar in distribution but is less extensive than in (A). A fibrillar structure in the form of a ring, designated as a single-stranded DNA accumulation site A, is identifiable in the center of the nucleus. (C) and (D) Localization of single-stranded viral DNA in the absence of NaOH treatment of sections. (From Puvion-Dutilleul, F. and Puvion, E., *J. Struct. Biol.*, 103, 280, 1990. With permission.) (C) After double digestion with protease and RNase, gold particles are abundant over the single-stranded DNA accumulation site A, which is here in the form of a ring. Some labeling also occurs over the fibrillo-granular network of the surrounding region, which is designated as the peripheral replicative zone (P). (D) Following RNase digestion alone, the single-stranded DNA accumulation sites A are identifiable by their higher electron-opacity. Gold labeling is similar in distribution to what it is in (C), but its intensity is markedly less.

It can even detect a single viral genome, since the nucleoids of the viruses are clearly labeled. Our procedure should prove fruitful in the study of other viral systems and also in the localization of repeated and amplified DNA sequences in noninfected cells.

VI. DETECTION OF SINGLE-STRANDED VIRAL DNA

A. CHARACTERISTICS OF THE HSV-1 AND Ad5 SYSTEMS

Both HSV-1 and Ad5 infected cells contain single-stranded portions of viral DNA. In HSV-1 infected cells, about 80% of the viral genomes of the mature viruses contained single-stranded portions which are randomly distributed along the genome. Their presence is without effect on the infectivity of such viruses.[29,36,37] We do not know if such single-stranded regions are present in the HSV-1 genomes prior to their insertion into viral capsids. In adenovirus infection, the semiconservative mode of replication of the viral genomes leads to the transient, but ineluctable, formation of single-stranded viral DNA. Indeed, DNA replication occurs by a continuous process which involves strand displacement.[38] Subsequently, the displaced single-stranded molecule becomes the template for complementary strand synthesis. Therefore, adenovirus DNA replication induces the formation of several types of replicative intermediates, one of which has a single-stranded DNA for template. We have no information on either the sites of the different replicative intermediates nor on the specific distribution of the viral single-stranded adenovirus DNA in the various structures which appear in the nucleus.

B. METHOD

In order to localize only the single-stranded viral DNA, hybridization is performed by incubating the denatured probe at the surface of Lowicryl sections (see Protocols 1 and 2) in which the DNA has not been denatured (Figure 7). The probe only reacts with the viral nucleic acid sequences of the section, which are normally single-stranded. A RNase treatment (see Protocol 4) prior to hybridization is, therefore, mandatory in order to eliminate the binding of the probe to the viral RNA sequences.

The method is similar to that described above, in Section V.B, for localizing viral genomes but, in the step 3, the alkali denaturation of the section is omitted.

C. RESULTS AND COMMENTS

When NaOH denaturation of protease-RNase digested sections is omitted, the distribution of the gold particles varies with the type of viral infection.

Following HSV-1 infection,[9] only a few gold particles are bound to the section. They are restricted to some, but not all, viral nucleoids at different stages of formation and location in the cell, and also to a few filaments in the virus-induced clear region of the nucleus. The data indicate that single-stranded portions of DNA in progeny viral genomes occur *in vivo,* at least in the fragment recognized by the probe (*Bam*H1 F fragment), both before and after viral genome encapsidation. The data also demonstrate that single-stranded viral DNA is present in parental viral DNA during its intracellular migration in the target cell.[5]

Following Ad5 infection, in early infected cells the labeling is concentrated over the small fibrillar masses.[8] In the intermediate stage of nuclear transformation (Figure 9C), gold labeling is intense over the compact fibrillar masses, which are in their phase of large size and varied in shape, and which we have designated as single-stranded DNA accumulation sites. The labeling is rarer over the surrounding fibrillo-granular network.[30] At the final stage of nuclear transformation, gold particles are restricted mainly to the now smaller, rounded, single-stranded DNA accumulation sites.[8] All other components of the cells, including viruses, are unlabeled. Our data reveals for the first time that a morphologically well-delineated structure results from the accumulation of viral single-stranded DNA, and that it is related to the replication of viral genomes. They also indicate that single-stranded viral DNA is almost entirely hidden by proteins, since labeling is minimal in the absence of protease digestion of sections (Figure 9D).

In conclusion, EM hybridization for labeling single-stranded DNA is suitable for detecting single-stranded viral DNA segments in infected cells, even when they are few in number, especially at the initial stage of herpetic infection. We interpret the existence of single-stranded DNA accumulation sites in adenovirus infected cells as being a result of the rapid segregation of large numbers of newly formed single-stranded DNA molecules, some of which have been released from the replicative intermediates. More generally, EM hybridization contributes to a better definition of nuclear-protein interactions and nuclear compartmentalization involved in the different stages in the replication of viral genomes.

VII. DETECTION OF VIRAL RNA

A. CHARACTERISTICS OF THE HSV-1 AND Ad5 SYSTEMS

Transcription of viral genomes had been extensively studied in biochemical experiments, which revealed that viral HSV-1 and Ad5 RNAs occur in the nucleus of lytically infected cells, as a result of transcription of regions of the viral genomes, and that they enter the cytoplasm after viral maturation processes. Ultrastructural studies are rarer due to the difficulty of revealing viral RNA only by *in situ* experiments such as autoradiography. To specifically identify viral RNA by EM hybridization, the application of the freshly denatured hybridization solution to sections in which DNA is not denatured by NaOH treatment is not enough. Because of the presence of portions of single-stranded HSV-1 and Ad5 DNA, a DNase digestion of the sections is mandatory. Although a protease treatment does not markedly improve the extent of hybridization, it has the advantage of suppressing the occasional labeling of cell and viral components other than newly transcribed viral RNA.

B. METHOD

The method which allows the detection of viral RNA (Figure 7) is roughly similar to that described above, in Section V, for localizing viral DNA. Only the specific details for detecting viral RNA are mentioned below.

1. *Biological material.* Both sections of formaldehyde- and glutaraldehyde-fixed cells can be used (see Protocol 1). Glutaraldehyde fixation also allows hybrid formation, although to a slightly lesser extent.[10] In addition, proteolytic enzymes and nucleases are effective on Lowicryl sections of glutaraldehyde-fixed material for the digestion of infected cells. We must mention, however, that the enzymatic digestion conditions we used in this protocol are without evident effect on the nucleolar ultrastructure of noninfected human HeLa cells.[24]
2. *Enzymatic digestions.* DNase digestion (see Protocol 5) is mandatory, and protease treatment (see Protocol 3) is recommended.
3. *DNA denaturation.* The alkali treatment of the sections prior to hybridization must be omitted. Only the DNA molecules of the hybridization solution should be denatured by heat immediately before use (see Protocol 7).
4. *Hybrid formation.* The incubation of the enzyme-digested sections in the presence of the denatured hybridization solution is the same as for detecting viral DNA (see Protocol 8). However, the 1 h incubation at 37°C, which gives an optimal hybridization signal in HSV-1 and Ad5 RNA labeling,[8,10,11] is not recommended for localizing ribosomal RNA in normal HeLa cells.[24] A longer incubation period (4 h) at higher temperature (up to 64°C) gives better ribosomal RNA detection.
5. *Hybrid revelation and staining.* The revelation of hybrids using anti-biotin antibodies, and the subsequent staining of sections, are unchanged (see Protocols 9 and 10).

C. RESULTS AND COMMENTS

The double digestion with protease and DNase made the identification of the different cell compartments difficult because marginated host chromatin is dissolved and no longer constitutes a visible layer between the cytoplasmic and intranuclear viral regions. Gold particles, isolated or grouped in clusters, are present over regions of the nucleus and cytoplasm which are known to contain viral RNA and may be considered as internal controls.

In HSV-1 infection,[9] the hybridization signal is not intense (Figure 10A). In the nucleus, gold particles are mainly restricted to the virus-induced clear regions, and especially to its fibrillo-granular network. In the cytoplasm, clustered gold particles are scattered over the areas which contain numerous ribosomes. The other cell and viral substructures are unlabeled. The moderate

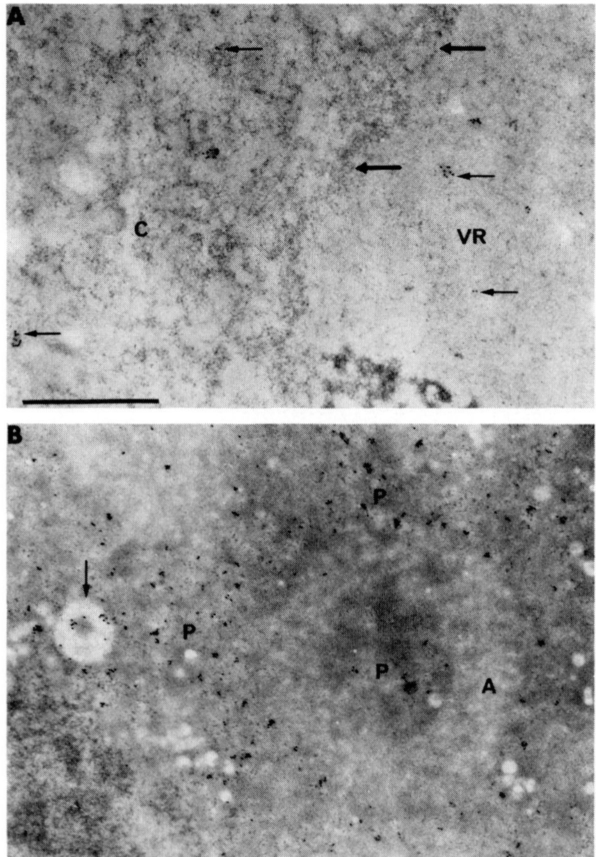

FIGURE 10. *In situ* hybridization using HSV-1 (A) and Ad5 (B) DNA probes on protease-DNase digested Lowicryl sections of HSV-1 and Ad5 infected cells respectively. Uranyl acetate staining. Bars = 1 μm. (A) Seventeen hours after HSV-1 infection. Glutaraldehyde fixation. (From Puvion-Dutilleul, F. and Puvion, E., *J. Electron Microsc. Tech.*, 18, 336, 1991. With permission.) The thick arrows point to the almost entirely digested condensed chromatin, which does not constitute a visible layer between the cytoplasm (C) and the intranuclear viral region (VR). Labeling (thin arrows) of the viral region (VR) and the ribosome-riçh area of the cytoplasm (C) is slight. (B) Seventeen hours Ad5 infection. Intermediate stage of virus-induced nuclear transformation. (From Puvion-Dutilleul, F. and Puvion, E., *Biol. Cell*, 71, 135, 1991. With permission.) Formaldehyde fixation. Part of nucleoplasm in which gold particles are most numerous over the fibrillo-granular network of the peripheral replicative zone (P). The ring-shaped single-stranded DNA accumulation site (A) is bleached by the double protease-DNase treatment of the section and is almost devoid of labeling. The digested smaller ring (on the left) (arrow) is decorated by gold particles, but is of unknown significance.

intensity of labeling is a result of the fact that the labeled viral RNA molecules were derived only from the transcribed sequence of the *Bam* H1 F fragment of HSV-1 genomes. Only the viral RNA, which originates from a short segment of the viral genomes, and only indeed that portion of it which is at the surface of the Lowicryl section, can bind the probe. We verified that when enzymatic digestions are omitted, the nucleoids of some enveloped and nonenveloped nucleocapsids are intensely labeled, and a few gold particles are scattered over mitochondria and rough endoplasmic reticulum, which indicates a certain binding of the viral probe to both cell and viral proteins. Protease treatment alone suppresses the occasional labeling of mitochondria and rough endoplasmic reticulum, and virtually eliminates the labeling of viral nucleoids. The latter, which result from the binding of the probe to single-stranded portions of encapsidated viral genomes, are completely suppressed by an additional DNase digestion.[8]

In adenovirus infection, the probe which we use spans the entire viral genome and permits the detection of all of the viral RNA molecules.[10,11] Therefore, the hybridization signal is markedly higher than that obtained above in HSV-1 infection. In the nucleus (Figure 10B), gold particles are present over the fibrillo-granular network which surrounds the very slightly labeled single-stranded DNA accumulation sites following protease-DNase digestion of sections prior to hybridization. They also are associated with some enlarged clusters of interchromatin granules and virus-induced structures resembling compact rings. In the cytoplasm, gold particles are numerous over regions rich in ribosomes. The other cell and virus components are unlabeled. The omission of the DNase treatment of the sections prior to hybridization results in an additional intense labeling of the single-stranded DNA accumulation sites, since the probe reacts with both viral RNA and the appropriate single-stranded sequences of the viral replicative intermediates. DNase treatment of the sections is imperative in order to localize Ad5 RNA exclusively. Although viral RNA-associated proteins do not constitute an impassable barrier to the probe, and do allow hybridization of the probe to viral RNA, a protease treatment increases the labeling of intranuclear Ad5 RNA.

VIII. EQUIPMENT AND REAGENTS

A. EQUIPMENT

EM hybridization experiments do not require elaborate material.

1. Grids — Gold grids are employed for recovering the ultrathin sections in order to eliminate the formation of precipitates on the sections during the different incubation steps. Mesh 200 gold grids are coated with a Formvar film strengthened by evaporation of a thin layer of carbon. The gold grids coated with Formvar and carbon, with or without sections of cells, may be stored for months at room temperature in sealed boxes. Uncoated gold grids, even of mesh 600, are not satisfactory because the sections crack under the electron beam when enzymatic digestions are made prior to hybridization.

2. Parafilm — Sheets of parafilm are used as a support for the drops of reagents and washing solutions, for several reasons. For one thing, the parafilm does not react with the chemicals used. Also, droplets of solutions do not flatten on this surface, a property which allows the use of only 1 to 2 μl of the probe (when the grid is added, a droplet of this volume flattens to form a continuous film under the section) and larger drops of enzyme and washing solutions. Another advantage is that the drops of reagents never mix, which means that numerous drops can be placed close together on a small surface and that the drops can be of different solutions (i.e., enzymes, washing solutions, and probes). This in turn permits a side-to-side disposition of grids from different biological samples (i.e., normal cells, infected cells) or samples of a given type undergoing different treatments (i.e., different digestions). For the detection of viral RNA the position in which a drop is to be placed is first cleaned with DEPC (see below in Section VIII.B).

3. Moist chamber — For both enzymatic digestions and the hybridization stage, the reactions are carried out in a moist chamber (Figure 1), as follows. All of the grids, including the controls, i.e., 16 to 20 grids per experiments, are floated on their individual droplets of reagent on parafilm lining the bottom of a 5 cm glass petri dish. This dish is placed in a 15-cm glass petri dish beside a piece of wet cellulose, so that evaporation of the H_2O from the latter maintains a humid atmosphere and prevents drying of the drops of reagent.

4. Incubator — An incubator at 37°C is used for both enzymatic digestions and hybridization. It must be mentioned that 37°C is not always the temperature which gives an optimum hybridization signal. For example, labeling of ribosomal RNA in normal and infected cells is better done at 64°C than at 37°C.[24]

B. REAGENTS

1. **Enzymes** — Highly purified enzymes, devoid of enzymatic contaminating activity, are dissolved immediately before use in appropriate solutions. Incubation of grids is performed at 37°C.

   a. Protease (Bacterial protease type VI, Sigma Chemical Comp., St. Louis, MO, U.S.) is dissolved at 0.2 mg/ml in distilled water (see Protocol 3).
   b. DNase 1 (Worthington Biochemical Corp., Freehold, NJ, U.S.) is dissolved at 1 mg/ml in 10 m*M* Tris-HCl buffer, pH 7.3, containing 5 m*M* $MgCl_2$, 2% RNasin ribonuclease inhibitor (Progema, Madison, WI, U.S.) and 2 m*M* dithiothreitol.[9-11] RNasin, the activity of which requires the presence of dithiotreitol in the solution, is used to suppress possible contaminating RNase activities (see Protocol 5).
   c. RNase A (Pancreatic RNase) (BDH Biochemical Ltd., Poole, U.K.) is dissolved at 1 mg/ml in 10 m*M* Tris-HCl buffer, pH 7.3 (see Protocol 4).

2. **Diethylpyrocarbonate (DEPC)** — DEPC (Sigma Chemical Comp., St. Louis, MO, U.S.) is dissolved at 0.1% in distilled water, autoclaved or heated at 37°C for 1 h, and then shaken vigorously in order to eliminate alcoholic vapors and CO_2 in the solution (see Protocol 10). The solution can be stored for up to 3 months at room temperature or at 4°C and should be discarded when a precipitate forms. All laboratory material (pipettes, glassware, the surface of parafilm where droplets of reagents will be applied) is rinsed in it before use and then allowed to air-dry. It is used in order to inhibit contaminating RNase activities.

PROTOCOL 10 — PREPARATION OF DEPC

1. Dissolve 100 mg DEPC in 100 ml distilled water
2. Autoclave the solution or heat it for 1 h at 37°C
3. Shake vigorously in order to eliminate alcoholic vapors and CO_2
4. Store at room temperature or at 4°C

3. **Phosphate-buffered saline (PBS)** — PBS, pH 7.3, consists of 137 m*M* NaCl, 3 m*M* KCl, 16 m*M* Na_2HPO_4, and 2 m*M* KH_2PO_4. It is sterilized and stored at 4°C, and used for the dilution of antibodies and for washing the grids after both hybridization and antibody incubations.
4. **Bovine serum albumin (BSA)** — BSA is diluted at 5% in PBS and stored at 4°C. It is used prior to the first antibody incubation in order to suppress the nonspecific binding of the antibodies to the section. The grids are exposed to a droplet of BSA for a second, then excess BSA on the grid is shaken off, not rinsed off, and the grid is immediately placed on the antibody.
5. **Antibodies** — Rabbit anti-biotin antibodies (Enzo) (stored at −20°C or at 4°C) and goat anti-rabbit IgG conjugated to gold particles 10 nm in diameter (Amersham Intern. plc, Amersham, U.K.) (stored at 4°C) are successively used for the indirect immunological detection of the hybrids (see Protocol 9). More recently, we have successfully employed direct immunological detection of hybrids using goat antibiotin immunoglobulins conjugated to gold particles 10 nm in diameter (Biocell Research Laboratories, Cardiff, U.K.) for localizing U3 RNA and ribosomal RNA in human HeLa and mouse 3T3 cells (see Protocol 10).[25]

IX. CONCLUSION

Post-embedding EM *in situ* hybridization reveals nucleic acid sequences on ultrathin Lowicryl sections without encountering problems of accessibility caused by the restriction of diffusion which occurs with pre-embedding techniques. The nucleic acid molecules, including small molecules such as messenger RNA, are retained in the Lowicryl sections even after removal of the associated proteins by a proteolytic treatment. The EM hybridization procedures described here have been used successfully to determine the chemical composition and function of intranuclear structures involved in virus replication. Provided that an appropriate protocol is adapted for each biological system, EM *in situ* hybridization can provide a broad field of applications in both animal and plant cells for localizing both cellular and viral nucleic acid sequences.

ACKNOWLEDGMENTS

The author is very indebted to Prof. E. H. Leduc (Brown University, Providence, RI, U.S.) for critical reading of the manuscript. This work is supported by general grants from the Centre National de la Recherche Scientifique and a special grant from the Association pour la Recherche sur le Cancer (ARC, Villejuif, France). F. P-D is a member of the Institut National de la Santé et de la Recherche Médicale.

REFERENCES

1. **Croissant, O., Duguet, C., Jeanteur, P., and Orth, G.,** Application de la technique d'hybridation moléculaire *in situ* à la mise en évidence au microscope électronique, de la réplication végétative de l'ADN viral dans les papillomes provoqués par le virus de Shope chez le lapin cottontail, *C. R. Acad. Sci., Paris,* 274, 614, 1972.
2. **De Zoeten, G. A., Powell, C. A., Gaard, G., and German, T. L.,** *In situ* localization of pea enation mosaic virus double-stranded ribonucleic acid, *Virology,* 70, 459, 1976.
3. **Geuskens, M. and May, E.,** Ultrastructural localization of SV40 viral DNA in cells during lytic infection, by *in situ* molecular hybridization, *Exp. Cell Res.,* 87, 175, 1974.
4. **Pautrat, G., Morel, G., Dihl, F., Allasia, C., Rey, F., Spire, J., Kourilsky, F., and Chermann, J. G.,** Distribution polaire des particules virales à la surface des cellules d'une lignée lymphocytaire infectées *in vitro* par le virus HIV: détection intracellulaire du génome viral, *Retrovirus,* 47, 511, 1988.
5. **Puvion-Dutilleul, F., Pichard, E., Laithier, M., and Puvion, E.,** Cytochemical study of the localization and organization of parental Herpes simplex virus type 1 DNA during initial infection of the cell, *Eur. J. Cell Biol.,* 50, 187, 1989.
6. **Puvion-Dutilleul, F. and Puvion, E.,** Ultrastructural localization of viral DNA in thin sections of Herpes simplex virus type 1 infected cells by *in situ* hybridization, *Eur. J. Cell Biol.,* 49, 99, 1989.
7. **Puvion-Dutilleul, F. and Puvion, E.,** Replicating single-stranded adenovirus type 5 DNA molecules accumulate within well-delimited intranuclear areas of lytically infected HeLa cells, *Eur. J. Cell Biol.,* 52, 379, 1990.
8. **Puvion-Dutilleul, F. and Puvion, E.,** Analysis by *in situ* hybridization and autoradiography of sites of replication and storage of single- and double-stranded Adenovirus type 5 DNA in lytically infected HeLa cells, *J. Struct. Biol.,* 103, 280, 1990.
9. **Puvion-Dutilleul, F. and Puvion, E.,** Ultrastructural localization of defined sequences of viral RNA and DNA by *in situ* hybridization of biotinylated DNA probes on sections of Herpes simplex virus type 1 infected cells, *J. Electron Microsc. Tech.,* 18, 336, 1991.

10. **Puvion-Dutilleul, F. and Puvion, E.**, Sites of transcription of adenovirus type 5 genomes in relation to early viral DNA replication in infected HeLa cells. A high resolution *in situ* hybridization and autoradiographical study, *Biol. Cell,* 71, 135, 1991.
11. **Puvion-Dutilleul, F., Roussev, R., and Puvion, E.**, Distribution of viral RNA during the adenovirus type 5 infectious cycle in HeLa cells, *J. Struct. Biol.*, 108, 209, 1992.
12. **Wolber, R. A., Beals, T. F., and Maassab, H. F.**, Ultrastructural localization of viral nucleic acid by *in situ* hybridization, *Lab. Invest.*, 59, 144, 1988.
13. **Wolber, R. A., Beals, T. F., and Maassab, H. F.**, Ultrastructural localization of Herpes simplex virus RNA by *in situ* hybridization, *J. Histochem. Cytochem.*, 37, 97, 1989.
14. **Bendayan, M.**, Protein-A gold electron microscopic immunocytochemistry: methods, applications, and limitations, *J. Electron Microsc. Tech.*, 1, 243, 1984.
15. **Carlemalm, E., Villiger, W., and Acetarin, J. D.**, Advances in specimen preparation for electron microscopy. I. Novel low temperature embedding resins and a reformulated Vestopal, *Experentia,* 36, 740, 1980.
16. **Roth, J., Bendayan, M., Carlemalm, E., and Villiger, W.**, Enhancement of structural preservation and immunocytochemical staining in low temperature embedded pancreatic tissue, *J. Histochem. Cytochem.*, 29, 663, 1981.
17. **Langer, P. R., Waldrop, A. A., and Ward, D. C.**, Immunological methods for mapping genes on *Drosophila* polytene chromosomes, *Proc. Natl. Acad. Sci. U.S.A.*, 79, 4381, 1981.
18. **Binder, M., Tourmente, S., Roth, J., Renaud, M., and Gehring, W. J.**, *In situ* Lowicryl K4M-embedded tissue using biotinylated probes and protein A-gold complexes, *J. Cell Biol.*, 102, 1646, 1986.
19. **Puvion-Dutilleul, F., Laithier, M., and Sheldrick, P.**, Ultrastructural localization of the herpes simplex virus major DNA-binding protein in the nucleus of infected cells, *J. Gen. Virol.*, 66, 15, 1985.
20. **Puvion-Dutilleul, F., Pédron, J., and Cajean-Feroldi, C.**, Identification of intranuclear structures containing the 72K DNA-binding protein of human adenovirus type 5, *Eur. J. Cell Biol.*, 34, 313, 1984.
21. **Puvion-Dutilleul, F.**, Localisation de séquences d'ADN viral par hybridation *in situ* de sondes biotinylées à la surface de coupes ultrafines de cellules enrobées dans le Lowicryl K4M, in *Techniques en Microscopie Electronique: Cryométhodes, Immunocytologie, Autoradiographie, Hybridation In Situ,* Morel, G., Ed., Editions INSERM, Paris, 1991, 553.
22. **Puvion-Dutilleul, F.**, Localisation d'ARN viraux par hybridation *in situ* de sondes biotinylées à la surface de coupes ultrafines de cellules enrobées dans le Lowicryl K4M, in *Techniques en Microscopie Electronique: Cryométhodes, Immunocytologie, Autoradiographie, Hybridation In Situ,* Morel, G., Ed., Editions INSERM, Press, 1991, 565.
23. **Puvion-Dutilleul, F.**, Localisation intracellulaire des génomes viraux et des ARN spécifiques dans des cellules permissives infectées par des virus à ADN humains, in *Hybridation In Situ: Méthodes Pratiques,* Calas, A., Bloch, B., Fournier, J. G., and Trembleau, A., Eds., Société Française de Microscopie Electronique, Ivry, France, 1992, 151.
24. **Puvion-Dutilleul, F., Bachellerie, J.-P., and Puvion, E.**, Nucleolar organization of HeLa cells as studied by *in situ* hybridization, *Chromosoma,* 100, 395, 1991.
25. **Puvion-Dutilleul, F., Mazan, S., Nicoloso, M., Christensen, M. E., and Bachellerie, J.-P.**, Localization of U3 RNA molecules in nucleoli of HeLa and mouse 3T3 cells by high resolution *in situ* hybridization, *Eur. J. Cell Biol.*, 56, 178, 1991.
26. **Puvion-Dutilleul, F., Pichard, E., Laithier, M., and Leduc, E. H.**, Effect of dehydrating agents on DNA organization in herpes viruses, *J. Histochem. Cytochem.*, 35, 635, 1987.
27. **Locker, H. and Frenkel, L.**, Bam I, Kpn I, Sal I enzyme maps of the DNAs of herpes simplex virus strains Justin and F: occurrence of heterogeneities in defined regions of the viral DNA, *J. Virol.*, 32, 429, 1979.

28. **Pellett, P. E., McKnight, J. L. C., Jenkins, F. J., and Roizman, B.,** Nucleotide sequence and predicted amino acid sequence of a protein encoded in a small herpes simplex virus DNA fragment capable of trans-inducing α genes, *Proc. Natl. Acad. Sci. U.S.A.*, 82, 5870, 1985.
29. **Batterson, W. and Roizman, B.,** Characterization of the herpes simplex virion-associated factor responsible for the induction of α genes, *Proc. Natl. Acad. Sci. U.S.A.*, 46, 371, 1983.
30. **Campbell, M. E. M., Palfreyman, J. W., and Preston, C. M.,** Identification of herpes simplex virus DNA sequences which encode a trans-acting polypeptide responsible for stimulation of immediate-early transcription, *J. Mol. Biol.*, 180, 1, 1984.
31. **Post, L. E., Mackem, S., and Roizman, B.,** Regulation of α genes of herpes simplex virus: expression of chimeric genes produced by fusion of thymidine kinase with α gene promoters, *Cell,* 24, 555, 1981.
32. **Galdieri, M. and Monesi, V.,** A possible artefact of *in situ* molecular hybridization technique in mouse spermatids, *Boll. Zool.*, 40, 411, 1973.
33. **Pardue, M. L. and Gall, J. G.,** Molecular hybridization of radio-active DNA to the DNA of cytological preparations, *Proc. Natl. Acad. Sci. U.S.A.*, 64, 600, 1969.
34. **Cogliati, R. and Gautier, A.,** Mise en évidence de l'ADN et des polysaccharides à l'aide d'un réactif de "type Schiff", *C. R. Acad. Sci.*, Paris, 276, 3041, 1973.
35. **Bernhard, W.,** A new staining procedure for electron microscopical cytology, *J. Ultrastruct. Res.*, 27, 250, 1969.
36. **Frenkel, N. and Roizman, B.,** Separation of the herpes deoxyribonucleic acid on sedimentation in alkaline gradients, *J. Virol.*, 10, 565, 1972.
37. **Wilkie, N. M.,** The synthesis and substructure of herpes virus DNA: the distribution of alkali labile single strand interruptions in HSV-1 DNA, *J. Gen. Virol.*, 21, 453, 1973.
38. **Lechner, R. L. and Kelly, T. J., Jr.,** The structure of the replicating adenovirus type 2 DNA molecules, *Cell,* 12, 1007, 1977.

Chapter 10

ELECTRON MICROSCOPIC *IN SITU* HYBRIDIZATION TO RNA OR DNA IN PLANT CELLS

Judith Brangeon and Lucienne Sossountzov

TABLE OF CONTENTS

0-8493-4414-X/93/$0.00 + $.50

I. INTRODUCTION

The *in situ* hybridization (ISH) technique makes possible the detection and localization of specific nucleic acid sequences within tissue sections or whole-mount preparations of single cells, organelles, or chromosomes. It is based on the principle that a labeled single-stranded fragment of DNA or RNA (probe), exogenously applied, will hybridize to a complementary sequence on the cellular or viral DNA or RNA (target), forming stable hybrids. The hybrids are located at a particular morphological site by using a system of detection (visible marker) which can be coupled to the labeled probe and visualized in the microscope.

If the probe is labeled with a radioisotope, the hybrid is detected by autoradiographic procedures, whereas nonradioactive labels such as biotin, 2-acetyl aminofluorine (AFF), bromo-deoxyuridine (BrdU), and digoxigenin are invariably detected by histochemical techniques (enzyme reactions, affinity, or immunocytochemical [ICC] methods).

In situ hybridization is a unique tool which allows an observer to determine the spatial sites where a particular DNA or RNA sequence is situated with respect to cell morphology. If used in a developmental context, it can provide insights into the temporal aspects of gene expression.

ISH was practiced initially at the light microscopic level (LM), but in recent years the attraction of the fine resolution offered by the electron microscope (EM) has extended the method to more routine use at the ultrastructural level.[1,2] Considerable literature on hybridizing to RNAs or DNAs in plant systems at the LM level is available, and it should serve as a convenient stepping stone to ultrastructural detection (see references in 2). In this report only methodologies and applications of ISH at the EM level will be considered.

This chapter will be concerned with certain characteristics specific to higher plant systems and the ways in which methodologies have been modified to overcome difficulties inherent in plant tissue. Applications of *in situ* hybridization to plant biology will be illustrated by recent advances in the ultrastructural identification of specific mRNA and DNA sequences. Methods for their detection will be given in the text in the form of outlined protocols.

II. HISTORICAL BACKGROUND

Since the first paper by Gall and Pardue,[3] the technique of *in situ* hybridization (ISH) has been developed at the light microscopic level and applied to cell smears,[4] chromosomes spreads,[5-8] cryostat and vibratome sections,[7,8] to sections of both paraffin-embedded samples, and, to a lesser extent, to resin-embedded samples.[5,9-12] More recently, ISH on sections obtained by ultracryotomy (see Chapter 6) has been achieved. Up until 1984, only radiolabeled probes and autoradiographic detection techniques were used. These early ISH experiments mainly made use of the so-called pre-embedding method,[6,7,13-19] whole-mount chromosomes, or cell spreads.[4-6,12] The nonradioactive post-embedding technique for the detection of DNA-DNA, DNA-RNA, or RNA-RNA hybrids in ultrathin sections is more recent.[20-26]

Binder et al.[27] were the first to test biotinylated probes for visualizing hybridized targets on ultrathin sections from Lowicryl K4M-embedded material in conjunction with protein A-gold labeling. To maintain good cellular integrity, they used phosphate-buffered saline washes in place of the more commonly used formamide-SSC buffers. Furthermore, their immunocytochemical system employed a two-step method with anti-biotin antibodies, combined with protein A-gold complexes to enhance the labeling. They obtained rapid signal detection and low background, with good cellular integrity.

In plant biology, only a few laboratories have tried ISH procedures at the ultrastructural level (see references in 2), although the method has been routinely used at the light microscopic level, mainly with ^{35}S- or ^{3}H-probes.[28-39] This approach has demonstrated the cellular heterogeneity of tissues producing specific mRNAs for a number of genes. But in this regard there is now a need for high resolution to identify the location of specific DNA or RNA sequences in cells and organelles from heterogeneous cell populations. Some laboratories have developed EM pre-embedding techniques[40-42] with radiolabeled or biotinylated probes; others worked on resin-embedded sections.[31,43] An elegant study was carried out by McFadden and his collaborators[2,43,44] on ultrathin resin sections using a double labeling system with biotin- and digoxigenin-tagged probes for the simultaneous detection of two specific rRNAs.

At the EM level, there is a major problem concerning the "thickness" of preparations in which only a small portion of the cell nucleic acids are present. Using probes with high specific activity, the number of hybrids formed at one site is necessarily low; hence the need for an efficient detection system. The second problem deals with properties of the fixative which should retain the maximum amount of target DNA or RNA and preserve the integrity of cellular structure for high resolution, while at the same time allowing sufficient penetration and diffusion of the labeled probes to their appropriate targets.[45,46]

III. PLANT TISSUE PARTICULARITIES

Plant and animal cells possess common functional features, in that they both have the cell molecular machinery to store genetic information, give instructions, and regulate the synthesis of molecules. For this reason, *in situ* hybridization methods that had initially been developed for animal systems have been later transposed to plant systems. However, plant cells are characterized by the presence of two unique and specific compartments: (1) the cell wall, an extracellular matrix consisting of cellulose and pectic substance, often impregnated with lignins and specialized proteinaceous material; it is a relatively thin envelope in young dividing cells (Figure 1a) but can become massive and elaborate in certain mature specialized cells. In addition to the cell and organelle boundary membranes, the cell wall constitutes a rigid semipermeable barrier; (2) the vacuole, a water-filled membrane-bound compartment which, in differentiated tissue, can occupy the greater part of the plant cell (Figure 1b). This centrally located compartment contains a high solute concentration and, as a result, maintains cell turgidity and positions the cytoplasm in a relatively thin film against the cell wall. Osmolarity must be maintained during tissue preparatory steps in order to avoid deleterious effects on cell structure. Another feature which sets plant systems apart from animal systems is the existence of permanently embryonic regions termed "meristems". These zones of continuously dividing cells are often distributed throughout tissue, and consequently young cells can be contiguous to "juvenile" and/or mature cells, thus contributing to a heterogeneous tissue pattern, with each form having particular characteristics, which necessitates a compromise for fixation/permeabilization procedures. In addition, numerous air spaces are interspersed among cells in certain types of tissue, and these interfere with or impede the free movement and penetration of solutions between cells. These morphological features constitute drawbacks for certain technical aspects of ISH and have undoubtedly contributed to its slower development in plant biology. It is more difficult to obtain from plant tissues than from animal samples a compromise between good cellular morphology, good retention of cellular nucleic acids (particularly RNA, DNA being more stable), and access of the corresponding probes to cellular targets. Modifications that have been adapted to plant material will be mentioned in the appropriate protocols.

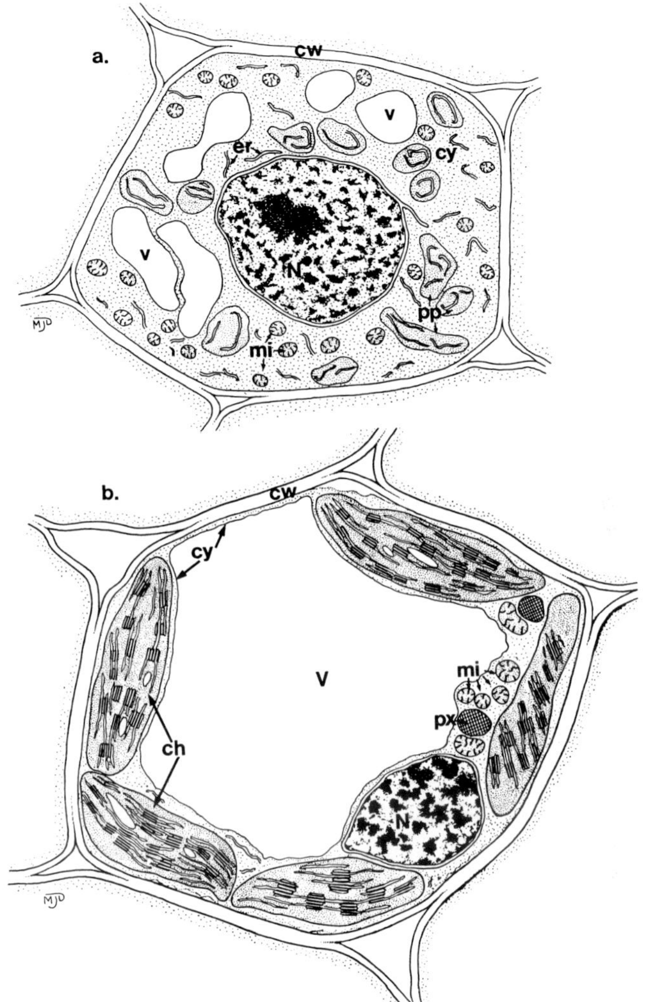

FIGURE 1. Characteristics of meristematic and differentiated cells. a., Meristematic leaf cell filled with ribosome-rich cytoplasm (cy) in which reside numerous proplastids (young chloroplasts, pp), mitochondria (mi), and endoplasmic reticulum (er); a thin cell wall (CW) surrounds the young cell; b., a differentiated leaf cell characterized by a large central vacuole (V) which positions the cytoplasm and organelles — chloroplasts (ch), mitochondria, peroxisomes (px), and nucleus (N) — against a thick cell wall.

IV. STRATEGY

Our laboratories have recently developed ISH at the EM level, with a view to: (1) localizing mRNAs coding for the large and small subunits of a chloroplastic enzyme, ribulose bisphosphate carboxylase;[40,41,47] (2) characterizing the modalities of replication of a plant virus, the cauliflower mosaic virus (CaMV);[48] and (3) localizing mRNA coding for lipid transfer protein (LTP) present in abundance in maize coleoptile.[49] In two of these cases, final gene products had been previously identified, and their intracellular distribution established, by immunocytochemical techniques using specific antibodies.[41,50] Once familiar with the ICC methods, we chose to use nonisotopic haptenized probes, detected either by affinity or ICC conjugates, for the ISH study (Figure 2).

Although most ISH procedures employ radiolabeled probes, we chose nonisotopic methods because recent developments have shown their increased sensitivity and resolution as well as their rapid use, which has led to the development of routine procedures at the LM and EM level in animal and medical research.[51] Biotinylated double-stranded DNA probes are the most commonly used. Biotinylated oligonucleotides and riboprobes have been used more frequently only during the last 3 to 4 years. According to Höfler,[52] and to Ozden et al.,[53] RNA probes have a higher specific activity than nick translated DNA probes and are now known to be as sensitive as radiolabeled probes. Furthermore, the possibility of using a posthybridization RNase treatment to digest the unhybridized probe leads to a high signal-to-noise ratio. However, one disadvantage is the stringent conditions of their use; all the steps of the procedure require precautions to be taken in order to prevent contamination by ubiquitous RNases.

To identify CaMV in turnip leaves and to detect specific mRNAs in tobacco leaves, we used cDNA and cRNA probes biotinylated by bio 11-dUTP or bio 11-UTP, respectively, because of the high sensitivity they can achieve. Due to the lability of cellular mRNA, there are problems of target retention, while at the same time good probe penetration into the tissues must be assured. For hybridization of LTP mRNA, we used a fairly new method of labeling: an *antisense* oligo(deoxy)nucleotide (single-stranded DNA probe) labeled by bromo-deoxyuridine according to Jirikowsky et al.[54] and Ramalho-Ortigao et al.[55] and known to give very stable hybrids with the target mRNA. DNA probes, oligonucleotide probes, and riboprobes have the advantage of being more stable than radiolabeled probes, remaining usable up to several months.[56] The coupled affinity detection system is less time-consuming than histoautoradiographic techniques: the chosen protocols take 2 days. The hybrid duplexes can be detected by affinity or ICC systems using specific, commercially available antibodies: polyclonal anti-biotin IgG or monoclonal anti-BrdU.

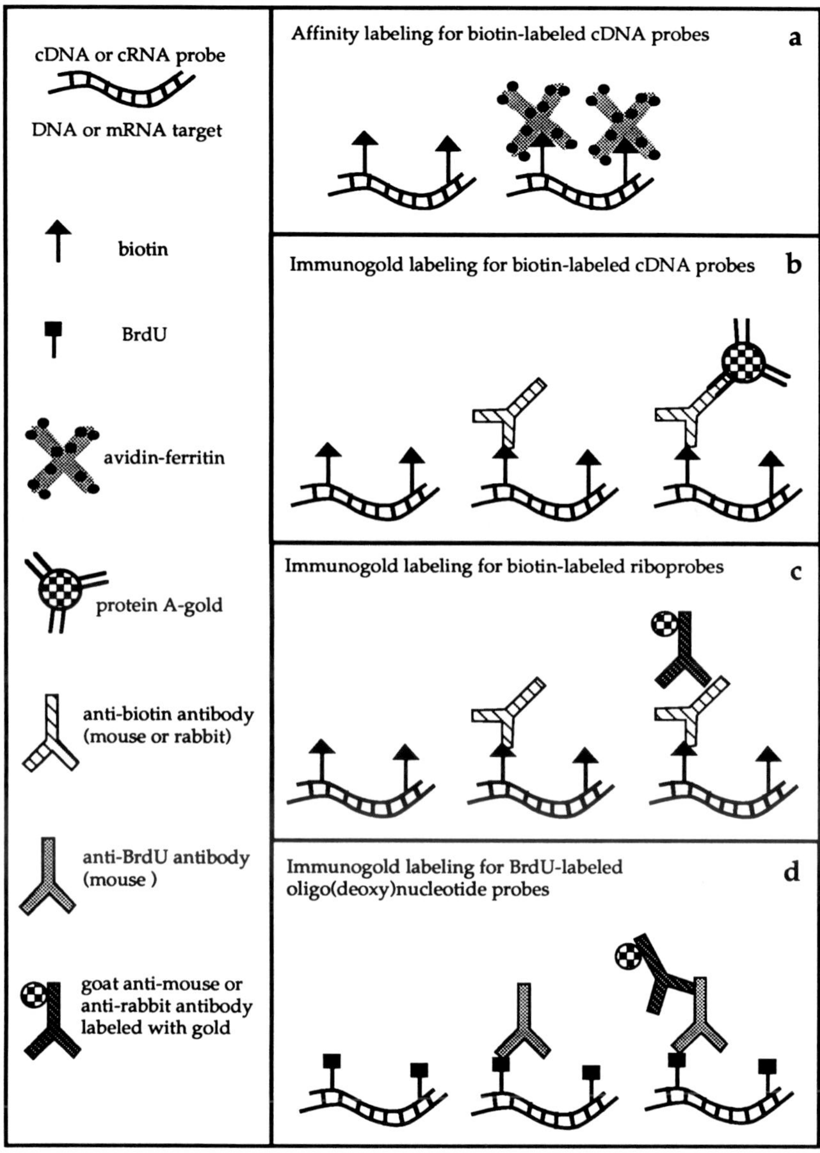

FIGURE 2. Schematic outline of procedures used for the EM detection of target DNA and mRNA by affinity (a) and immunocytochemical techniques (b, c, and d). These procedures were applied on plant block samples according to the pre-embedding technique (a), and to ultrathin sections "on grid" after pre-embedding in LR White or Lowicryl K4M acrylic resins (b, c, and d), depending on the probe used (cDNA probes, riboprobes, and oligoprobes) and their labeling (biotin and bromodeoxyuridine).

In this chapter, we give examples of ISH procedures obtained with resin-embedded plant organs for ultrastructural observations. We shall describe in detail the use of:

1. Biotinylated cDNA probes to localize mRNAs coding for the small subunit and large subunit composites of a chloroplastic enzyme, ribulose bisphosphate carboxylase in tobacco leaves.[40,41,47]
2. Biotinylated riboprobes to localize viral DNA of *minus* and *plus* polarity in cells of systemically infected turnip leaves by a DNA virus, the CaMV.[48]
3. BrdU-labeled *antisense*- and *sense*- oligoprobes to localize and to verify the location of mRNA coding for a lipid transfer protein in maize coleoptile.[49]

Our goal is to provide a practical guide for ISH at the EM level using embedded tissues, to show the variety of approaches that can be used to attain these objectives, and, hopefully, to stimulate further interest in this method for our colleagues in plant research.

V. METHODOLOGY

There are two ISH approaches that have been successfully employed in plant systems to detect endogenous cell nucleic acid targets: the *pre-embedding* method and the *post-embedding* method, both of which have their strengths and limitations. The choice of method is governed by the starting living material (fixation and impermeabilization requirements), the type and quantity (copy numbers) of targets, and the ease with which probes and/or tags (i.e., probe length, marker sizes) can enter and diffuse within preparations.

The pre-embedding approach — This method starts out with small samples (blocks) of prefixed tissue, which must be permeabilized by a ''pore-forming'' treatment (enzymatic or detergent treatment) in order to facilitate the penetration of the probe and its accessibility to targets throughout the volume of the tissue block. Once inside the cells the probe drives the reaction, and virtually all specific targets would be bound, which greatly increases the signals. The labeled probes are detected either by autoradiography or by the subsequent diffusion of a visible marker into the tissue block. The tissue is then embedded in a resin for sectioning and observation. The pre-embedding method has been extensively used in animal systems[14-16,19,24,57] with both chemically fixed and frozen preparations. It has been shown that probe and marker systems can be diffused into the tissue with relative ease and with minimum disruption of cell morphology. Given that cell barriers in most plant systems are relatively sturdy, the treatment for rendering cells porous without cell damage can be delicate and, often, a compromise must be reached.[40,42]

The post-embedding approach — This technique also starts out with a light fixation procedure; samples are then processed and embedded in a resin, and ultrathin sections are prepared. As the term implies, probe-target formation (hybridization) takes place on resin-embedded sectioned material. Sensitivity can be rather low given the probability that only targets exposed at the surface of the section are accessible for hybridizing. This drawback may be reduced by the use of hydrophilic acrylic resins such as LR (London Resin) White or Gold; it has been suggested that these resins are semipermeable and allow some probe penetration, and thus access to targets deeper within the section. Hybridization on ultrathin sections has been applied in both animal and plant systems using low-temperature embedding in Lowicryl K4M[20,23,24,27] and room temperature embedding in LR Gold[43,44,58] or LR White.[25,45]

VI. EXAMPLES OF ISH METHODS IN PLANT MATERIAL

A. INTRACELLULAR LOCALIZATION OF RIBULOSE-1,5-BISPHOSPHATE CARBOXYLASE SUBUNIT mRNAS IN TOBACCO LEAVES

We have carried out a pre-embedding EM *in situ* hybridization using paraformaldehyde-fixed leaf material. The endogenous mRNAs were hybridized with biotin-labeled DNA probes and subsequently visualized using avidin-ferritin conjugates. As our initial goal was to establish a reliable and reproducible technology to locate cell RNAs, we chose as a test system an enzyme/mRNA system in which targets are abundant and localization to specific subcellular compartments is well documented.

Ribulose 1,5-bisphosphate carboxylase, a chloroplastic enzyme, is a key enzyme of CO_2 fixation in C3 plants. It is composed of two types of subunit, the small subunit (SSU) being nuclear-encoded and synthesized as a precursor on free cytoplasmic polysomes whereas the large subunit (LSU) is encoded and synthesized inside the chloroplasts, where final enzyme assembly also occurs.[59]

The implication of both nuclear- and plastid-coding genes in the synthesis of the enzyme constitutes a rather unique system of coordinated interactions between cell compartments. Prior to undertaking the ISH study, immunogold cytochemistry was used to visualize target proteins confined exclusively to the stroma matrix of wild type tobacco chloroplasts; a totally immunonegative enzyme-deficient tobacco mutant was used as a specificity control.[42]

1. Preparation and Biotinylation of DNA Probes

The SSU probe from tobacco was kindly provided by Dr. J. Fleck (Clone ps TV 34). The hybridization was performed with the insert (543 bp) excised by CFoI restriction enzyme. The LSU probe from tobacco chloroplast DNA was cloned (clone pt B1) by Shinozaki and Sugiura;[60] the *Bam* HT fragment of the 3′ coding region was used for hybridization. Control hybridizations were carried out with a pBR322 probe.

The nick translation with biotinylated dUTP (11 atom linkers) was performed with a Bethesda Research Laboratories (BRL, Bethesda, MD, U.S.) kit following the instruction manual with slight modifications (Figure 3). The biotinylated nucleotide analogs are substrate for DNA polymerases and can be incorporated into nucleic acids by *in vitro* polymerization; the biotin moiety is attached to the nucleotide analog at the C-5 position in deoxyuridine (see Protocol 1).

PROTOCOL 1 — INCORPORATION OF BIOTIN 11-dUTP INTO DNA

1. Prepare reaction mixture composed of 5 μg DNA, 10 units of DNA polymerase I, 100 pg of DNase, 0.4 m*M* each of dATP, dGTP, dCTP, and biotin 11-dUTP in a final volume of 50 μl of 100 m*M* Tris-HCl, 2 m*M* $MgCl_2$, pH 7.8, and 10 μg/ml bovine serum albumin (BSA) according to Harris and Croy.[42] Incubate at 16°C for 90 min.
2. Load reaction mixture onto a 5- to 10-ml Sephadex G50 superfine column equilibrated with 1× SSC (0.15 *M* sodium chloride, 0.015 *M* sodium citrate, pH 7.0), containing 0.1% (v/v) SDS. A plastic pipette fitted with siliconized glass wool is a satisfactory column. Collect excluded fractions in 1.5-ml polypropylene microcentrifuge tubes. The presence of 0.1% SDS in elution buffer will reduce nonspecific binding of the biotin-labeled DNA to the plastic tubes.
3. To verify biotin incorporation, spot 1 to 2 μl of each fraction onto nitrocellulose paper and visualize by streptavidin coupled to alkaline phosphate as described in the DNA detection instructions of the BRL kit.
4. Store aliquots of biotinylated probes in DNA dilution buffer (provided in the BRL kit) at −20°C until use.

2. Pre-Embedding Method for ISH in Leaf Tissue

Tobacco plants *(Nicotiana tabacum)* were grown in a growth chamber under a 16-h day/8-h night regime. Samples from young developing leaves and from fully expanded leaves were excised from lateral apices.

The operations to be carried out are presented in Protocols 2, 3, and 4.

PROTOCOL 2 — PREFIXATION AND PERMEABILIZATION[a]

1. Cut thin slices (about 200 μm) of leaf material in a droplet of freshly made 4% paraformaldehyde in Na-phosphate-sucrose buffer (P-S: 0.05 *M* phosphate buffer, 0.05 *M* sucrose, pH 7) for 5 h at room temperature. The addition of sucrose to the buffer maintains osmolarity during fixation, prevents cell plasmolysis, and improves preservation of structure. Change fixation solution twice during 5 h. If samples float, place under gentle vacuum to improve penetration of fixative.
2. Wash samples in same P-S buffer for 4 × 15 min.

Nick translation

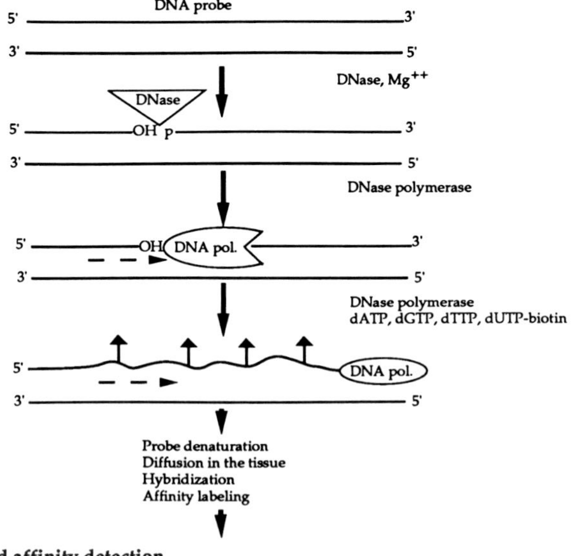

Hybridization and affinity detection

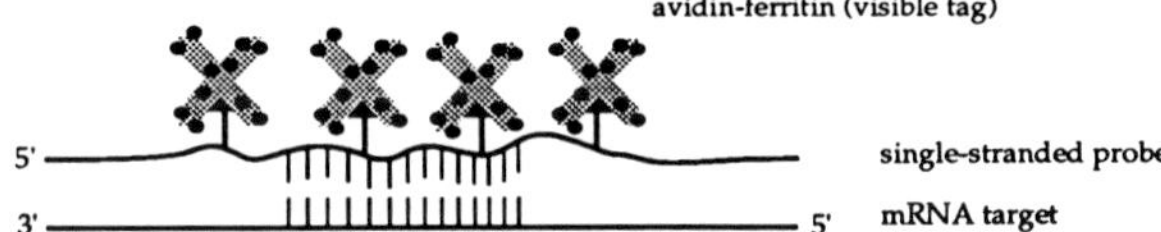

Intra-cellular targets-single labeling

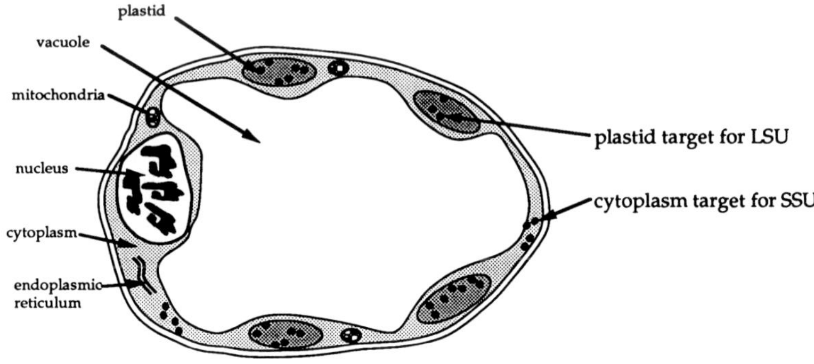

FIGURE 3. Schematic representation of nick translation and labeling procedure of two DNA probes for the ISH detection of mRNAs coding for large subunit (LSU) and small subunit (SSU) composites of ribulose bisphosphate carboxylase. See text for further details and Figure 2a.

3. Soak samples in a preparation of Pronase B at 25 μg/ml in P-S buffer containing 5 m*M* EDTA for 15 min at 18°C.
4. Stop digestive (pronase) activity by incubating samples in 0.5 m*M* phenyl methylsulphonyl fluoride in P-S/EDTA buffer containing 4 mg/ml glycine to quench free aldehyde groups for 1 h at 18°C.

[a] This method has also been used to locate mRNA for the storage protein legumin in pea cotyledons. After prefixation, tissue pieces were recut to approximately 100-μm slices using a Mickle tissue chopper.[42]

PROTOCOL 3 — PREHYBRIDIZATION AND HYBRIDIZATION

1. Place individual samples in siliconized Eppendorf tubes and equilibrate with 50% formamide in 0.3 *M* NaCl/0.03 *M* Na citrate, 5 m*M* EDTA, 1% Denhardt's solution, pH 7.0, (hybridization buffer) for 2 × 15 min at room temperature.
2. Denature cDNA biotinylated probes prior to use by plunging in boiling water for 3 min, then place on ice.
3. Incubate tissue samples in Eppendorf tubes containing 150 μl of 50% formamide: 50% hybridization buffer with 0.5 μg/ml of denatured biotinylated probe cDNA (either SSU probe, LSU probe, or control pBR322 probe) plus 250 μg/ml of salmon sperm DNA for 40 h at 37°C. Another control consisted of incubation in the formamide-hybridization buffer with no probe, followed by avidin-ferritin incubation (see Protocol 4, step 2).

PROTOCOL 4 — POSTHYBRIDIZATION STEPS AND EMBEDDING

1. Wash thoroughly with several changes of phosphate-buffered saline (PBS) over a 4 to 5 h period.
2. Incubate in 5 μg/ml avidin-ferritin (purchased from Sigma, St. Louis, MO, U.S.) in PBS for 2 h at 37°C (Figure 2a).
3. Wash tissue samples in PBS 4 × 15 min and leave overnight in PBS with gentle agitation.
4. Post-fix in 1% osmium tetroxide in distilled H_2O for 1 h, followed by several washes in H_2O.
5. Dehydrate in a graded series of alcohols, infiltrate, and embed in Spurr resin. Since hybrid duplexes are stable in tissue samples, conventional hydrophobic resins (Spurr, Epon) may be used.
6. Cut pale-gold ultrathin sections and place on 200-mesh copper grids. Do not poststain as this can yield electron-dense cytoplasmic backgrounds which interfere with ferritin particle visualization. If tissue has high endogenous levels of biotin, it is possible to use a blocking step before hybridizations using an avidin-biotin blocking system (Vector Laboratories Inc., Burlingame, CA, U.S.).[61]

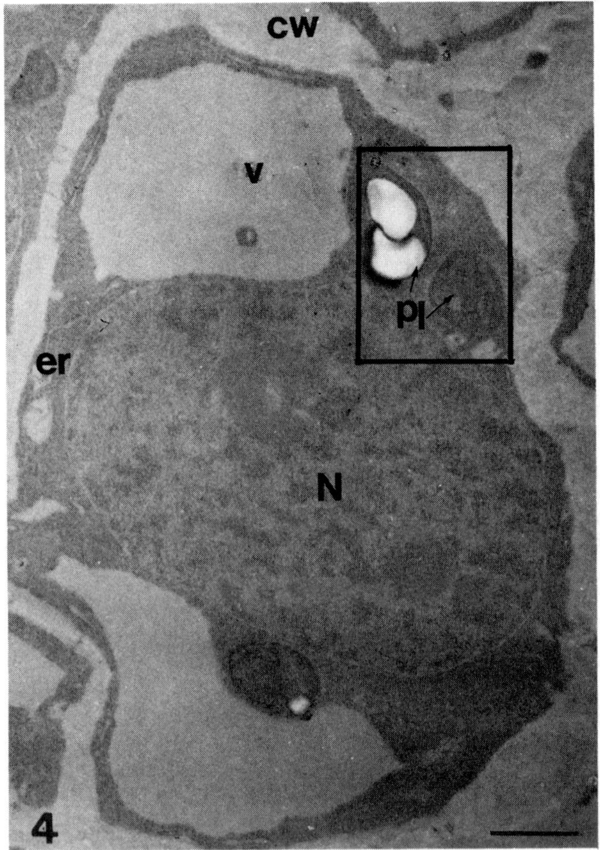

FIGURE 4. Young tobacco leaf cell incubated with biotinylated LSU probe and subsequently with avidin-ferritin markers. Cell components — nucleus (N), plastids (pl), endoplasmic reticulum (er), vacuole, (v), and cell wall (cw) — are readily discernible.

We have used this protocol for protoplast (leaf cells devoid of cell walls) preparations and isolated organelles. After the prefixation step, the samples were centrifuged at low speed to form a pellet and embedded in 1% agar in H_2O. The agar blocks were then carried through the subsequent steps as were the tissue pieces.

3. Results

Figure 4 illustrates a young leaf cell incubated with the biotinylated LSU probe, and subsequently with avidin-ferritin. The cell architecture is relatively well preserved, and cell components, nucleus (N), plastids (pl), and endoplasmic reticulum (er) are readily recognizable. These positively labeled cells, which are located near the outer surface of the leaf slice, show no visible sign of structural disruption. The two enclosed plastids are shown at a higher

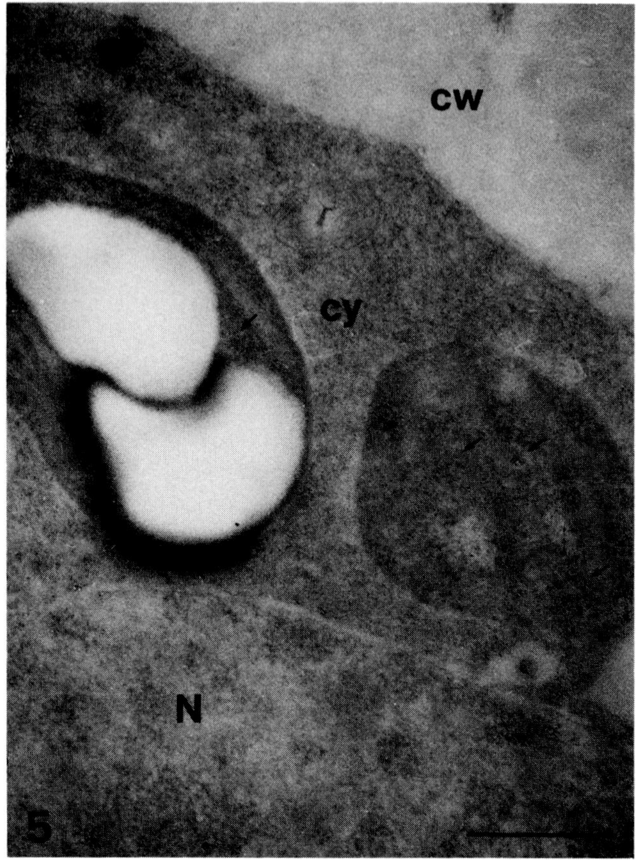

FIGURE 5. High magnification of boxed-in plastids shows ferritin markers (arrows), indicating that LSU mRNAs are located exclusively in plastids. Cytoplasm and nucleus are untagged.

magnification in Figure 5; ferritin markers (arrows) indicating the target LSU mRNAs are exclusively located in the plastid stroma, while the surrounding cytoplasm (cy), nucleus (N), and cell wall (CW) are unlabeled. This specificity was observed in all neighboring cells, although some internal cells of thicker leaf slices were unlabeled, undoubtedly due to low probe penetration. The sections were unstained, since contrasting with either uranyl acetate or lead citrate led to electron-dense cytoplasmic or stroma backgrounds (mostly caused by ribosomes), which interfered with viewing the ferritin particles. Although these markers have a low electron density and are not easily visible in the electron microscope, their small size (6 nm) is advantageous for cell penetration. A portion of an SSU-probed cell is shown in Figure 6. Probe penetration and avidin-ferritin tagging were excellent in these SSU-probed cells; a homogeneous distribution of ferritin throughout the cytoplasm indicated the

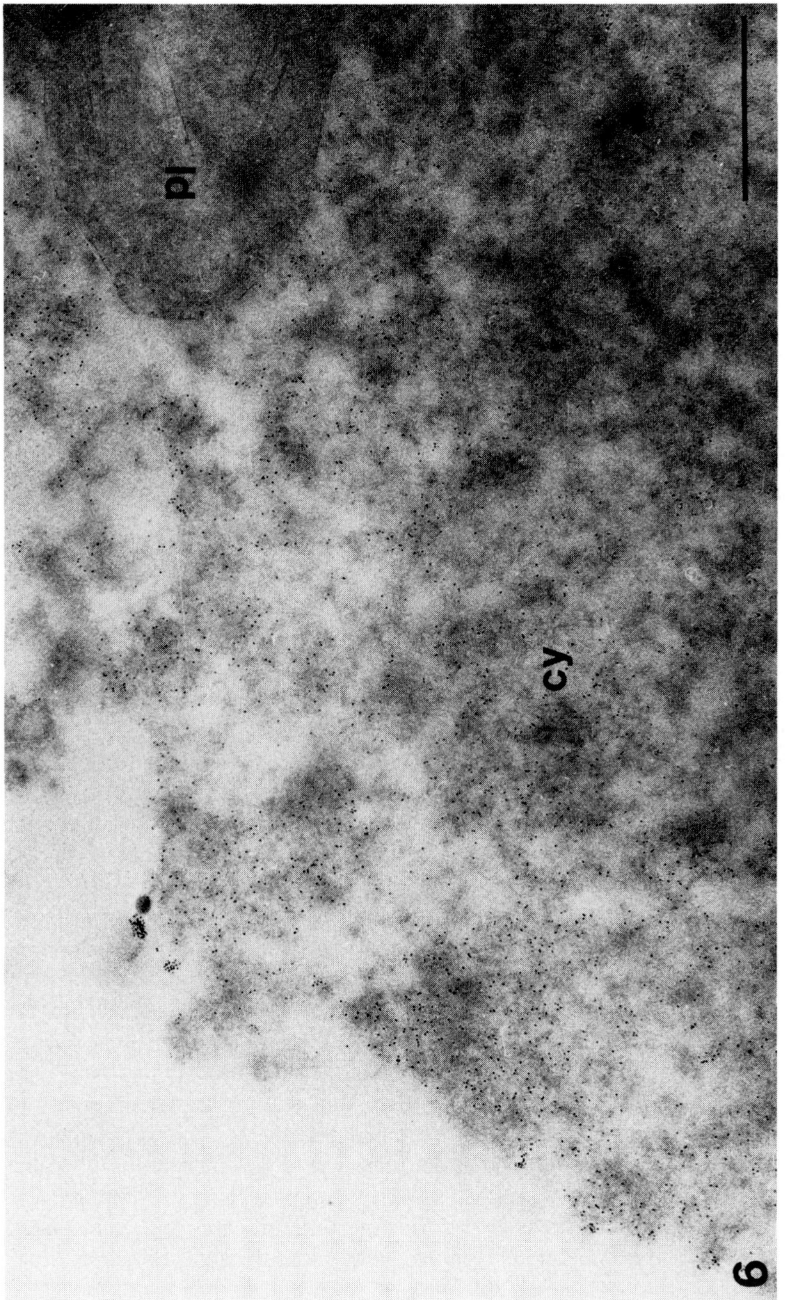

FIGURE 6. Portion of tobacco cell incubated with biotinylated SSU probe and avidin-ferritin. Cytoplasm is densely labeled, indicating SSU mRNAs; the plastid is ferritin-free.

TABLE 1
Evaluation of the Pre-Embedding Method in Plant Tissue

Advantages	Disadvantages
Target molecules distributed throughout cell volume. Access to probe increased.	Low contrast of ferritin: EM observation difficult.
High affinity between avidin-biotin: stable bridge.	Risk of extraction of cell components by enzyme pretreatment can lead to poor morphological detail.
Dense particulate labeling per biotinylated site.	Variable probe penetration in some tissue slices: heterogeneous labeling.
High sensitivity and specificity at intracellular level.	Immunodetection not possible due to large molecular size of anti-biotin.
Low background in controls and nontarget areas.	

SSU messengers. No specific tagging accumulated in the plastids or in any other cell organelles (unlabeled vacuoles, peroxisomes, mitochondria); background labeling was minimal and comparable to that in controls.

Control samples, hybridized with biotinylated pBR322 probe and incubated with avidin-ferritin or treated with avidin-ferritin only, revealed no preferential labeling and a low level of nonspecific tagging.

These results demonstrate the application of *in situ* hybridization to the detection of cellular mRNA molecules and, in particular, the accumulation of these molecules in subcellular cytoplasmic or plastidial compartments. It is noteworthy that in some SSU-probed cells, labeling was associated with the nuclei; ferritin particles were distributed uniformly and densely throughout the nucleoplasm, suggesting that SSU premessenger RNA was targeted. The nucleoli, sites of ribosomal RNA synthesis, were ferritin-free. To confirm this observation, isolated nuclear preparations were incubated with either LSU or SSU probes. Again ferritin tagging was dense within SSU-probed nuclei; LSU probing of both nuclei and cells gave negative results, consistent with the fact that the large subunit is chloroplast-encoded. Although these are preliminary results, it is possible that ISH can be used to detect unprocessed mRNA within nuclei.

For the pre-embedding method (Table 1), permeabilization is essential in order to assure the access of the probe to the target nucleotides. The rate of enzymatic digestion and the amount of nonionic detergents depend on the type of tissue.[61] The thick cell walls of many plant tissues constitute serious barriers. A major drawback which we encountered with this approach was the variability of probe penetration or diffusion, leading to heterogeneity in labeling, particularly in highly differentiated leaf tissue. If deproteinization was increased, this inevitably led to deterioration of cell morphology. On the other hand, hybridization of pronase-treated young leaf tissue, protoplast (leaf cells with cell wall removed by enzymatic digestion) preparations, and isolated organelles yielded homogenous labeling, with no detrimental effect on structures.

Attempts to replace the avidin-ferritin tags with streptavidin-gold complex (Sigma, St. Louis, MO, U.S.) were unsuccessful. It may well be that the large molecular size of the streptavidin-gold complex impeded its movement throughout the tissue, or that steric hindrance makes the binding of this complex to biotin impossible.[62]

The use of small oligoprobes[14-16,57] (see also Section VI.C) in conjunction with small colloidal gold markers (1 to 5 nm), now commercially available, would certainly improve probe/tag diffusion and reduce the need for rather severe permeabilization procedures, thus extending the pre-embedding method to more widespread use.

4. Post-Embedding Method for ISH in Leaf Tissue

All plant material will not be easily amenable to this pre-embedding approach. As an alternative method, we have recently experimented with a post-embedding procedure, using the same material and probes as above. This method is based on previously published procedures on resin-embedded sectioned material,[24,25,43,44] (see Section VI.B and C). Thin sections of aldehyde-fixed, LR White embedded tobacco leaves were probed with biotinylated probes; the hybrids were subsequently detected by an anti-biotin, protein A-gold complex as described in the following Protocols 5, 6, and 7. Leaf material was sampled as described for the pre-embedding method.

PROTOCOL 5 — FIXATION AND EMBEDDING

1. Cut small samples (2 mm^2) of leaf in a droplet of freshly made 4% paraformaldehyde-0.5% glutaraldehyde in Na-phosphate-sucrose buffer (PS: 0.1 *M* phosphate buffer, 0.04 *M* sucrose, pH 7.2). Place samples in fixation solution for 5 h at 4°C using intermittent vacuum in a desiccator until samples sink. Change fixation solution three times.
2. Wash samples in same P-S buffer for 4 × 15 min.
3. Dehydrate in a graded ethanol series (30%, 50%, 70%, 80%, 90%, 95%) for 15 min each, 100% ethanol for 2 × 10 min).
4. Orient leaf (or other tissue) samples as desired (e.g., for x sections) in 1 to 2% agar in H_2O, cut into cubes, and continue dehydration (between 50 to 70% step).
5. Infiltrate LR White resin (LRW) by transferring specimens to a 3:1 mixture of ethanol/LRW, a 1:1 mixture, a 1:3 mixture for 2 h, and rotate overnight in pure LRW at room temperature. Change several times the following day prior to embedding in pure LRW. Use tightly capped gelatin capsules to exclude oxygen during polymerization at 50°C for 10 h.
6. Prepare ultrathin light-gold sections and place on 200-mesh gold grids.
7. Place sectioned material in a covered petri dish and store in a desiccator until use.

PROTOCOL 6 — PREHYBRIDIZATION AND HYBRIDIZATION

1. Place sections in an oven at 60°C for 1 h prior to hybridization.
2. Prepare probe-hybridization buffer (0.5 μg of biotinylated probe: LSU, SSU, or pBR322) in 50% formamide per 50% 0.3 *M* NaCl and 0.03 *M* Na citrate, pH 7.0, and salmon sperm DNA in 250 μg/ml to block nonspecific binding. Denature double-stranded probes prior to use by boiling for 2 min, immediately cool to 0°C in ice where kept until use.
3. Line a glass petri dish with Parafilm and place in a larger recipient containing moist filter paper. Carefully place a number of 2 to 5-μl drops of probe-mixture on parafilm, taking care not to spread them.
4. Place grids on droplets, seal the recipient with tape, and hybridize overnight (16 to 18 h) at selected temperature (38°C → 40°C). For controls, place sections on pBR322 probe mixture or on hybridization buffer only, and treat as in following steps.
5. Rinse by dipping grids in PBS buffer and then float on buffer droplets 4 × 10 min.

PROTOCOL 7 — IMMUNOGOLD DETECTION

1. Preincubate grids by floating on droplets of 0.1 *M* Tris-HCl, 0.1 *M* NaCl, 1% BSA, and 1% Tween-20, pH 7.2 for 30 min. Drain off excess solution.
2. Incubate with rabbit anti-biotin (Enzo Biochemical Inc., NY, U.S.) diluted 1:100 in preincubation buffer containing 0.1% BSA and 0.2% Tween-20 (Figure 2b).
3. Rinse in Tris-HCl (as above) without BSA for 4 × 5 min. Drain off excess buffer.
4. Incubate in Protein A-gold conjugate (10-nm particles, Amersham Int. Pic, Amersham, U.K.), diluted to 1:50 in PBS for 1 h (Figure 2b). Wash with PBS for 3 × 5 min.
5. Rinse by dipping grids several times in distilled H_2O.
6. Counterstain in saturated aqueous uranyl acetate for 30 min.

5. Results

Lightly fixed leaf material, embedded in LRW acrylic resin, showed good ultrastructural preservation of cellular architecture. However, the delimiting membranes, such as tonoplast, plasmalemma, and chloroplast envelopes, appeared bleached and were not clearly resolved. Occasional leaching of cytoplasm was noted, compared to that obtained by conventional fixations. Organelle ultrastructure was sufficiently good to permit unequivocal identification.

The compartmentation of the large subunit messenger RNAs to the plastids and the small subunit mRNAs to the cytoplasm was confirmed using the post-embedding method. Gold particles were readily visible, exhibiting a random

TABLE 2
Evaluation of the Post-Embedding Method in Plant Tissue

Advantages	Disadvantages
Good fine morphology: no or limited protease digestion, hence little detrimental effects on structure	Only targets on section surface are tagged, which reduces signals (low copy targets difficult to detect).
Immunogold detection rapid, and gold markers readily visible in EM.	Hybrid duplex less stable at section surface.
Mild dehydration for LRW resin.	Low sensitivity: ultrathin section contains small number of cell mRNAs compared to tissue blocks.
Anti-biotin antibodies: high specificity.	''Stickiness'' of resin can increase noise level.
LRW suitable for use when ICC and ISH required for same material for both LM and EM.	

distribution in their respective compartments; the specificity of the labeling pattern was identical to that described using the pre-embedding method. However, the concentration of gold particles was always less in plastids than in cytoplasm. We are unable to determine whether this was a consequence of differences in probe size, probe penetration, or accessibility of endogenous RNAs, or if indeed there were real differences in the abundance of mRNA molecules between the two compartments (Table 2).

Preliminary tests showed relatively poor morphological preservation, with no significant differences in signal detection after pronase B or proteinase K treatment (see Protocol 2, step 3); hence, we eliminated the pretreatment digestion from the protocol.[63] These preliminary results show that a relatively simplified ISH technique can yield sensitive and specific hybridization with acceptable low background. Admittedly, we are dealing with high copy nucleic acids; this protocol has not been tested on less abundant targets. Single gold particles sometimes associated with the LRW resin result in a higher background with this method than with the pre-embedding technique.

B. REPLICATION OF CAULIFLOWER MOSAIC VIRUS: DETECTION AND INTRACELLULAR LOCALIZATION OF SINGLE-STRANDED INTERMEDIATES

Applications of ISH to animal virology is widespread.[64] This is not the case for plant viruses because of the difficulties encountered in using ISH at the EM level with plant tissues. ISH has made it possible to localize sites of active viral infection at the subcellular level and/or to elucidate the mode of viral replication. In these respects, ISH was used to study CaMV, a DNA virus[65] which accumulates in cytoplasmic inclusion bodies or *viroplasms* in infected cells, and which replicates by reverse transcription from an RNA intermediate (35S RNA). Various data indicate that CaMV DNA replication could occur in viroplasms,[66-68] including the reverse transcription step.[69,70] Moreover, experiments using a subcellular fraction obtained from CaMV-infected turnip leaves,[71] purified CaMV replication complexes,[72] and purified viroplasms able to support viral DNA synthesis[69,70] gave results indicating an asymmetric replication of CaMV DNA. Such an asymmetry was shown *in vivo* by ISH at the LM level for Hepatitis B virus,[73-75] another retroid virus.

In order to analyze the *in vivo* situation for CaMV, free single-stranded DNA intermediates of defined polarity were studied by ISH on ultrathin sections of infected turnip leaves. Indeed, if the replication of CaMV DNA proceeds asymmetrically by reverse transcription from 35S RNA of *plus* polarity, large amounts of free-single stranded DNA of *minus* polarity should be accumulated. For this purpose, single-stranded RNA probes were used because of their known advantages: high hybridization efficiency, strand-specific probes of defined size, and the absence of competing strands.[2,75-79] The nonisotopic method using bio 11-UTP to label the probes gives low background level,[64] and the detection with anti-biotin antibodies and immunogold, used in this study, is more sensitive than with streptavidin-gold.[79]

1. Preparation of Probes

a. In vitro *Transcription Vectors*

In order to prevent hybridization with a small DNA intermediate of *minus* polarity (sa-DNA of 600 nucleotides, located counterclockwise from the zero of the genome map), present in large amounts in CaMV-infected tissues, and which could mask the hybridization signal of sequences of the same polarity originating from other parts of the viral genome, we chose a probe with the sa-DNA region deleted. In this experiment, the probe corresponded only to the coat protein gene which is located opposite the sa-DNA region on the viral circular genome of 8 kb.[65] The probe did not overlap the CaMV 19S RNA messenger for the major viroplasm protein. A DNA fragment of 1.6 kb containing the coat protein gene was obtained from a cloned CaMV genome[80-82] by digestion with the restriction enzyme *Bam* H1 and *Acc* I (Figure 7).

PROTOCOL 8 — PREPARATION OF *PLUS* AND *MINUS* PROBES

1. Fill in the ends of the DNA fragment with Klenow fragment of DNA polymerase and ligate with *Bam* H1 linkers.
2. Insert the resulting fragment into the transcription plasmid Bluescript KS (Stratagene, La Jolla, CA, U.S.) at the *Bam* H1 site of the polylinker in either orientation with regard to the promoter of the phage T7. The experimental conditions of the different techniques are described elsewhere.[83]
3. Linearize the recombinant plasmids by treatment with *Sma* I, transcription by T7 DNA-dependent RNA polymerase allowing the synthesis of RNA of either positive (*plus* probe) or negative (*minus* probe) polarity (Figure 7).

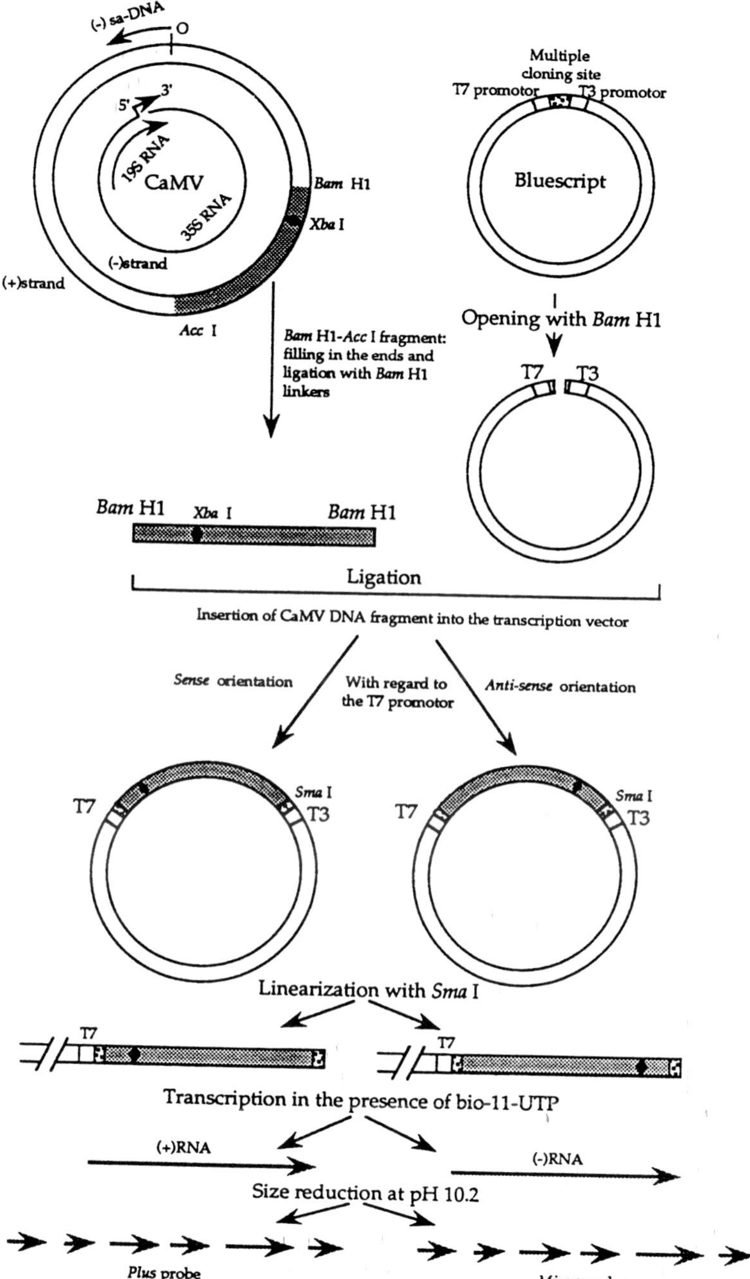

FIGURE 7. Schematic representation of cauliflower mosaic virus genome, and the molecular techniques for the preparation and biotinylation of *plus* and *minus* cRNA probes used for ISH applied to infected turnip leaves.

b. Labeling of the Probes

Protection against RNase contamination is crucial. All the experiments are performed under RNase-free conditions. Gloves are always worn while preparing or using RNA probes; glassware is autoclaved at 220°C for 4 h and plastic tubes or tips, at 120°C for 20 min. Water for buffers is treated with diethylpyrocarbonate (DEPC) at 0.1% and autoclaved.

The transcription protocol is essentially the same as that described by Folsom et al.[84] for the incorporation of a nucleotid analog, N6-(6-aminohexyl)ATP, with just a few modifications: addition of bovine serum albumin (BSA)[83] and biotin 11-UTP diluted with UTP; preincubation at 37°C for 5 min before adding the template DNA and half of the total amount of RNA polymerase, and addition of the second half of the enzyme after 1 h of incubation.

PROTOCOL 9 — INCORPORATION OF BIOTIN 11-UTP INTO RNA

1. For preparative purpose, scale up the final volume to 250 μl. It is made up of: 2.5 μl of BSA at 10 mg/ml; 50 μl of 5× transcription buffer (200 m*M* Tris-HCl, pH 8.0, 40 m*M* $MgCl_2$, 10 m*M* spermidine-HCl, 125 m*M* NaCl; 10 μl of 10 m*M* ATP; 10 μl of 10 m*M* GTP; 10 μl of 10 m*M* CTP; 5 μl of 10 m*M* UTP; 7.5 μl of 10 m*M* bio 11-UTP (Sigma); 12.5 μl of 100 m*M* DTE; 5 μl of RNasin (human placental RNase inhibitor, 40 U/μl, Promega, Biotech, Madison, WI, U.S.); 130.5 μl of DEPC-treated water. Preincubate this mixture at 37°C for 5 min and then add 5 μl of linearized template DNA (5 μg of 4.6-kb plasmid) and 1 μl of T7 RNA polymerase (50 U, Strategene).
2. After 1 h of incubation, add 1 μl of RNA polymerase and continue the reaction for another hour.
3. At the end of the transcription, add 2.5 μl of RNase-free DNase at 30 U/μl (Boehringer, Mannheim, Germany) and incubate at 37°C for 15 min.
4. Treat the mixture three times with one volume of phenol/chloroform (v/v).
5. Precipitate RNA in the final aqueous phase by addition of 25 μl of 3 *M* Na acetate, pH 6, and 600 μl of ethanol, and wash the RNA pellet (collected by centrifugation) twice with ethanol at 70%; dry it and dissolve in 90 μl of DEPC-H_2O.
6. For size reduction of RNA,[85] add 10 μl of 400 m*M* $NaHCO_3$, 600 m*M* Na_2CO_3, pH 10.2, to the solution. Hydrolysis is performed at 60°C for 10 min. It is checked by 2% agarose gel electrophoresis carried out in the presence of RNA markers (BRL; 0.24 to 9.5-kb RNA ladder). After migration, the gel is treated with ethidium bromide (2 μg/ml) to reveal the position of the markers and the synthetized RNA.

7. Transfer RNA to nitrocellulose filter and detect biotinylated RNA with streptavidin and biotinylated phosphatase, as described in the DNA detection system instruction manual of BRL. Average length of the probes is 250 bases. RNA amount is estimated by dotting on nitrocellulose using the same BRL system and biotinylated denatured DNA as standard.
8. Precipitate the biotinylated RNA in the presence of Na acetate and ethanol, spin down, and wash RNA twice with ethanol at 70%, in which it is kept until use.

2. Post-Embedding Method for ISH of CaMV

a. Tissue Processing

The experiments were carried out on turnip plants infected with CaMV Cabb B-PV 147 (American Type Culture Collection, Rockville, MD, U.S.) by abrasion at a four-leaf stage. After 4 weeks, the young leaves, systemically infected, as checked by morphological changes, were sampled for ISH procedure. For leaf tissue processing, fixation and embedding procedures were previously optimized for immunocytochemical studies.[86] The entire protocol is schematized in Figures 2c and 8. See also Protocols 10 and 11.

PROTOCOL 10 — FIXATION AND EMBEDDING

1. Prepare a 4% paraformaldehyde fixative in 0.1 *M* Na phosphate buffer (PB), pH 7, and add, just prior to use, 0.1 to 0.5% glutaraldehyde from a 25% solution, electron microscopy grade. Use within 1 h.
2. Immerse the whole leaf in the fixative and slice it into 4- to 5-mm thick pieces. The sliced samples are immersed in a new bath of fixative; penetration is improved by gentle pumping for 1 h. Renew the fixative and fix for 3 h at room temperature.
3. Wash extensively with PB containing 0.1 *M* glycine for 30 min to quench the free aldehyde groups, and then with PB (avoid postfixation with OsO_4 as the acrylic resin used is incompatible with osmium).
4. After dehydration in a graded ethanol series, carry out the infiltration in Lowicryl K4M for 3 to 4 d at −25°C (infiltration time is longer than with animal tissues). Embed the leaf pieces with a new bath of resin in gelatin capsules. Polymerize by UV irradiation at −25°C for 2 to 3 d and at room temperature for an additional 2 d. Keep the embedded samples in a desiccator until sectioned.

There are several ways of processing the grids for ISH incubations. Gold grids are fragile, hence we chose to use the Multistain chambers (Polaron equipment, Biorad, Watford, U.K.), making it practical to manipulate a large series of grids simultaneously: grids remain in their original place during the

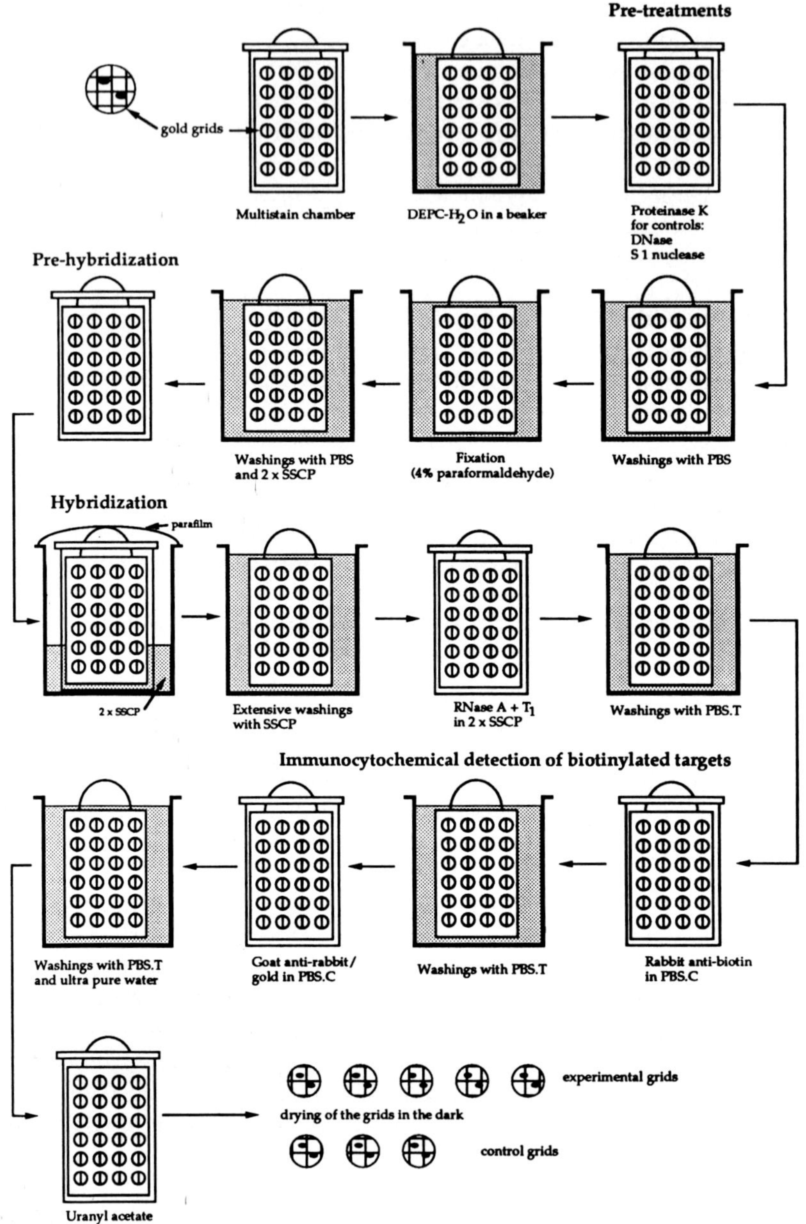

FIGURE 8. Schematic outline of the ISH technique used for the immunogold detection of *plus* and *minus* nucleic acids of cauliflower mosaic virus in infected turnip leaves, with biotinylated riboprobes (see also Figure 2c). Application of all steps of the ISH procedure using Multistain chambers for good hybridization and low background.

whole ISH procedure (see Figure 8). However, it is possible to transfer grids to different incubations and rinses by the "floating drop" method or in microtest plates.

PROTOCOL 11 — SECTION PREPARATION

1. Coat 200-mesh gold grids with Formvar film (0.3% in chloroform [w/v]) and a thin layer of carbon.
2. Prepare serial pale-gold ultrathin sections with a diamond knife.
3. Dry sectioned material in an oven for 1 h at 40°C and keep in a desiccator for 2 or 3 days to prevent unsticking. They can then be used, or stored in the dark at room temperature.

b. Stock Solutions

During the time elapsed for infiltration and polymerization of the block samples, prepare all the solutions needed for ISH procedure. As for probe preparation, protection against RNase contamination is crucial. Use sterile glassware and DEPC-treated water. To obtain low background, large volumes of buffer are necessary for rinses.

PROTOCOL 12 — BUFFERS

1. Prepare ultrapure deionized water (DEPC-H_2O), filtered on 0.22-μm Millipore filters, pretreated by 0.1 and 0.2% DEPC. Stir overnight at room temperature in a hood. Sterilize for 30 min at 130°C to decompose the residual DEPC.
2. Prepare *bottles* with 0.1 *M* PBS with DEPC-H_2O and autoclave.
3. Prepare in *baked colored glassware:* PBS + 5 m*M* EDTA; PBS + 0.2% glycine; 4% paraformaldehyde in PBS; 20× SSCP (173 g NaCl + 88.2 g sodium citrate + 31.2 g NaH_2PO_4 with DEPC-H_2O to make 1 l). Sterilize solutions by autoclaving. From the 20× SSCP buffer, prepare 4× SSCP, 2 × SSCP, and 1× SSCP. All these solutions are kept in a cold chamber at 4°C before use.
4. Prepare in *prebaked polyethylene tubes:*
   - 1 *M* Tris (12.1 g in 100 ml DEPC-H_2O; adjust at pH 7.5 with HCl).
   - 0.5 *M* EDTA (18.6 g for 100 ml DEPC-H_2O; adjust at pH 7.5 with NaOH).
   - 1% sodium dodecyl sulfate (SDS) in DEPC-H_2O. Dissolve by stirring at 50°C.
   - ×50 Denhardt's solution (1 g Ficoll + 1 g Polyvinylpyrrolidone + 1 g BSA (fraction V, Sigma). Add DEPC-H_2O to make 100 ml.
   - Deionized formamide (take care in preparing this solution and use only if it remains clear).

   Store these solutions at −25°C until used.

5. Prepare in *baked Eppendorf tubes* solutions for nucleic acid and enzyme treatments.

Nucleic acids are added to reduce nonspecific binding of the probes, i.e., background: yeast tRNA, RNase-free (type V, Sigma) at 10 mg/ml, and salmon sperm DNA, RNase-free (Sigma) at 10 mg/ml, dissolved in sterile DEPC-H_2O. For enzymes used in pre- and posthybridization steps, prepare:

- Proteinase K (Boehringer) at 10 mg/ml, in 20 m*M* Tris + 2 m*M* $CaCl_2$, pH 7.4 (2.4 g Tris in 100 ml DEPC-H_2O; adjust to pH 7.4 with HCl and add 22 mg anhydrous $CaCl_2$).
- DNase I, RNase-free (Boehringer) at 30 U/ml, in 20 m*M* Tris + 10 m*M* $MgCl_2$, pH 7.4 (add 203 mg $MgCl_2$ to the Tris buffer as above).
- RNase A (Boehringer) at 10 mg/ml + RNase T1 at 100.000 U/ml, in 2× SSCP (96.8 ml) + 10 m*M* Tris (1 ml of 1 *M* Tris) + 1 m*M* EDTA (200 μl of 0.5 *M* EDTA). Add RNase A at 40 μg/ml, and RNase T1 at 30 U/ml.
- S1 nuclease (Boehringer) at 400 U/μl, in 33 m*M* Na acetate + 50 m *M* NaCl + 0.03 m*M* $ZnSO_4$, pH 4.5 (prepare first a solution of $ZnSO_4$ at 1 mg/ml; add 272 mg Na acetate to 100 ml DEPC-H_2O; stir and then add 292 mg NaCl and 0.5 ml of the $ZnSO_4$ solution).

Store all these solutions at −25°C until use.

PROTOCOL 13 — HYBRIDIZATION BUFFER

1. In sterile centrifuge tubes, prepare 100 ml of buffer with:

50 ml deionized formamide	50% in final solution
15 ml 20× SSCP	3× SSCP
2 ml × 50 Denhardt's solution	1%
1 ml of 1 *M* Tris	10 m*M*
1 ml of 0.5 *M* EDTA	5 m*M*
10 ml SDS 1%	0.1%
21 ml DEPC-H_2O	

2. Add to this buffer 10 g Dextran sulfate (10% in final solution). Vortex to mix well and stir for 30 min at 45°C until completely dissolved.
3. Filter through 0.45-μm Millipore filters before adding salmon sperm DNA at 250 μg/ml and tRNA at 250 μg/ml. Place aliquots of hybridization buffer into the Multistain chambers, in which the grids are further treated without possibility of drying.

This buffer was adapted from a number of published procedures and tested in preliminary experiments performed on turnip leaves. Formamide is used to facilitate hybridization at lower temperature, dextran sulfate to accelerate the rate of hybrid formation, and 3× SSCP to maintain an ionic strength compatible with a rapid rate of hybridization; Denhardt's solution, yeast tRNA, and sperm DNA are used to saturate nonspecific binding sites. Our experience has shown that the use of SDS improves the access of the probes to the cellular targets.

c. Pretreatments of Ultrathin Sections

Because of the problems of probe access into plant tissues, several kinds of pretreatments have been tried.[87] We have found proteinase K to be the most efficient. Pretreatment with 0.2 *N* HCl, often recommended in order to remove basic proteins which may bind nucleotides nonspecifically, was omitted because it did not improve the signal and was detrimental to cellular resolution.[79]

PROTOCOL 14 — EXPERIMENTAL AND CONTROL TREATMENTS

1. Place 5 (controls) to 10 (experimental sections) in Multistain chambers according to the enzymatic treatments and number of probes. Wet them for 5 min in DEPC-H_2O. This wetting step was found to reduce the background and precipitates on the sections.
2. Treat all the grids with 10 μg/ml proteinase K for 15 min at 37°C. It is advisable to preheat the protease solution.
3. Wash rapidly in PBS and fix with 4% paraformaldehyde solution for 10 min to stop the protease reaction. Wash briefly with PBS and twice for 5 min with 0.1 *M* glycine in PBS. Wash again 3 × 5 min in PBS and in 2× SSCP for 5 min: the grids are now ready to be hybridized.
4. For control sections, two enzymatic treatments were used:
   a. *DNase pretreatment* for a set of grids in 100 μg/ml DNase for 1 h at 37°C. Follow the steps described in 3.
   b. *S1 nuclease:* treat the selected grids with 625 U/ml of the enzyme for 2 h at room temperature. Follows the steps described in 3.

Conditions for hybridization are described in Protocol 15.

PROTOCOL 15 — HYBRIDIZATION WITH THE TWO RIBOPROBES

1. After centrifugation, evaporate the ethanol traces (see Protocol 9, step 8) in Speed-Vac to obtain the probes as a dry precipitate. Dissolve this precipitate in 1-ml hybridization buffer. Close tightly and vortex to mix well. Add 11 ml of the buffer to obtain a probe concentration of 100 ng/ml.

2. Incubate the grids for 20 h at 40°C. To prevent evaporation, the Multistain chambers are placed in containers humidified with 50-ml 2× SSCP and tightly capped with Parafilm.

d. Posthybridization Treatments

The washing steps (see Protocol 16) are crucial, partly due to the presence of viscous dextran sulfate used to reduce the nonspecific binding of the probes, but with reduced damage to the tissues. Furthermore, since cRNA probes often stick to mononucleic acid constituents, treatment with RNase A and T1 digest nonhybridized probes was used (see Protocol 16, step 3).[38,76,88,89]

For the ICC detection of hybrids (see Protocol 17), all treatments are carried out at room temperature. Washings are done with PBS + 0.1% Triton X-100 (PBS.T) to ensure a low background. All sera are diluted in PBS containing 0.2% Triton X-100 and 0.2% Tween-20 to facilitate antibody penetration; 0.2% Carrageenan gelatin (type IV, Sigma) and 1 mg/ml nonimmunized goat IgG are added to saturate protein binding sites (PBS.C).

PROTOCOL 16 — WASHINGS

1. Rinse in large volumes (250 ml for each Multistain rack) of 4× SSCP at 45°C with gentle agitation. Care must be taken not to bend the flexible gold grids by excessively energetic agitation.
2. Rinse twice for 2 min in 2× SSCP at room temperature and once for 10 min in 1× SSCP.
3. Treat for 30 min at 37°C with RNase A and RNase T1 solution to digest any unhybridized single-stranded RNA that may have stuck to tissue. This step greatly reduces nonspecific labeling.
4. Rinse with 2× SSCP and with 1× SSCP for 5 min each.
5. Wash in PBS for 3 × 5 min, prior to the ICC detection of the biotinylated hybrids.

PROTOCOL 17 — IMMUNOGOLD HYBRID DETECTION

1. Treat all the grids with PBS.C for 30 min.
2. Without rinsing, drain off excess PBS.C and apply rabbit anti-biotin IgG (Enzo Diagnostics, New York, NY, U.S.) diluted in PBS.C at 6 μg/ml for 5 h (Figure 2c).
3. Wash in PBS.T for 4 × 5 min.
4. Treat again with PBS.C for 30 min, and incubate with secondary goat anti-rabbit IgG conjugated with 30-nm gold particles (EM GAR G30, Amersham) for 90 min, diluted 1:30 in PBS.C.
5. Wash under gentle agitation with PBS.T, 4 × 5 min, and with ultrapure deionized water, 4 × 5 min.
6. To prevent the loss of gold label, postfix with 2.5% glutaraldehyde in ultrapure water for 15 min, and rinse extensively.
7. Counterstain with saturated uranyl acetate in water for 1 h.

3. Results

Ultrathin sections of CaMV-infected leaves of turnip were used without prior denaturation of DNA, as viral DNA should be single-stranded *in vivo*. Incubation was carried out with *plus*- or *minus*-strand biotinylated RNA.

With the *plus* probe, hybridization was mainly associated with viroplasms, i.e., inclusion bodies where viral particles accumulate in infected cells. Irregularly shaped viroplasms (Figure 9) were more heavily labeled than spherical dense viroplasms. No significant labeling appeared elsewhere, in particular over the cytoplasm, plastids, mitochondria, cell walls, or nuclei: the few background gold particles appeared in singlets. The positive reaction was higher in mesophyll cells than in the phloem and xylem cell tissues which constitute the vascular strands irrigating the leaf blades. Gold particles appeared in clusters over the viroplasms. Free viral particles found in the cytosol[86] were not labeled. After pretreatment of the sections with DNase (Figure 10) or S1 nuclease, most of the viroplasms was unlabeled. In the absence of a biotinylated RNA probe, no gold particles appeared, or only few single gold grains randomly distributed over the whole section. All these data indicate that free single-stranded DNA of *minus* polarity is present in viroplasms.

With the *minus* probe, labeling was also observed at the level of irregular viroplasms, especially in the mesophyll cells, and at a greater density than with the *plus* probe (Figure 11), probably in part due to the presence of ^{35}S RNA.[70] Indeed, the hybridization signal was greatly reduced after DNase pretreatment and was completely eliminated with S1 nuclease (Figure 12). From these experiments, one can deduce that free single-stranded DNA of *plus* polarity exists also in viroplasms. One might expect that the newly synthesized DNA of *plus* polarity should exist as a duplex, i.e., associated with its template of *minus* polarity.

Although an accumulation of free *minus*-strand DNA due to an asymmetric replication *in vivo* of the CaMV DNA was not detected under our experimental conditions, we showed that ISH can be used to look for and localize specific forms of viral replication intermediates (DNA and/or RNA) in plant cells. In the case of CaMV, these intermediates accumulated in irregular viroplasms with a less dense matrix than the spherical ones. They are more frequent in young leaves than in adult leaves, and in mesophyll cells than in phloem and xylem tissues.[86] These viroplasms are most probably the site of replication of CaMV, while rounded dense viroplasms may correspond to mature forms where viral particles are stored.

It is interesting to note that no hybridization was observed with RNA probes prepared by incorporation and subsequent modification of N6-(6-aminohexyl)ATP.[84,90] Allyl-UTP (BRL) is difficult to use in ISH techniques, because of background problems.[52]

From a technical point of view, our results demonstrated that *in situ* detection of viral nucleic acid sequences in plant tissue sections is feasible, with relatively good preservation of cellular details. Using a hydrophilic resin and biotinylated riboprobes, we have compared for the first time the specific

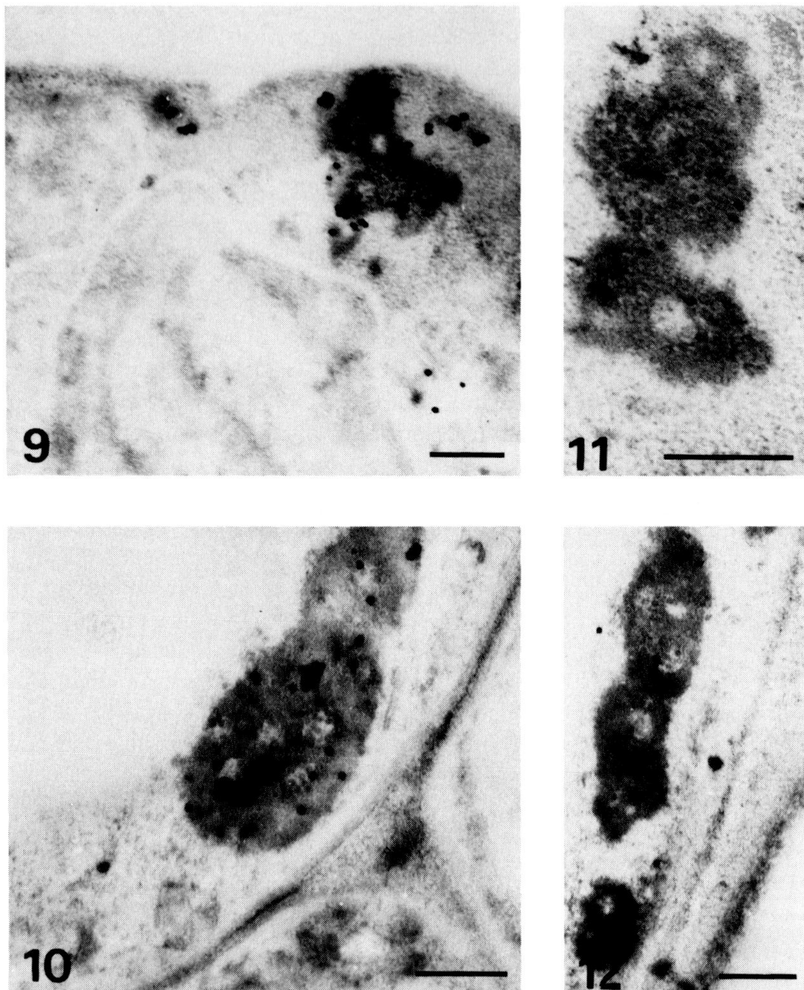

FIGURES 9 to 12. *In situ* hybridization with biotinylated riboprobes on proteinase K digested-Lowicryl sections of turnip leaves infected with cauliflower mosaic virus. Signal detection with rabbit anti-biotin and goat anti-rabbit IgG coupled to 30-nm gold particles. Comparison of viroplasm labeling after hybridization with probe of *plus* polarity without (9), or after DNase treatment (10), and with probe of *minus* polarity without (11), or after S1 nuclease treatment (12).

location of viral DNA of *plus* or *minus* polarity in virus-infected plant cells. Although care has to be taken with the use of riboprobes, as well as the detection with antibodies which may stick to some plant constituents,[91] such as phenolic compounds, lignin, and cell wall matrix, the background can be kept very low. The technique we used was reproduced with no variations in two experiments on the same plant material. The method, similar to that used by Binder[20] and Binder et al.,[27] appears well adapted to plant tissue sections. Recent reports have shown localization of viral DNA or RNA at the EM level, using pre-embedding ISH.[19,26] As demonstrated by specificity controls, there is a large disparity of labeling between cells treated with DNase and S1 nuclease, and untreated cells: specific hybrids between cRNA and DNA of *plus* and *minus* polarity are formed under our experimental conditions, with a high signal-to-noise ratio.

ISH has become a standard method at the LM level for analyzing viral infections.[19,26,73-75,77,87,88,92-96] However, there are very few reports on this technique at the EM level. Puvion-Dutilleul and Puvion[22] have localized herpes simplex virus, a DNA virus, in Lowicryl-embedded cells with biotinylated double-stranded DNA probes, and Troxler et al.,[79] using single-stranded RNA probes, have detected poliovirus on Lowicryl-embedded cells. The latter have observed that ss-RNA probes are more sensitive and give lower background than DNA probes. In preliminary experiments, we obtained poor results, and poor morphological resolution with ds-DNA biotinylated probes, as already mentioned.[92] Furthermore, anti-biotin antibodies visualized indirectly using IgG-gold also gave better results than streptavidin-gold, which is in agreement with Troxler et al.[79] and Wenderoth and Eisenberg.[25] The signal was high, with a minimum of nonspecific background. An advantage of using ss-RNA probes is that they are strand-specific, and whatever their labeling (^{35}S or ^{3}H,[77,88]), they permit the detection of each strand of the viral genome during replication. In the plant field, viruses have been mainly localized on protoplast squashes,[97] on isolated organelles,[98] or on pollen by dot-hybridization.[99] For plant virology, we believe that the method we present could be helpful in the future in studies of DNA and RNA viruses.

C. DISTRIBUTION OF LIPID TRANSFER PROTEIN mRNA IN MAIZE COLEOPTILE USING BROMO-DEOXYURIDINE-LABELED OLIGONUCLEOTIDES[49]

LTPs, which are widely distributed in animals, bacteria, and yeast, are also abundant in plants.[100] They ensure the transport of lipids between membranes and seem to play an important role during early seedling growth. The use of ICC methods with specific anti-LTP antibodies at both light[50] and electron microscopic (unpublished results) level has demonstrated that LTP is mainly cytoplasmic and is heterogeneously distributed in the complex tissues constituting the maize coleoptile and its enrolled leaves. This distribution raises the question as to the differential expression of LTP genes. *In situ* hybridization on Spurr semithin sections with biotinylated cDNA probes[101] and immunogold detection confirmed that LTP gene expression is also cell-specific.[50] Therefore, at the LM level, the LTP gene is expressed as both LTP mRNA and the end product of the gene in the same cells, suggesting that the site of synthesis and the site of use are the same. The presence of both mRNA and LTP translation products suggests zonal intracellular synthesis even where accumulation or secretion later occur.[102] The application of ISH at the EM level introduces the possibilities of using the localization of genomic regulation of the LTP biogenesis to answer the question: are there significant differences in the heterogeneous tissues of a maize coleoptile between the gene coding for LTP mRNA and the localization of the corresponding protein? To reach this goal, we replaced biotinylated cDNA probes which gave negative results with oligoprobes and subsequent immunogold detection.

Knowing the advantages of oligo(deoxy)nucleotides as probes, recently cited by Baldino et al.,[103,104] and Dirks et al.,[105,106] we used an *antisense* (complementary to the cellular mRNA) and a *sense* (identical to the cellular mRNA) oligoprobes. Traditionally, oligoprobes have been labeled with ^{32}P, ^{35}S, and ^{3}H[107-111] or bio 11-UTP.[104-106,112-114] They have generally been applied to whole cells,[108,110] to vibratome, and cryostat sections[54,55,103,107,109,111,115] or to paraffin-embedded tissues.[55,105,108,111,113] Elegant examples of the use of ^{3}H-labeled and bio 11-UTP oligoprobes are to be found in EM studies by Penshow et al.[18] and by Guitteny et al.,[15,16] respectively, using pre-embedding techniques.

Here we describe: (1) the use of oligonucleotide probes complementary to LTP mRNA on ultrathin sections; (2) the labeling of these probes by 5′-bromo-deoxyuridine (BrdU, Sigma). This novel technique was first initiated by Jirikowsky et al.[54] and Ramalho et al.,[55] the detection method using monoclonal anti-BrdU and anti-mouse IgG labeled with gold permitting a rapid and sensitive visualization of the hybrids.[116]

1. Preparation of Oligoprobes

LTP mRNA was detected using a 22-base synthetic nucleotide complementary to bases 212 to 233 of the LTP 9C2 cDNA.[101] This probe will be referred to as the *antisense* probe. Control hybridizations were performed with a 22-base synthetic oligonucleotide overlapping the sequence 212 to 233 of cDNA bases. This *sense* probe will be referred to as the control probe (Figure 13).

3′ terminal labeling was performed according to the method of Ramalho-Ortigao et al.[55]

PROTOCOL 18 — TERMINAL LABELING OF THE PROBES WITH BrdU

1. Mix the following components:
   - 15 pmol of oligonucleotide
   - 100 m*M* of sodium cacodylate, pH 7.0

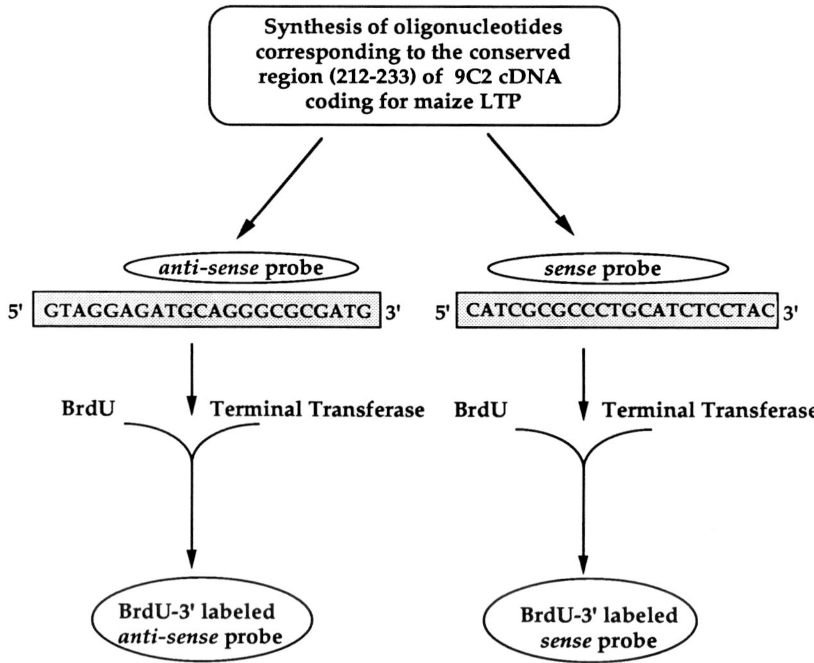

FIGURE 13. Schematic representation of the preparation of single-stranded *antisense-* and *sense-* oligo(deoxy)nucleotide probes and their labeling with bromo-deoxyuridine for the detection of lipid tranfer protein mRNA in maize coleoptile (see also Figure 2d).

- 1 m*M* $CoCl_2$
- 1 m*M* b-mercaptoethanol
- 100 μg/ml BSA (nuclease free, Boehringer)
- 500 m*M* 5-bromo-2′-deoxyuridine-triphosphate (Sigma)
- 39 U of terminal deoxynucleotidyltransferase (Pharmacia, Uppsala, Sweden).

2. Incubate for 30 min at 37°C, and stop the reaction by boiling for 2 min.
3. Separate the unincorporated BrdU by chromatography in a Sephadex G50 column equilibrated with water. Store at −40°C.

2. ISH of LTP mRNA in Maize Coleoptile

The experiments were carried out on coleoptile grown from seeds of W 64A inbred line of maize *(Zea mays),* developed on moist vermiculite in the dark at 30°C for 3 d.

Tissue processing was adapted from optimized immunocytochemical studies.[50] The entire protocol is schematized in Figure 14.

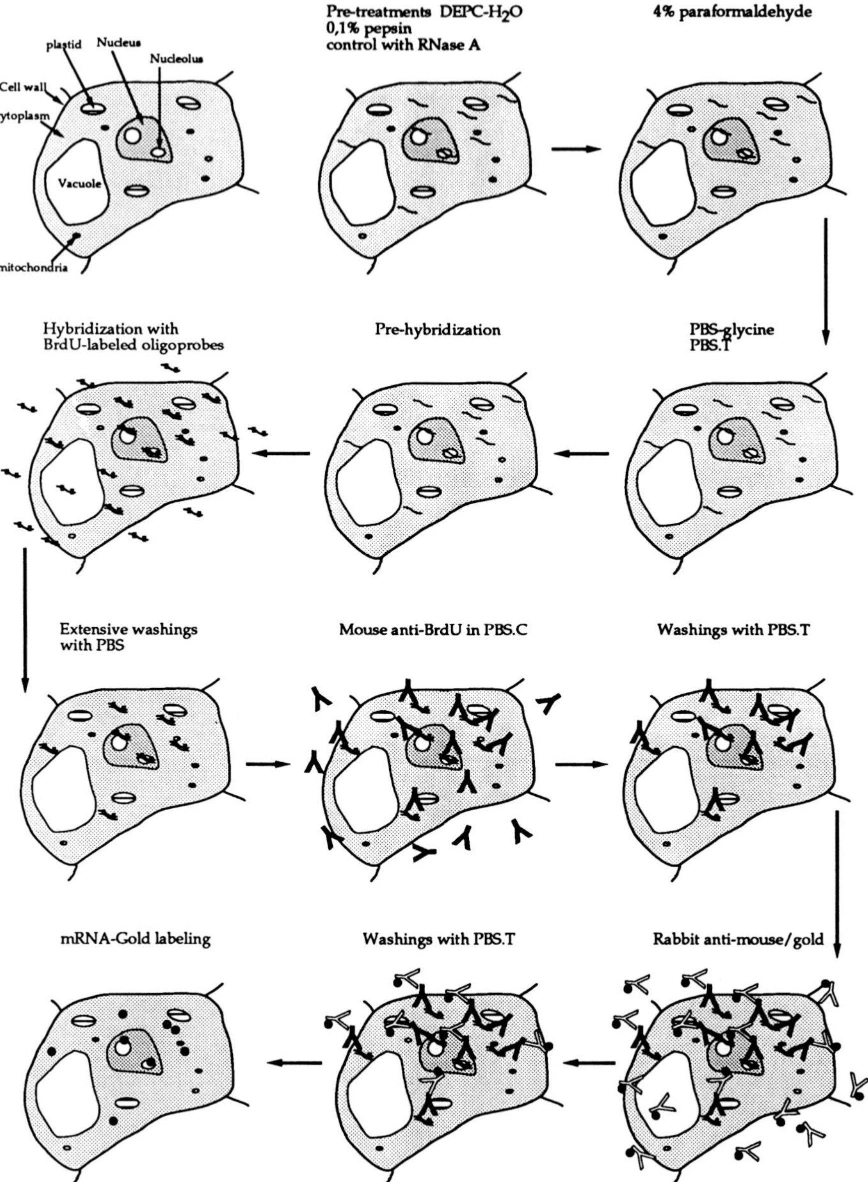

FIGURE 14. Schematic illustration of the use of *antisense-* and *sense-* oligoprobes labeled with BrdU on maize coleoptile, and the immunocytochemical detection of the hybrids formed.

PROTOCOL 19 — TISSUE AND SECTION PREPARATION

1. Slice the coleoptiles in a fixative containing 4% paraformaldehyde and 0.5% glutaraldehyde in a 0.5 *M* Na phosphate buffer, pH 7.
2. Fix for 6 h at 4°C.
3. Wash, dehydrate, and embed in Lowicryl K4M as described above (see Protocol 10, steps 3 and 4).
4. For sectioning, use the same protocol as described for the second example (see Protocol 11).

a. Stock Solutions

The same precautions as those described in the second example for riboprobes are used to preserve tissue mRNA and to prevent its degradation by RNases. All RNase-free solutions are prepared with DEPC-H_2O and sterilized. Chemicals are of the highest purity. Glassware, tips, Eppendorf cones, and tubes are baked at 180°C. Gloves are worn throughout the whole procedure.

PROTOCOL 20 — PREPARATION OF BUFFERS AND SOLUTIONS

1. Prepare a large number of bottles with 6× SSCP, and PBS with and without Triton X-100.
2. Prepare the hybridization buffer. It contains for 10 ml:

- 5 ml deionized formamide
- 3.5 ml 6× SSCP
- 1 ml × 50 Denhardt's solution
- 250-μl sheared, denatured salmon sperm DNA at 1 mg/ml
- 250-μl sonicated yeast tRNA at 1 mg/ml.

The buffer contains: 50% deionized formamide, 5% Denhardt's, 2.1× SSCP (0.3 *M* NaCl, 30 m*M* sodium citrate, 40 m*M* sodium phosphate), 250-μg salmon sperm DNA and tRNA. Store all these solutions at −25°C until use. The hybridization buffer does not contain dextran sulfate, and so posthybridization washing steps are less stringent than in the above example. The grids are washed only with PBS, resulting in an improved cellular integrity, as previously noted.[27]

3. Buffers for pretreatments: Prepare a 0.1% pepsin (Boehringer) in 0.2 *N* HCl with DEPC-H_2O.

After preliminary experiments, pepsin was used in place of proteinase K, giving a better cellular preservation.

Prepare RNase A, 1 mg/ml, in 20 m*M* Tris-HCl and 10 m*M* EDTA from the 1 *M* Tris and 0.5 *M* EDTA stock solutions (see Protocol 12, step 4).

Prior to the ICC detection of hybrids (Figure 2d), ultrathin sections were treated according to Protocols 21 and 22, and Figure 14.

PROTOCOL 21 — PRETREATMENTS OF ULTRATHIN SECTIONS

1. Wet the gold grids bearing 2 sections and place in Multistain chambers with DEPC-H_2O for 5 min.
2. Treat half the grids with preheated 0.1% acidic pepsin solution for 20 min at 37°C.
3. Stop the protease reaction by rinsing twice with 0.1 *M* glycine in PBS. The treatment by pepsin to improve probe penetration by removing proteins and ribosomes was found to be preferable to proteinase K for the cellular integrity.[96,105,117] However, we found that this protease treatment did not increase the number of hybrid signals.[27]
4. For control sections, treat some grids (at least 4) with RNase A for 1 h at 37°C in order to eliminate the specific signal and to evaluate the background level.
5. After these two pretreatments, wash the grids with PBS; fix with 4% paraformaldehyde fixative for 15 min to stabilize cellular mRNA and to prevent its loss; wash with PBS-glycine to eliminate free aldehyde groups, and finally with PBS.T (10 min, 2 baths).
6. Prehybridize for 2 h at 37°C, i.e., the same temperature as the hybridization temperature. Prehybridization with excess salmon DNA, yeast tRNA, and BSA was used to block the nonspecific binding of probes to cellular constituents.

The available volume of oligoprobes per grid was 5 μl. Therefore, the grids were removed with prebaked tweezers from the hybridization buffer in Multistain chambers and rapidly immersed in sterilized conic Beem capsules containing the probes.

PROTOCOL 22 — HYBRIDIZATION

1. Add the *antisense* and the *sense* probes adjusted at a concentration of 6 pmol/ml to the hybridization buffer.
2. Incubate the grids for 3 h at 37°C.

b. Posthybridization Treatments

The washing step is not as crucial in this case as it is for riboprobes used for viral genome detection, since hybridization occurs here without the viscous dextran sulfate. We followed the technique used by Binder et al.[27] and washed the grids in large volumes of PBS under moderate agitation. Then the grids were again placed in Multistain chambers, taking great care to prevent their drying.

PROTOCOL 23 — WASHINGS AND IMMUNOGOLD HYBRID DETECTION

1. Wash for 4 × 5 min with PBS.
2. Wash for 15 min with PBS containing normal goat serum at 1 mg/ml protein concentration.
3. To detect BrdU-labeled hybrids, treat with monoclonal (mouse) anti-BrdU antibody (Becton Dickinson, Mountain View, CA, U.S.) diluted 1:50 in PBS.C (PBS + 0.2% Triton X-100 and Tween-20 + 0.2% Carrageenan gelatin, 1 mg/ml nonimmune goat IgG) to saturate protein binding sits and to facilitate the antibody penetration. Incubate for 18 h at room temperature, in the dark.
4. The following day, wash with PBS for 3 × 5 min, and with PBS.T 3 × 2 min.
5. Treat with PBS containing normal goat serum, 1 mg/ml.
6. For gold labeling, incubate the grids for 2.5 h at room temperature with goat anti-mouse IgG labeled with 15-nm gold particles (Ultra-GAM-Gold-15, BHL18, British Bio-technology, Oxford, U.K.) diluted 1:30 in PBS.C g. Wash with PBS.T, 3 × 10 min, and with ultrapure water, 3 × 5 min.
7. Postfix with 2.5% glutaraldehyde in ultrapure water to reduce the loss of gold.
8. Counterstain the grids with a saturated aqueous uranyl acetate solution for 30 min and with lead citrate (according to Reynolds) diluted with boiled water (v/v) for 1 min.

c. *Specificity Controls*

All the control experiments were performed twice on serial ultrathin sections.

PROTOCOL 24 — NEGATIVE AND POSITIVE CONTROLS

1. Put aside some grids that receive only the hybridization buffer.
2. Probe with labeled oligoprobe complementary to mRNA *(antisense)* and labeled oligoprobe homologous to the mRNA (*sense,* negative control).
3. Treat some grids, pretreated with RNase A, with the two probes.
4. Check by ICC with anti-LTP antibodies that the tissues contain both the mRNA and the coded LTP.[50]

3. Results

In agreement with the immunocytochemical data obtained with anti-LTP antibodies,[50] labeling of mRNA transcripts detected by the *antisense* probe was observed in the same cells as LTP. Gold grains were mainly associated with the cytoplasm of the highly differentiated cells of the outer coleoptile epidermis (Figure 15). In the inner epidermis, only a very few clusters of gold grains were noted. In the enrolled leaves, gold particles were rare in the epidermal cells, but some cells in the vascular strands of leaves and coleoptiles were immunopositive: the LTP transcripts appeared more numerous, and gold particles rarely occurred as singlets. In these positive cells, grains appeared mainly as clusters randomly distributed over the cytoplasm. The majority of cells showed predominant cytoplasmic localization, but in some cells, particularly in young differentiating cells of conducting tissue, gold labeling was visible in nucleoli (Figure 16). As estimated by Weiss and Chen,[114] using a oligonucleotide poly d(T) probe at the LM level, most of the cellular mRNA is located in the nucleus. Such high levels of mRNA targets around the nucleoli might result from an enhanced transcription of the LTP gene, or from the fact that the transcribed mRNA might be protected from degradation, as suggested by Fontaine et al.[89] Interestingly, we noted that not all nucleoli are labeled with the *antisense* probe. The distribution observed suggests that factors such as cell cycle phase are likely to be the cause of the heterogeneity in nucleolar labeling. This nucleolar labeling was eliminated when the sections were pretreated with RNase A or with the *sense* probe (Figure 17).

Background was very low on sections without probes and those treated with the *sense* strand, where there were very few single gold particles. RNase A-treated sections hybridized with the *antisense* probe were almost devoid of gold grains. The comparison of *antisense* and *sense* probes in hybridization permits easy measurement of the signal-to-noise ratio. The use of the *sense* oligonucleotide is one of the best tests of specificity: it removes the signal and shows that there is no possibility of artifacts or nonspecific binding.

In conclusion, we can state that mRNA is localized in cells containing the protein product of the encoding gene. The use of ICC[50] and ISH establishes that some cells are involved in LTP biosynthesis and that the protein remains at its site of synthesis.

From a technical point of view, the labelings obtained indicate that the aldehyde mixture used is adequate for the retention of mRNA and good preservation of cells. We have noted that the degree of cellular resolution was higher than that observed with turnip leaves. The main reason for this might be the small size of the oligoprobe, which penetrates more easily into tissues, without the absolute need for protease treatment, always detrimental to cellular integrity.[18,113] We observed no differences in the intensity of labeling between pepsin-treated and untreated sections, as observed by Steinert et al.[10] Though pepsin pretreatment was shown to be unnecessary with the short-length probes, the beneficial effect of protease, on the contrary, was clear in that it removed, at least partly, dense proteinaceous matrix of viroplasms in the turnip leaves infected by CaMV, undoubtedly because the formaldehyde fixation traps viral DNA and RNA in a dense protein shell. As a result, structural integrity is not as well preserved in turnip as in maize. Some researchers[27,63] prefer to avoid these pretreatments. If we compare the results of preliminary experiments at the LM level obtained with protease treatment,[50] it appears that at the EM level the intensity of the mRNA signal is substantially lower than that found on semithin sections. It is possible that mRNA targets are not completely accessible to the probe after mild treatment

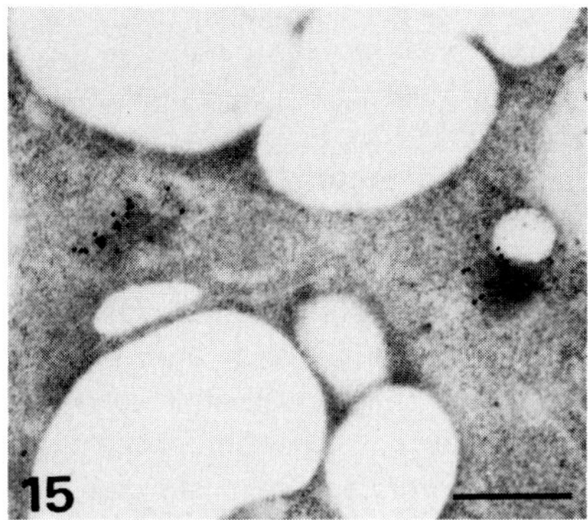

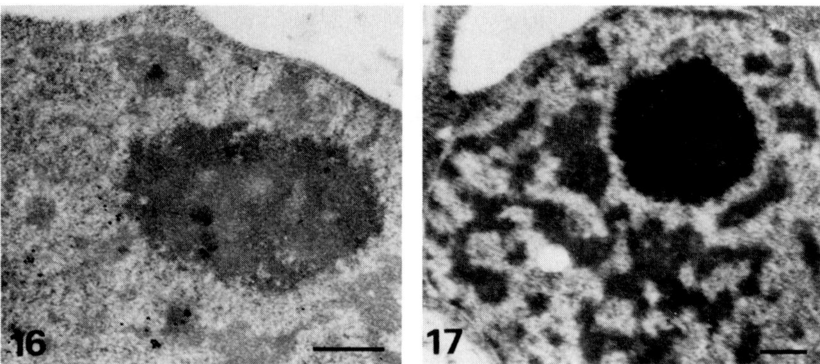

FIGURES 15 to 17. *In situ* hybridization with oligo(deoxy)nucleotide probes labeled with BrdU. Signal detection with mouse anti-BrdU and goat anti-mouse IgG coupled with 15-nm gold particles. Localization of mRNA coding for lipid transfer protein in maize coleoptile after Lowicryl embedding. With the *antisense* probe, labeling of the cytosol (15) and of the nucleolus (16). With the *sense* probe, all cell compartments were unlabeled; an unlabeled nucleolus is shown in (17), as an example.

with pepsin, or in its absence. Etched Spurr semithin sections[50] were submitted to treatment by 0.4% in place of 0.2% pepsin which had no marked effect.[50] The mRNA targets may be potentially blocked by associated proteins, ribosomes, or other constituents. A loss of mRNA from tissues was improbable because the same tissue sample was used for the LM and EM studies. In this type of experiment, one has to choose between high labeling and good resolution. The fact that dextran sulfate is absent, which can lead to a sparser labeling of mRNA transcripts,[63,118] means that less stringent posthybridization washes are necessary, hence leading to better ultrastructural morphology.[27] We have demonstrated the feasibility of using oligo(deoxy)nucleotide probes labeled with BrdU at the EM level to detect cDNA-RNA hybrids formed in a heterogeneous cell population. The method we described for embedded tissues is fairly new. This method gives a high level of cellular integrity, and good resolution in comparison with electron microscope radiography. Though synthetic oligoprobes have been more popular[113] in the last 5 years, they have essentially been used for LM level work, mostly with radiolabeling or conjugation with biotin.[15,16,103,104,106-112,115,119] We found BrdU-labeled oligonucleotide probes to be superior to biotinylation, but the reason for this difference is unclear. It may be due to the disturbing effect of biotin on base pair formation, as suggested by Jirikowsky et al.[54] Whatever their labeling, oligoprobes are advantageous when compared to cDNA probes: they are single-stranded and easily labeled at the 3′ or 5′ end; their synthesis and labeling is rapid and easy to perform; they are of short size (20 to 50 bases), which facilitates their access to their corresponding targets; the hybrids formed are stable; many specificity controls can be performed to verify the specificity of the signal; as with the use of riboprobes, oligoprobes with a complementary sequence (*sense* probe) can be compared to the *antisense* probe (this negative control, run in parallel, is very useful as a way to measure nonspecific labeling); BrdU-oligoprobes are stable for at least 9 months when stored in aqueous solution at −25°C (we used the two probes described here 1 and 11 months after their preparation, with no effective differences in the results).

From a biological point of view, ISH observations can solve some problems relating to the transcription of the LTP gene and the outcome of the encoded protein. With anti-LTP antibodies, we have detected the cells in which LTP is present and used ISH to identify the cells actually synthetizing LTP mRNA. The present data show that gene transcripts match well with the presence of the protein translation product. These data appeared inconsistent with the hypothesis of a collection of LTP mRNA in the outer epidermis, since all the cells, mainly those of the minor veins irrigating the leaves, clearly showed clusters of gold grains, i.e., mRNA targets. The amount of mRNA detected appeared lower than the amount of protein detected by ICC;[50] it is possible that the amount of transcripts revealed is slightly below the sensitivity of our experiments, as mentioned above. Further experiments using stronger proteolytic treatments may clear up this problem. Other reasons may be considered: LTP may accumulate in some cells after being synthesized *de novo* in other cells, or mRNA turnover could be high, LTP being more stable.

VII. CONCLUDING REMARKS

In situ hybridization will certainly continue to develop in plant biology to study differentiation and development mechanisms, and in particular, to analyze expression patterns of foreign genes in transgenic plants. The possibility of relating cellular and/or intracellular ultrastructure to molecular aspects of gene expression on a cell-to-cell basis is a powerful tool and will complement other molecular biological techniques to further the understanding of plant developmental systems. The ultimate usefulness of this technique depends on the accuracy with which markers reflect local concentrations of target nucleotides, so that the gene products can be topologically and chronologically evaluated.

With this goal in mind, we have attempted to present a practical guide to ISH at the EM level, giving some advantages and drawbacks of the approach as applied to plant material. A brief evaluation of our experiments is given below:

- Chemical fixation with paraformaldehyde along with a low concentration of glutaraldehyde is suitable for the retention of nucleotide targets and gives good structural preservation.[120] The fixative may inactivate endogenous RNases.
- RNA and oligonucleotide probes give a lower background of nonspecific bound label than do DNA probes and also better signal intensity.
- Immunochemical methods of visualizing labeled probes are superior to affinity reaction methods.
- Enzymatic digestion to allow probe access is not always necessary (and probably depends on target accessibility); preliminary trials should be carried out to determine the degree of necessity.
- The use of *sense*- and *antisense* strands is the best control system for confirming true targets; nonspecific reactions of probes or labels to tissue components can lead to false positive results. A combination of specificity controls can validate the ISH approach.
- The post-embedding method is more widely applicable in plant biology; however, methods giving increased sensitivity via amplification of the signals must be developed for dealing with single copy genes.

ACKNOWLEDGMENTS

We wish to thank Mrs. Defoug for cell drawings and Mr. R. Maldiney for computer-schematic drawings.

REFERENCES

1. **Wilkinson, D. G.,** mRNA *in situ* hybridization and the study of development, in *In Situ Hybridization Principles and Practice,* Polak, J. M. and McGee, J. O'D., Eds., Oxford University Press, New York, 1990, chap. 8.
2. **McFadden, G. I.,** *In situ* hybridization techniques: molecular cytology goes ultrastructural, in *Electron Microscopy of Plant Cells,* Hawes, Ed., London, 1991, chap. 6.
3. **Gall, J. G. and Pardue, M. L.,** Formation and detection of RNA-DNA hybrid molecules in cytological preparations, *Proc. Natl. Acad. Sci. U.S.A.,* 63, 378, 1969.
4. **Singer, R. H., Langevin, G. L., and Lawrence, J. B.,** Ultrastructural visualization of cytoskeletal mRNAs and their associated proteins using double-label *in situ* hybridization, *J. Cell Biol.,* 108, 2343, 1989.
5. **Manuelidis, L. and Ward, D. C.,** Chromosomal and nuclear distribution of the *Hin*d III 1.9 kb human DNA repeat segment, *Chromosoma,* 91, 28, 1984.

6. **Hutchinson, N. J.,** Hybridization histochemistry: *in situ* hybridization at the electron microscopic level, in *Immunolabelling for Electron Microscopy,* Polak, J. M. and Varndell, I. M., Eds., Elsevier, New York, 1984, chap. 23.
7. **Manuelidis, L., Langer-Safer, P. R., and Ward, D. C.,** High resolution mapping of satellite DNA using biotin-labeled DNA probes, *J. Cell Biol.,* 95, 619, 1982.
8. **Manuelidis, L.,** Indications of centromere movement during interphase and differentiation, *Ann. Acad. Sci. N.Y.,* 450, 205, 1985.
9. **Jamrich, M., Mahon, K. A., Gavis, E. R., and Gall, J. G.,** Histone mRNA in amphibian oocytes visualized by *in situ* hybridization to methacrylate-embedded tissue sections, *EMBO J.,* 3, 1939, 1984.
10. **Steinert, G., Felsani, A., Kettmann, R., and Brachet, J.,** Presence of rRNA in the heavy bodies of sea urchin eggs. An *in situ* hybridization study with the electron microscope, *Exp. Cell Res.,* 154, 203, 1984.
11. **Pardue, M. L.,** *In situ* hybridization, in *Nucleic Acid Hybridization,* Hames, B. D. and Miggins, S. J., Eds., IRL Press, Oxford, 1987, 179.
12. **Narayanswami, S. and Hamkalo, B. A.,** Hybridization to chromatin and whole chromosome mounts, in *Electron Microscopy in Molecular Biology. A Practical Approach,* Rickwood, D. and Hames, B. D., Eds., IRL Press, Oxford, 1987, chap. 10.
13. **Geuskens, M. and May, E.,** Ultrastructural localization of SV 40 viral DNA in cells, during lytic infections, by *in situ* molecular hybridization, *Exp. Cell Res.,* 87, 175, 1974.
14. **Guitteny, A. F. and Bloch, B.,** Ultrastructural detection of the vasopressin messenger RNA in the normal and Brattleboro rat, *Histochemistry,* 92, 277, 1989.
15. **Guitteny, A. F., Fouque, B., Mougin, Ch., Teoule, R., and Bloch, B.,** Histological detection of messenger RNAs with biotinylated synthetic oligonucleotide probes, *J. Histochem. Cytochem.,* 36, 563, 1988.
16. **Guitteny, A. F., Fouque, B., Teoule, R., and Bloch, B.,** Vasopressin gene expression in the normal and brattleboro rat: a histological analysis in semi-thin sections with biotinylated oligonucleotide probes, *J. Histochem. Cytochem.,* 37, 1479, 1989.
17. **Pelletier, G., Tong, Y., Simard, J., Zhao, H. F., and Labrie, F.,** Localization of peptide gene expression by *in situ* hybridization at the electron microscope level, *Methods Neurosci.,* 1, 197, 1989.
18. **Penschow, J. D., Haralambidis, J., and Coghlan, J. P.,** Location of glandular kallikrein mRNAs in mouse submandibular gland at the cellular and ultrastructural level by hybridization histochemistry using ^{32}P and ^{3}H oligodeoxyribonucleotide probes, *J. Histochem. Cytochem.,* 39, 835, 1991.
19. **Wolber, R. A., Beals, T. F., and Maassab, H. F.,** Ultrastructural localization of herpes simplex virus RNA by *in situ* hybridization, *J. Histochem. Cytochem.,* 37, 97, 1989.
20. **Binder, M.,** *In situ* hybridization at the electron microscope level, *Scan. Microscopy,* 1, 331, 1987.
21. **Le Guellec, D., Frappart, L., and Desprez, P. Y.,** Ultrastructural localization of mRNA encoding for the EGF receptor in human breast cell cancer line BT20 by *in situ* hybridization, *J. Histochem. Cytochem.,* 39, 1, 1991.
22. **Puvion-Dutilleul, F. and Puvion, E.,** Ultrastructural localization of viral DNA in thin sections of herpes simplex virus type 1 infected cells by *in situ* hybridization, *Eur. J. Cell Biol.,* 49, 99, 1989.
23. **Thiry, M. and Thiry-Blaise, L.,** *In situ* hybridization at the electron microscopic level: an improved method for precise localization of ribosomal DNA and RNA, *Eur. J. Cell Biol.,* 50, 235, 1989.
24. **Webster, H. de F., Lamperth, L., Favilla, J. T., Lemke, G., Tesin, D., and Manuelidis, L.,** Use of biotinylated probe and *in situ* hybridization for light and electron microscopic localization of P_o mRNA in myelin-forming Schwann cells, *Histochemistry,* 86, 44, 1987.

25. **Wenderoth, M. P. and Eisenberg, B. R.,** Ultrastructural distribution of myosin heavy chain mRNA in cardiac tissue: a comparison of frozen and LR White embedment, *J. Histochem. Cytochem.*, 39, 1025, 1991.
26. **Wolber, R. A. and Beals, T. F.,** Streptavidin-gold labeling for ultrastructural *in situ* nucleic acid hybridization, in *Colloidal Gold: Principles, Methods and Applications,* Academic Press, London, 1989, chap. 19.
27. **Binder, M., Tourmente, S., Roth, J., Renaud, M., and Gehring, W. J.,** *In situ* hybridization at the electron microscope level: localization of transcripts on ultrathin sections of Lowicryl K4M-embedded tissue using biotinylated probes and protein-A gold complexes, *J. Cell Biol.*, 102, 1646, 1986.
28. **Ambros, P. F., Matzke, M. A., and Matzke, A. J. M.,** Detection of 17 kb unique sequence (T-DNA) in plant chromosomes by *in situ* hybridization, *Chromosoma,* 94, 11, 1986.
29. **Bergey, D. R., Stelly, D. M., Price, H. J., and McKnight, T. D.,** *In situ* hybridization of biotinylated DNA probes to cotton meiotic chromosomes, *Stain Technol.*, 64, 25, 1989.
30. **Cox, K. H. and Golberg, R. B.,** Analysis of plant gene expression, in *Plant Molecular Biology: a Practical Approach,* Shaw, C. H., Ed., IRL Press, Oxford, 1988, 1.
31. **Harris, N., Grindley, H., Mulchrone, J., and Croy, R. R. D.,** Correlated *in situ* hybridisation and immunochemical studies of leguminin storage protein deposition in pea (*Pisum sativum* L.), *Cell Biol. Int. Rep.*, 13, 23, 1989.
32. **Kelly, A. J., Zagotta, M. T., White, R. A., Chang, C., and Meeks-Wagner, D. R.,** Identification of genes expressed in the tobacco shoot apex during the floral transition, *Plant Cell,* 2, 963, 1990.
33. **Langdale, J. A., Zelitch, I., Miller, E., and Nelson, T.,** Cell position and light influence C4 versus C3 patterns of photosynthetic gene expression in maize, *EMBO J.*, 7, 3643, 1988.
34. **Mansfield, M. A. and Raikhel, N. V.,** Abscisic acid enhances the transcription of wheat-germ agglutinin mRNA without altering its tissue-specific expression, *Planta,* 180, 548, 1990.
35. **Perez-Grau, L. and Golberg, R. B.,** Soybean seed protein genes are regulated spatially during embryogenesis, *Plant Cell,* 1, 1095, 1989.
36. **Perrot-Rechenmann, C., Joannes, M., Squalli, D., and Lebacq, P.,** Detection of phosphoenolpyruvate and ribulose 1,5-bisphosphate carboxylase transcripts in maize leaves by *in situ* hybridization with sulfonated cDNA probes, *J. Histochem. Cytochem.*, 37, 423, 1989.
37. **Raghavan, V.,** Changes in Poly(A) + RNA concentrations during germination of spores of the fern, *Onoclea sensibilis, Protoplasma,* 140, 55, 1987.
38. **Raikhel, N. V., Bednarek, S. Y., and Lerner, D. R.,** *In situ* hybridization in plant tissues, in *Plant Molecular Biology Manual,* Section B9, Gelvin, S. B. and Schilperoort, R. A., Eds., Kluwer, The Netherlands, 1989, 1.
39. **Yokoyama, M., Ogawa, M., Nozu, Y., and Hashimoto, J.,** Detection of specific RNAs by *in situ* hybridization in plant protoplasts, *Plant Cell Physiol.*, 31, 403, 1990.
40. **Brangeon, J., Prioul, J. L., and Forchioni, A.,** Localization of mRNAs for the small and large subunits of rubisco using electron microscope *in situ* hybridization, *Plant Physiol.*, 86, 990, 1988.
41. **Brangeon, J., Nato, A., and Forchioni, A.,** Ultrastructural detection of ribulose-1,5-bisphosphate carboxylase protein and its subunit mRNAs in wild-type and holoenzyme-deficient *Nicotiana* using immunogold and *in situ* hybridization techniques, *Planta,* 177, 151, 1989.
42. **Harris, N. and Croy, R. R. D.,** Localization of mRNA for pea leguminin: *in situ* hybridization using a biotinylated cDNA probe, *Protoplasma,* 130, 57, 1986.
43. **McFadden, G. I.,** *In situ* hybridization in plants: from macroscopic to ultrastructural resolution, *Cell Biol. Int. Rep.*, 13, 3, 1989.

44. **McFadden, G., Bönig, I., and Clarke, A.,** Double label *in situ* hybridization for electron microscopy, *Trans. R. Micros. Soc.*, 1, 683, 1990.
45. **Leitch, A. R., Mosgöller, W., Schwarzacher, T., Bennett, M. D., and Heslop-Harrison, J. S.,** Genomic *in situ* hybridization to sectioned nuclei shows chromosome domains in grass hybrids, *J. Cell Sci.*, 95, 335, 1990.
46. **Motte, P. M., Loppes, R., Menager, M., and Deltour, R.,** Three-dimensional electron microscopy of ribosomal chromatin in two higher plants: a cytochemical, immunocytochemical, and *in situ* hybridization approach, *J. Histochem. Cytochem.*, 39, 1495, 1991.
47. **Brangeon, J. and Forchioni, A.,** Détection ultrastructurale des ARNm de tissus foliaires par hybridation *in situ* à l'aide de sondes biotinylées, révélées à la ferritine, in *Techniques en Microscopie Électronique, Cryométhodes, Immunocytologie, Autoradiographie, Hybridation In Situ,* Morel, G., Ed., Paris, Editions INSERM, 1991, chap. 52.
48. **Dabos, P., Mazzolini, L,. Yot, P., and Sossountzov, L.,** Replication of cauliflower mosaic virus: detection and intracellular localization of single-stranded intermediates by *in situ* hybridization, unpublished data.
49. **Arondel, V., Kader, J. C., and Sossountzov, L.,** Distribution of lipid transfer protein mRNA in maize coleoptile using bromo-deoxyuridine labeled-oligonucleotides, unpublished data.
50. **Sossountzov, L., Ruiz-Avila, L., Vignols, F., Jolliot, A., Arondel, V., Tchang, F., Grosbois, M., Guerbette, F., Miginiac, E., Delseny, M., Puydomenech, P., and Kader, J. C.,** Spatial and temporal expression of a maize lipid transfer protein gene, *Plant Cell,* 3, 923, 1991.
51. **Chan, V. T. W. and McGee, J. O'D.,** Non-radioactive probes: preparation, characterization and detection, in *In Situ Hybridization Principles and Practice,* Polak, J. M. and McGee, J. O'D., Eds., Oxford University Press, New York, 1990, chap. 4.
52. **Höfler, H.,** What's new in ''*In situ* hybridization'', *Pathol. Res. Pract.*, 182, 421, 1987.
53. **Ozden, S., Aubert, C., Gonzales-Dunia, D., and Brahic, M.,** Simultaneous *in situ* detection of two mRNAs in the same cell using riboprobes labeled with biotin and ^{35}S, *J. Histochem. Cytochem.*, 38, 917, 1990.
54. **Jirikowski, G. F., Ramalho-Ortigao, J. F., Lindl, T., and Seliger, H.,** Immunocytochemistry of 5-bromo-2′-deoxyuridine labeled oligonucleotide probes. A novel technique for *in situ* hybridization, *Histochemistry,* 91, 51, 1989.
55. **Ramalho-Ortigao, J. F., Jirkowski, G. F., and Seliger, H.,** 5-Bromouridinated oligonucleotide for hybridization analysis of DNA and RNA on membranes and *in situ, Nucleosides and Nucleotides,* 8, 805, 1989
56. **Deniyn, M., De Weger, R. A., Berends, M. J. H., Compier-Spies, P. I., Jansz, H., Van Unnik, J. A. M., and Lips, C. J. M.,** Detection of calcitonin-encoding mRNA by radioactive and non-radioactive *in situ* hybridization: improved colorimetric detection and cellular localization of mRNA in thyroid sections, *J. Histochem. Cytochem.*, 38, 351, 1990.
57. **Trembleau, A., Ferre-Montagne, M., and Calas, A.,** Ultrastructural visualization of oxytocin mRNA by *in situ* hybridization. A high resolution radioautographic study using a tritiated oligonucleotide probe, *C. R. Acad. Sci., Paris,* III, 307, 869, 1988.
58. **McFadden, G. I., Bönig, I., Cornish, E. C., and Clarke, A. E.,** A simple fixation and embedding method for use in hybridization histochemistry on plant tissues, *Histochem. J.*, 20, 575, 1988.
59. **Ellis, R. J.,** Chloroplast proteins: synthesis, transport and assembly, *Annu. Rev. Plant Physiol.*, 32, 111, 1981.
60. **Shinozaki, K. and Sugiura, M.,** The nucleotide sequence of tobacco chloroplast gene for the large subunits of ribulose-1,5-bisphosphate carboxylase/oxygenase, *Gene,* 20, 91, 1982.
61. **Bresser, J. and Evinger-Hodges, M. J.,** Comparison and optimization of *in situ* hybridization procedures yielding rapid, sensitive mRNA detections, *Gen. Anal. Tech.*, 4, 89, 1987.

62. **Jackson, P., Lewis, F. A., and Wells, M.,** *In situ* hybridization techniques using an immuno-gold silver staining system, *Histochem. J.*, 21, 425, 1989.
63. **Lawrence, J. B. and Singer, R. J.,** Quantitative analysis of *in situ* hybridization methods for the detection of actin gene expression, *Nucleic Acids Res.*, 13, 1777, 1985.
64. **Teo, C. G.,** *In situ* hybridization in virology, in *In Situ Hybridization Principles and Practice,* Polak, J. and McGee, J. O'D., Eds., Oxford University Press, New York, 1990, chap. 9.
65. **Shepherd, R. J.,** Biochemistry of DNA plant viruses, in *The Biochemistry of Plants,* Marcus, A., Ed., Academic Press, New York, 1989, chap. 15.
66. **Favali, M. A., Bassi, M., and Conti, G. G.,** A quantitative autoradiographic study of intracellular sites for replication of cauliflower mosaic virus, *Virology,* 53, 115, 1973.
67. **Kamei, T., Rubio-Huertos, M., and Matsui, C.,** Thymidine-H^2 uptake by X-bodies associated with cauliflower mosaic virus infection, *Virology,* 37, 506, 1969.
68. **Modjtahedi, N., Volovitch, M., Sossountzov, L., Habricot, Y., Bonneville, J. M., and Yot, P.,** Cauliflower mosaic virus-induced viroplasms support viral synthesis in a cell-free system, *Virology,* 133, 289, 1984.
69. **Mazzolini, L., Bonneville, J. M., Volovitch, M., Magazin, M., and Yot, P.,** Strand-specific viral DNA synthesis in purified viroplasms isolated from turnip leaves infected with cauliflower mosaic virus, *Virology,* 145, 293, 1985.
70. **Mazzolini, L., Dabos, P., Constantin, S., and Yot, P.,** Further evidence that viroplasms are the site of cauliflower mosaic virus genome replication by reverse transcription during viral infection, *J. Gen. Virol.*, 70, 3439, 1989.
71. **Thomas, C. M., Hull, R., Bryant, J. A., and Maule, A. J.,** Isolation of a fraction from cauliflower mosaic virus-infected protoplasts which is active in the synthesis of (+) and (−) strand viral DNA and reverse transcription of primed RNA templates, *Nucleic Acids Res.*, 13, 4557, 1985.
72. **Marsh, L., Kuzj, A., and Guilfoyle, T.,** Identification and characterization of cauliflower mosaic virus replication complexes—analogy to hepatitis B virus, *Virology,* 143, 212, 1985.
73. **Blum, H. E., Stowring, L., Figus, A., Montgomery, C. K., Haase, A. T., and Vyas, G. N.,** Detection of hepatitis B virus DNA in hepatocytes, bile duct epithelium, and vascular elements by *in situ* hybridization, *Proc. Natl. Acad. Sci. U.S.A.*, 80, 6685, 1983.
74. **Blum, H. E., Haase, A. T., Harris, J. D., Walker, D., and Vyas, G. N.,** Asymmetric replication of hepatitis B virus DNA in human liver: demonstration of cytoplasmic-minus strand DNA by blot analysis and *in situ* hybridization, *Virology,* 139, 87, 1984.
75. **Jiang, X., Estes, M. K., and Metcalf, T. G.,** Detection of hepatitis A virus by hybridization with single-stranded RNA probes, *Appl. Environ. Microbiol.*, 53, 2487, 1987.
76. **Hoefler, H., Childers, H., Montminy, M. R., Lechan, R. M., Goodman, R. H., and Wolfe, H. J.,** *In situ* hybridization methods for the detection of somatostatin mRNA in tissue sections using antisense RNA probes, *Histochem. J.*, 18, 597, 1986.
77. **Schneider, A., Oltersdorf, T., Schneider, V., and Gissmann, L.,** Distribution pattern of human papilloma virus 16 genome in cervical neoplasia by molecular *in situ* hybridization of tissue sections, *Int. J. Cancer,* 39, 717, 1987.
78. **Simmons, D. M., Arriza, J. L., and Swarson, L. W.,** A complete protocol for *in situ* hybridization of messenger RNAs in brain and other tissues with radiolabelled single-stranded RNA probes, *J. Histotechnol.*, 12, 169, 1989.
79. **Troxler, M., Pasamontes, L., Egger, D., and Brenz, K.,** *In situ* hybridization for light and electron microscopy: a comparison of methods for the localization of viral RNA using biotinylated DNA and RNA probes, *J. Virol. Methods,* 30, 1, 1990.
80. **Howell, S. H., Walker, L. L., and Dudley, R. K.,** Cloned cauliflower mosaic virus DNA infects turnips *(Brassica rapa), Science,* 208, 1265, 1990.
81. **Howarth, A. J., Gardner, R. C., Messing, J., and Shepherd, R. J.,** Nucleotide sequence of naturally occurring deletion mutants of cauliflower mosaic virus, *Virology,* 112, 678, 1981.

82. **Gardner, R. C., Howarth, A. J., Hahn, P., Brown-Luedi, M., Shepherd, R. J., and Messing, J.,** The complete nucleotide sequences of an infectious clone of cauliflower mosaic virus by M13mp7 shot-gun sequencing, *Nucleic Acids Res.*, 9, 2871, 1981.
83. **Sambrook, J., Fritsch, E. F., and Maniatis, T.,** *Molecular Cloning. A Laboratory Manual,* Vol. 1, Cold Spring Harbor Laboratory Press, Cold Spring, NY, 1989, chaps. 1, 4, 5, and 6.
84. **Folsom, V., Haces, A., and Gaskill, M.,** N6-(6-aminohexyl) ATP for synthesis of ligand-modified RNA probes, *Focus,* 12, 45, 1990.
85. **Cox, K. H., DeLeon, D. V., Angerer, L. M., and Angerer, R. C.,** Detection of mRNAs in sea urchin embryos by *in situ* hybridization using asymmetric RNA probes, *Develop. Biol.*, 101, 485, 1984.
86. **Sossountzov, L., Habricot, Y., Dabos, P., and Yot, P.,** An ultrastructural and immunogold analysis of the short- and long-distance spread of cauliflower mosaic virus in turnip leaves, unpublished data.
87. **Brahic, M. and Haase, A. T.,** Detection of viral sequences of low reiteration frequency by *in situ* hybridization, *Proc. Natl. Acad. Sci. U.S.A.*, 75, 6125, 1978.
88. **Alexandersen, S., Bloom, M. E., Wolfinbarger, J., and Race, R. E.,** *In situ* molecular hybridization for detection of Aleutian Mink disease *Parvovirus* DNA by using strand specific probes: identification of target cells for viral replication in cell cultures and in mink kits with virus-induced interstitial pneumonia, *J. Virol.*, 61, 2407, 1987.
89. **Fontaine, B., Sassoon, D., Buckingham, M., and Changeux, J. P.,** Detection of the nitotinic acetylcholine receptor-subunit mRNA by *in situ* hybridization at neuromuscular junctions of 15-day-old chick striated muscles, *EMBO J.*, 7, 603, 1988.
90. **Giaid, A., Hamid, Q., Adams, C., Springall, D. R., Terenghi, G., and Polak, J. M.,** Non-isotopic RNA probes. Comparison between different labels and detection systems, *Histochemistry,* 93, 191, 1989.
91. **Knox, R. B.,** Methods for locating and identifying antigens in plant tissues, in *Immunocytochemistry,* Vol. 1, Bullock, G. R. and Petrusz, P., Eds., Academic Press, London, 1982, 205.
92. **Brigati, D. J., Myerson, D., Leary, J. J., Spalholz, B., Travis, S. Z., Fong, C. K. Y., Hsiung, G. D., and Ward, D. C.,** Detection of viral genomes in cultured cells and paraffin-embedded tissue sections using biotin-labeled hybridization probes, *Virology,* 126, 32, 1983.
93. **Croen, K. D., Ostrove, J. M., Dragovic, L. J., Smialek, J. E., and Straus, S. E.,** Latent herpes simplex virus in human trigeminal ganglia. Detection of an immediate early gene "anti-sense" transcript by *in situ* hybridization, *New Engl. J. Med.*, 317, 1427, 1987.
94. **Jilbert, A. R., Freiman, J. S., Gowans, E. J., Holmes, G. M., Cossart, Y. E., and Burrell, C. J.,** Duck hepatitis B virus DNA in liver, spleen, and pancreas: analysis by *in situ* and southern blot hybridization, *Virology,* 158, 330, 1987.
95. **Minnigan, H. and Moyer, R. W.,** Intracellular location of rabbit *Poxvirus* nucleic acid within infected cells as determined by *in situ* hybridization, *J. Virol.*, 55, 634, 1985.
96. **Raap, A. K., Geelen, J. L., van der Meer, J. W. M., van de Rijke, F. M., van den Boogaart, P., and van der Ploeg, F. M.,** Non-radioactive *in situ* hybridization for the detection of cytomegalovirus infections, *Histochemistry,* 88, 367, 1988.
97. **Steffensen, D. M., Wilson, H. J., and Goodman, R. M.,** *In situ* detection of early replication phases of a gemini virus in legume protoplasts, *Plant Cell Rep.*, 6, 462, 1987.
98. **Gröning, B. R., Abouzid, A., and Jeske, H.,** Single-stranded DNA from Abutilon mosaic *virus* is present in the plastids of infected *Abutilon sellovianum, Proc. Natl. Acad Sci. U.S.A.*, 84, 8996, 1987.
99. **Pesic, Z. and Hiruki, C.,** Comparison of ELISA and dot-hybridization for detection of alfalfa mosaic virus in alfalfa pollen, *Can. J. Plant Pathol.*, 10, 116, 1988.
100. **Kader, J. C.,** Lipid-binding proteins, *Plant Chem. Phys. Lipids,* 38, 51, 1985.

101. **Tchang, F., This, P., Stiefel, V., Arondel, V. Morch, M. D., Pages, M., Puidomenech, P., Grellet, F., Delseny, M., Bouillon, P., Huet, J. C., Guerbette, F., Beauvais-Conte, F., Duranton, H., Pernollet, J. C., and Kader, J. C.,** Phospholipid transfer protein: full-length cDNA and amino acid sequence in maize. Amino acid sequence homologies between phospholipid transfer proteins, *J. Biol. Chem.*, 263, 16849, 1988.
102. **Sterk, P., Booij, H., Schellekens, G. A., Van Kammen, A., and De Vries, S. C.,** Cell-specific expression of the carrot EP2 lipid transfer protein gene, *Plant Cell.*, 3, 907, 1991.
103. **Baldino, F. Jr., Chesselet, M. F., and Lewis, M. E.,** High-resolution *in situ* hybridization histochemistry, *Methods Enzymol.*, 168, 761, 1989.
104. **Baldino, F., Jr., Ruth, J. L., and Davis, L. G.,** Nonradioactive detection of vasopressin mRNA with *in situ* hybridization histochemistry, *Exp. Neurol.*, 104, 200, 1989.
105. **Dirks, R. W., Raap, A. K., Van Minnen, J., Vreugdenhil, E., Snut, A. B., and Van der Ploeg, M.,** Detection of mRNA molecules coding for neuropeptide hormones of the pond snail *Lymnaea stagnalis* by radioactive and non-radioactive *in situ* hybridization. A model study for mRNA detection, *J. Histochem. Cytochem.*, 37, 7, 1989.
106. **Dirks, R. W., Vangijlswijk, R. P. M., Tullis, R. H., Snut, A. B., Van Minnen, J., Van der Ploeg, M., and Raap, A. K.,** A simultaneous detection of different mRNA sequences coding for neuropeptide hormones by double *in situ* hybridization using FITC- and biotin-labeled oligonucleotides, *J. Histochem. Cytochem.*, 38, 467, 1990.
107. **Miller, M. A., Urban, J. H., and Dorsa, D. M.,** Quantification of mRNA in discrete cell groups of brain by *in situ* hybridization histochemistry, *Methods Neurosci.*, 1, 164, 1989.
108. **Ozden, S., Aubert, C., Gonzalez-Dunia, D., and Brahic, M.,** *In situ* analysis of proteolipid protein gene transcripts during persistent Theiler's virus infection, *J. Histochem. Cytochem.*, 39, 1305, 1991.
109. **Sutin, E. L., Montpied, P., and Jacobowitz, D. M.,** A synthetic oligonucleotide probe encoding for atrial natriuretic peptide detects specific mRNA transcripts in rat heart but not brain using *in situ* hybridization histochemistry, *Histochemistry,* 93, 519, 1990.
110. **Taneja, K. and Singer, R. H.,** Use of oligodeoxynucleotide probes for quantitative *in situ* hybridization to actin mRNA, *Anal. Biochem.*, 166, 389, 1987.
111. **Uhl, G. R.,** Messenger RNA localization with the microscope, in *Methods in Neurotransmitter Reception Analysis,* Yamamura, H. et al., Eds., Raven Press, New York, 1990, 219.
112. **Emson, P.C., Arai, H., Agrawal, S., Christodoulou, C., and Gait, M. J.,** Non radioactive methods of *in situ* hybridization: visualization of neuroendocrine mRNA, *Methods Enzymol.*, 168, 753, 1989.
113. **Larsson, L. I., Christensen, T., Dabboge, H.,** Detection of pro-opiomelanocortin mRNA by *in situ* hybridization using a biotinylated oligodeoxynucleotide probe and avidin-alkaline phosphatase histochemistry, *Histochemistry,* 89, 109, 1988.
114. **Weiss, L. M. and Chen, Y. Y.,** Effects of different fixatives on detection of nucleic acids from paraffin-embedded tissues *in situ* hybridization using oligonucleotide probes, *J. Histochem. Cytochem.*, 39, 1237, 1991.
115. **Kiyama, H. and Emson, P.,** An *in situ* hybridization histochemistry method for the use of alkaline phosphatase-labeled oligonucleotide probes in small intestine, *J. Histochem. Cytochem.*, 39, 1377, 1991.
116. **Stroobants, C., Sossountzov, L., and Miginiac, E.,** DNA synthesis in excised tobacco leaves after bromodeoxyuridine incorporation: immunohistochemical detection in semi-thin Spurr sections, *J. Histochem. Cytochem.*, 38, 641, 1990.
117. **Mullink, H., Walboomers, J. M. M., Tadema, T. M., Jansen, D. J., and Meijer, C. J. L. M.,** Combined immuno- and non-radioactive hybridocytochemistry on cells and tissue sections: influence of fixation, enzyme pre-treatment, and choice of chromogen on detection of antigen and DNA sequences, *J. Histochem. Cytochem.*, 37, 603, 1989.

118. **Chesselet, M. F., Weiss, L., Wuenschell, C., Tobin, A. J., and Affolter, H. U.,** Comparative distribution of mRNAs for glutamic and decarboxylase, tyrosine hydroxylase and tachykinins in the basal ganglia: an *in situ* hybridization study in the rodent brain, *J. Comp. Neurol.*, 262, 125, 1987.
119. **Denny, P., Hamid, Q., Krause, J. E., Polak, J. M., and Legon, S.,** Oligo-riboprobes. Tools for *in situ* hybridization, *Histochemistry,* 89, 481, 1988.
120. **Singer, R. H., Lawrence, J. B., and Villnave, C.,** Optimisation of *in situ* hybridization using isotopic and non-isotopic detection methods, *Biotechnology,* 4, 230, 1986.

INDEX

A

B

E

F

G

H

I

R

S

T

U